MICROWAVE TECHNIQUES
TRANSMISSION LINES

A. Kumar
Emeritus Scientist
Deptt. of Electrical and Communication Engineering
Indian Institute of Science
Bangalore

PUBLISHING FOR ONE WORLD
NEW AGE INTERNATIONAL (P) LIMITED, PUBLISHERS
New Delhi • Bangalore • Chennai • Cochin • Guwahati • Hyderabad
Jalandhar • Kolkata • Lucknow • Mumbai • Ranchi
Visit us at www.newagepublishers.com

Published by New Age International (P) Ltd.. Publishers
First Edition: 1998
Reprint: 2008

Branches:

- 36, Malikarjuna Temple Street, Opp. ICWA, Basavanagudi, **Bangalore**. © (080) 26677815
- 26, Damodaran Street, T. Nagar, **Chennai**. © (044) 24353401
- Hemsen Complex, Mohd. Shah Road, Paltan Bazar, Near Starline Hotel, **Guwahati**. © (0361) 2543669
- No. 105, 1st Floor, Madhiray Kaveri Tower, 3-2-19, Azam Jahi Road, Nimboliadda, **Hyderabad**. © (040) 24652456
- RDB Chambers (Formerly Lotus Cinema) 106A, 1st Floor, S.N. Banerjee Road, **Kolkata**. © (033) 22275247
- 18, Madan Mohan Malviya Marg, **Lucknow**. © (0522) 2209578
- 142C, Victor House, Ground Floor, N.M. Joshi Marg, Lower Parel, **Mumbai**. © (022) 24927869
- 22, Golden House, Daryaganj, **New Delhi**. © (011) 23262370, 23262368

ISBN: 978-81-224-1076-1

Rs. 185.00

C-08-07-2783

Printed in India at Mohanlal Printers, Delhi.
Typeset at Aptech, Delhi.

PUBLISHING FOR ONE WORLD
NEW AGE INTERNATIONAL (P) LIMITED, PUBLISHERS
4835/24, Ansari Road, Daryaganj, New Delhi-110002
Visit us at **www.newagepublishers.com**

To my wife Chandrika
and my son Niraj Sachdeva
and also to the fond memory of
my mother Shrimati Shivdevi Sachdeva,
my father Shri Ram Labhaya Sachdeva and
my brother Sqn Ldr Ramesh Chander Sachdeva, Vr.C.

PREFACE

The importance of microwave engineering is very well recognised by communication engineers/scientists of today. Radar requirements during World War II provided the initial spur for the development of microwave engineering, and the fact that microwaves have certain unusual and unique features that may be profitably exploited, gave a new dimension to the growth of the subject. Since then microwave engineering has had a phenomenal proliferation and today it is a discipline by itself. Its applications have pervaded almost all walks of life ranging from remote sensing from satellite-based platforms to food processing. However, the techniques that one is compelled to employ at these frequencies are entirely different from those used in the lower parts of the electromagnetic spectrum. Hence the need for such a book, which attempts to discuss and explain the basic microwave techniques.

It is encouraging to note that in our country microwave engineering occupies a prominent place in the undergraduate engineering curricula of almost all universities and institutions of higher learning. This book is an outgrowth of the classroom lectures that the author has been giving at the Indian Institute of Science, Bangalore, India, on the subject for over three decades. This book is, indeed, an edited version of his frequently updated class notes, toned up, wherever felt necessary, by the author's interaction with his students. The author likes to thank his students for the numerous classroom discussions they have had with him and for giving him, on many occasions, food for thought.

This book deals with Transmission Lines which form the foundation on which microwave circuits rest. It is intended to serve as a textbook for a first course in Microwave Engineering, that is, for the undergraduates in electrical communication engineering, electronics engineering and electrical engineering. It would be equally useful to postgraduates in applied physics. Engineering postgraduates, research students and practising research and development engineers would find this textbook useful for self study as well as for reference to acquire a better understanding of the fundamentals of microwave engineering.

Throughout the book emphasis has been laid on physical principles. An attempt has been made to use simple language and to make analytical treatments also simple but without losing rigour, so that even an average student may grasp the ideas easily. The reader would easily note from the text and the solved examples and exercises given that there is a slight leaning towards theoretical analysis. This has been done intentionally to build-up the analytical skills of the student and to train him better for research and development work.

The present volume is divided into ten chapters, the first being an introductory one. Chapter 2, presenting the transmission line theory, forms the core of the subject. Discussed in Chapter 3 are the variations of the line constants with the frequency of operation. Signal distortion that creeps in because of these variations has been discussed in Chapter 4. Chapters 2 and 3 are of interest to low frequency or telephone and telegraph engineers. If the reader is interested in very high or microwave frequencies only, he may skip these two chapters without any loss of continuity of the subject. Chapter 5 deals with the concepts of voltage and current waves and reflections on a transmission lines. This is an important chapter and should be paid the attention it deserves. A transmission line may, after all, be regarded as a two-port network and its terminal behaviour described in terms of two-port network parameters. This is done in Chapter 6 after recapitulating definitions and inter-relationships of the network parameters. Discussed in Chapter 7 are lossless line or the low loss lines that may be regarded as lossless. The remaining chapters deal with only lossless lines that all practical lines approximate to, when operating at high or microwave frequencies. Chapter 8 discusses the use of lossless lines for realisation of pure reactances or susceptances. Chapter 9 gives the theory and applications of the transmission line charts. Most of it, however, is devoted to the most popular and versatile one - the Smith Chart. Discussed in the last chapter is the very practical and important subject, namely, impedance matching.

"There is only one satisfactory way to study a theory, and that is by application to specific examples." Keeping this in mind, a set of problems requiring numerical or analytical solutions, has been given at the and of each chapter, except the first one. Some problems have also been given to substantiate, or to supplement some point discussed in the text, or else to drive home some point which could not be done in the text. Complete solutions for a few typical of these are given to teach the student how to solve problems, and the remaining are meant to be exercises for the student to solve. However, to let the student know whether he has done the exercises correctly or not, answers to each of the exercises have also been given. The students would do well to solve the problems given. But if they do not have the time to do so, they are advised to go through the solved examples at least. Doing so would clarify any doubts that they may have and would make their understanding of the subject more sound. While solving a problem, if a need is felt to discuss a point of practical interest, the solution is interrupted, the point discussed fully, and the solution resumed. The author has solved the illustrative examples and exercises by running on his 486 Personal Computer a large number of computer programmes that he has developed with the assistance of his son, Niraj Sachdeva, research student, Indian Institute of Science, Bangalore. The author likes to place on record his appreciation to him for his assistance. For further details about the softwares developed for solving a variety of transmission line problems, the interested readers may contact the author.

A distinctive feature of this book is that all the drawings and graphs/curves are computer-generated using data of some typical practical lines. However, for some curves, to bring out or accentuate some particular characteristic, fictitious data has been used.

The author likes to express his gratitude to Prof. G. Padmanaban, Director, Indian Institute of Science, Bangalore, Prof. M.A.L. Thathachar, Chairman, Electrical Sciences Division, I.I.Sc, and Prof. A Selvarajan, Chairman, Electrical Communication Engineering Department, I.I.Sc., Bangalore, for providing necessary facilities during the preparation of the manuscript. He is thankful to the All India Council for Technical Education for awarding him Emeritus Fellowship. He is indebted to the Council of Scientific and Industrial Research for awarding him Emeritus Scientistship. The author is grateful to Prof. T.S. Mruthyunjaya, Chairman, Centre for Continuing Education, Indian Institute of Science, Bangalore, for granting him the financial assistance for the preparation of the manuscript.

The author likes to thank his colleagues, Prof. Miss T.S. Vedavathy, E.C.E. Department; Prof. N. Balakrishnan, Chairman, Supercomputer Education and Research Centre, I.I.Sc, Bangalore, and Prof. K.C. Gupta of the University of Colorado at Boulder, Visiting Professor (1993-1994), E.C.E. Department, I.I.Sc, for the fruitful discussions that he has had with them and also for the interest evinced in his endeavour, the encouragement given and the support extended by them to him during the preparation of the manuscript. He gratefully acknowledges the assistance in the preparation of the manuscript that he has received from Messrs S.R. Nagesh, Niraj Sachdeva and P. Venkatesh Rao, formerly research students of this Department. In conclusion, the author takes this opportunity to express his sincere gratitude to Chandrika Kumar for the patience she evinced and for her constant and unstinted encouragement during the preparation of the manuscript.

ANAND KUMAR

CONTENTS

Chapter 1

INTRODUCTION

Microwaves are electromagnetic waves of wavelengths ranging from about 30 cm down to about 0.3 mm corresponding to the frequency range of 10^9 Hz to 10^{12} Hz. Microwave engineering deals with systems operating at these frequencies. The most important characteristic of this branch of engineering is the short wavelength involved. At microwave frequencies, the wavelength of operation is of the same order of magnitude as the components and interconnections of conventional electronic circuits. As the frequency of operation is increased, these components and interconnections radiate out or throw away from the circuit more and more amount of power, and thus become increasingly inefficient. Therefore, new circuit techniques for handling signals in the microwave frequency range are called for. Another consequence of increasing the frequency of operation to the microwave frequency range is the fact that the propagation time for signals from one point in a circuit to another becomes comparable with the time period of the signals. This fact coupled with the difficulty of uniquely defining voltages and currents in a microwave circuit renders the conventional low-frequency circuit analysis techniques based on Kirchhoff's laws and voltage-current concept inapplicable to microwave circuits. It becomes necessary, therefore, to carry out the analysis of microwave circuits in terms of the electric and magnetic fields.

1.1 APPLICATIONS OF MICROWAVES

Microwave engineering may be traced to have its origin in the early '40s. The war-time need for high-resolution radars capable of detecting and locating enemy planes and ships provided a great stimulus for the development of microwave systems. At that time, microwave engineering was almost synonymous with radar engineering. It is true that even today radar, in its varied forms like fire-control radar, missile-tracking radar, missile-guidance radar, airport traffic-control radar, highway collision-avoidance radar, weather-detecting radar, represents a major use of microwave frequencies. But today microwaves are applied to a variety of areas ranging from remote sensing to food processing. All these applications are possible because microwaves possess certain unusual and unique features which may profitably be exploited. For want of space, a complete account of all the applications of microwaves cannot be given here. However, a few of the applications of microwaves are described briefly below just to indicate their varied nature.

A tracking radar requires that the electromagnetic energy be radiated in a narrow pencil-like beam. This, in turn, calls for an antenna the relative size of whose radiating aperture is very large in terms of the wavelength. In practice, such an antenna may conveniently be obtained only at microwave frequencies. For instance, a typical radar antenna at 10 GHz is a 1-metre diameter paraboloid. To get the same performance at 10 MHz, a paraboloid of 1000-metre diameter would be required, a size much too big to be manoeuvred!

In communication systems, the bandwidth of a modulated carrier that can conveniently be handled or processed is only a few per cent of the carrier frequency, thus restricting the number of signals that a given carrier may handle. Microwaves, therefore, have the inherent potential of offering very large bandwidths and, hence, of meeting the ever-increasing demand for communication channels. For this reason, microwaves are finding widespread use in both the terrestrial and the satellite communication links.

The fact that the Earth's atmosphere has a transparent *window* in the lower part of the microwave band and the availability of highly sensitive, low-noise microwave receivers permit radio astronomic research of space in this band and also the establishment of communication between ground stations and spacecrafts.

Microwave techniques have also been applied to the area of basic research. A waveguide loaded periodically with shunt susceptances is used in a linear accelerator which produces high-energy beams of charged particles for use in atomic and nuclear research. Some molecular, atomic and nuclear systems exhibit resonance phenomena in the microwave band. Microwave spectroscopy may, therefore, be used for structure analysis. Microwave plasma is used for studying certain chemical processes, direct labelling and also for the synthesis of chemical compounds. Microwave plasma has also been used as an alternative to sputtering in the low-temperature preparation of a wide variety of non-crystalline films.

Another property of microwaves for which they are used is their dielectric heating effect. In wood industry, microwave heating is used for bonding of plywood veneers, destruction of dry rot fungus in wood and woodworms. Some more applications of microwave heating are: curing of concrete, breaking of concrete, sealing of plastics, drying, processing of dielectric sheets, freeze drying of foods, finish drying of potato-chips, thawing of frozen foods, heating of precooked food, cooking and baking of food, conditioning of flour and sterilization. Microwave heating provides uniform heating having fast temperature build up without any smoke, heat and odour. It does not change the shape, colour and flavour of the food, and it also preserves the heat-sensitive vitamins and proteins in the food. Microwave ovens are governed by the FFC regulations to operate in the ISM (Industrial, Scientific, Medical) bands which are (915 ± 13) MHz, (2450 ± 50) MHz and (5800 ± 75) MHz. Nearly all microwave ovens in the world operate in the 2450 MHz band. The demand for microwave ovens for domestic use is increasing rapidly and manufacture of these ovens has been taken up in our country also.

Microwave techniques have also been applied in steel rolling mills for an accurate control of the thickness of steel sheets. Microwaves have been employed in agriculture and medical diathermy too.

1.2 MICROWAVE RADIATION HAZARDS

While the previous section makes one feel that there is apparently no area where microwaves have not been useful, it may now be pointed out that microwaves can be hazardous too! Exposure to strong microwave fields can be harmful. Weight loss, changes in blood composition and a variety of other physiological effects are found to occur by exposure to strong microwave fields. It has been found that a body temperature rise, produced by exposure to microwaves, of more than about 5 to 10 degrees Celsius, may produce irreversible effects in the body. Eyes are particularly sensitive to microwave irradiation. Cataract formation is found to occur for radiation fluxes in excess of 100 mW/cm^2. Though more scientific work is required for a complete study of the effects of microwaves on human body yet sufficient information does exist to lay down safe limits for microwave irradiation. However, there are no universal standards, the USA standards are:

Power Density: 10 mW/cm^2 for period of 0.1 hour or more.

Energy Density: 1 mWh/cm^2 during any 0.1 hour period.

The standards followed by the former USSR were:

0.01 mW/cm^2 for an exposure duration of an entire day,

0.10 mW/cm^2 for an exposure of upto 2 hours, and

1.00 mW/cm^2 for an exposure of no more than 20 minutes per work day.

It may be mentioned that the power density inside the waveguide in a microwave laboratory is of the order of 20 mW/cm^2. Therefore, looking into a waveguide with the microwave power may be avoided.

Chapter 2

TRANSMISSION LINE THEORY

2.1 INTRODUCTION

In an electrical communication system, electromagnetic signal power is required to be transported from one point in the system to another without much loss. At low frequencies, this may conveniently be accomplished by simply connecting the two points by a conducting wire. But as the frequency of operation is increased, these interconnections tend to dissipate more and more of the signal power as radiations. The radiation loss increases, initially, as the square of the frequency. The tendency of the interconnecting wire to radiate may be curbed to some extent by making the *return current* flow through another wire placed very close and parallel to the interconnecting wire. These two conductors form a transmission system, shown in Fig. 2.1, called the ***Open-wire***, ***Two-wire or Two-conductor Transmission Line***. From the field point of view, the transmission line forms the simplest structure for guiding the electromagnetic wave along it in the form of a transverse electromagnetic field (TEM-mode) as given in Fig. 2.2.

The open-wire transmission line may be used satisfactorily upto about 300 MHz. At frequencies higher than this, the radiation losses become far too much to be tolerated. The radiation losses may be made negligible by employing a circular cylindrical conducting tube, placed coaxially with an inner conductor [Ref. Fig. 2.3] as the conductor for carrying the ***return currents***. The resulting wave guiding structure, shown in Fig. 2.3, is called the ***coaxial line*** The fields in the coaxial line remain confined to the annular region between the inner and the outer conductors and, hence, no radiation takes place. The field configuration inside the coaxial line also corresponds to the TEM mode as depicted in Fig. 2.4.

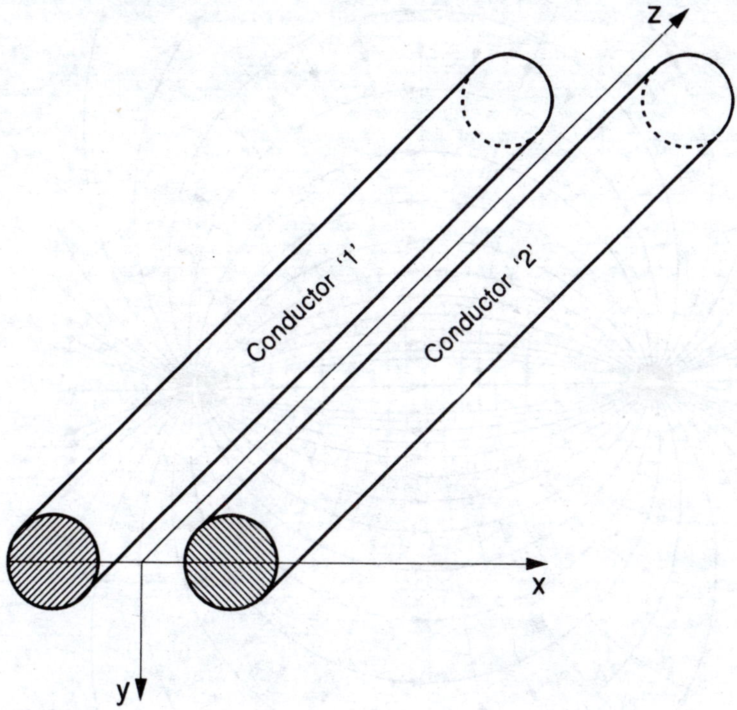

Fig. 2.1 Geometry of a Two-conductor or Open-wire Transmission Line.

The coaxial line is quite acceptable as a transmission structure upto about 3000 MHz above which, losses due to imperfections of the dielectric and the conductor, become appreciable and cannot be neglected. These losses may be reduced considerably by dispensing with the centre conductor and using merely a hollow metal tube of circular or rectangular (or theoretically, any other) cross-section [Ref. Fig. 2.5] — the rectangular cross-section being very commonly employed. Such a system is called the ***Waveguide***. A waveguide differs from the former two transmission or waveguiding structures in its construction in that it is a single-conductor system whereas the transmission line and the coaxial line are two-conductor systems. The single conductor is folded to form a tube of rectangular or any other geometrical shape cross-section. If a field analysis of a waveguide is carried out, it is found that the boundary conditions to be satisfied by the fields inside a hollow metal waveguide, are such that the TEM-mode cannot exist. Instead, a longitudinal component of either the electric field or the magnetic field, as shown in Fig. 2.6, or of both fields, must exist. If the field configuration is such that only the electric field has a longitudinal component, but the magnetic field is entirely in the transverse plane, it corresponds to a ***TM-mode** or **E-mode***. Propagation through the waveguide is said to be in a ***TE-mode** or **H-mode***, if only the magnetic field has a longitudinal component but the electric field is entirely in the transverse plane. Comparatively, the attenuation of these waveguides is, extremely low. Although

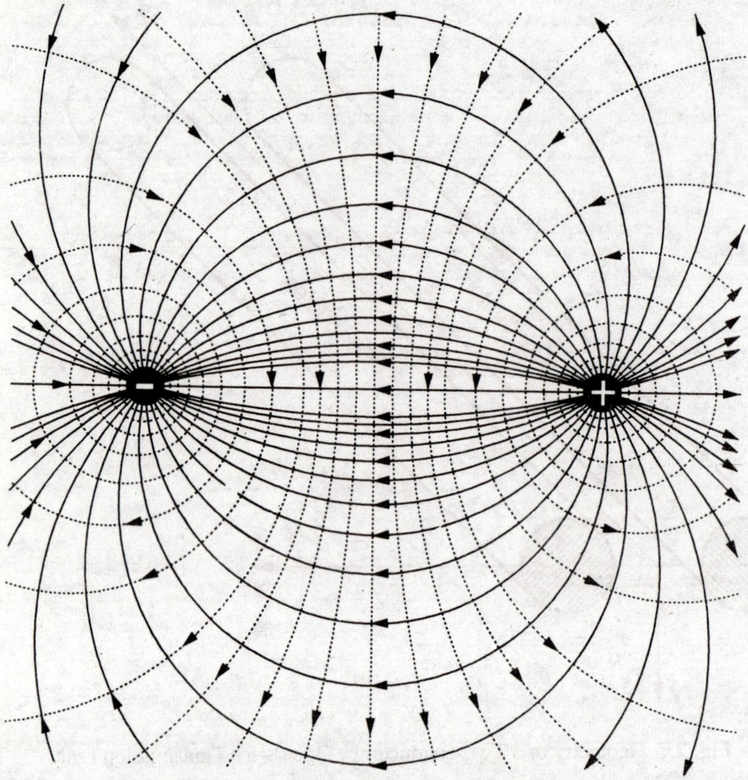

Fig. 2.2 Transverse Plane Field Configuration (Two-conductor Transmission Line).

waveguides are the best in so far as the efficiency of power transmission is concerned yet they cannot practically be used at lower frequencies because, it turns out that the cross-sectional dimensions of such pipes or waveguides have to be directly related to the wavelength of operation in order that transmission of power at all takes place through them. It is, therefore, only when the frequency of operation is a few giga hertz that the cross-sectional dimensions are small enough for waveguides to be convenient for practical use. In so far as propagation of waves along the waveguiding structures is concerned, in contrast to the two-conductor transmission line and the coaxial line, these waveguides possess a low frequency ***cut-off*** and exhibit a high-pass property.

Discussed in this chapter is the theory of two-conductor transmission lines. Propagation of high frequency electromagnetic waves along a two-conductor transmission line cannot be analysed by the application of lumped-element circuit analysis, wherein it is assumed that all points of an ideal connecting wire are at the same potential. It is well-known that the velocity with which electromagnetic disturbances travel along the line is but finite. Therefore, at high frequencies, the potential at various points even on an ideal connecting wire cannot be the same.

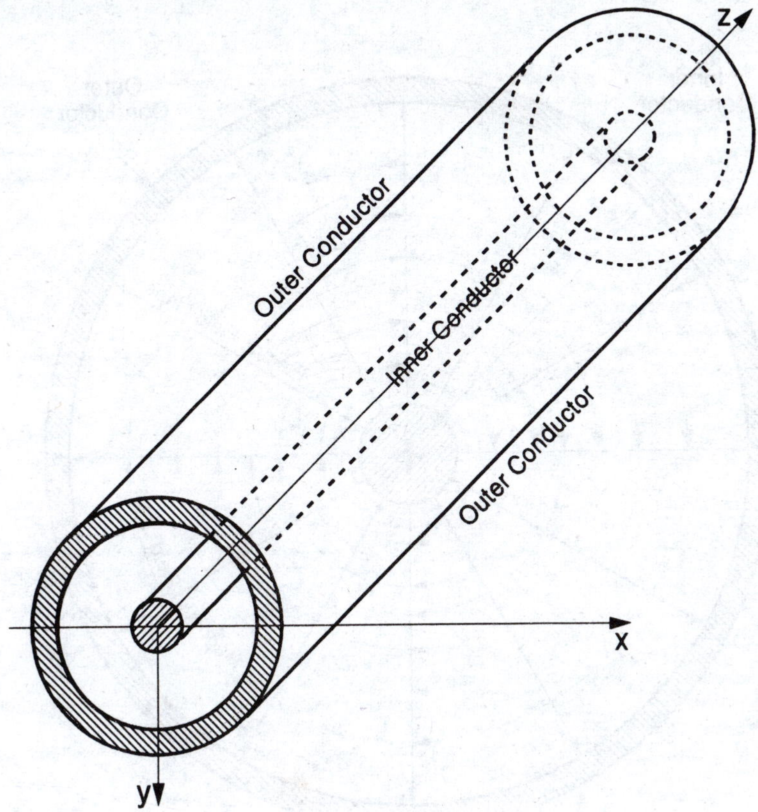

Fig. 2.3 Geometry of Coaxial Line.

Hence, there is a need for studying, from the basic principles, the variation of the voltage and current along the length of a transmission line and also with respect to time which amounts to studying the propagation of electromagnetic waves along the transmission line. It may be pointed out that the transmission line theory for a two-conductor line, developed and discussed in this chapter, may be applicable to waveguides and other transmission waveguiding structures also, though only under certain restricted conditions.

2.2 TRANSMISSION LINE THEORY

Consider a two-conductor line like the transmission line or the coaxial line as shown in Fig. 2.7(a). A two-conductor line is symbolically represented by two straight lines drawn parallel to each other [Ref. Fig. 2.7(b)]. The propagation of electromagnetic waves along this line may be studied by carrying out a field analysis, but a simpler treatment of the subject consists in extending the circuit or network theory concepts to cover the transmission line which may be regarded as a ***distributed parameter*** circuit. The latter is considered here. Although the transmission line theory is applicable to all two-conductor transmission systems yet it would be better

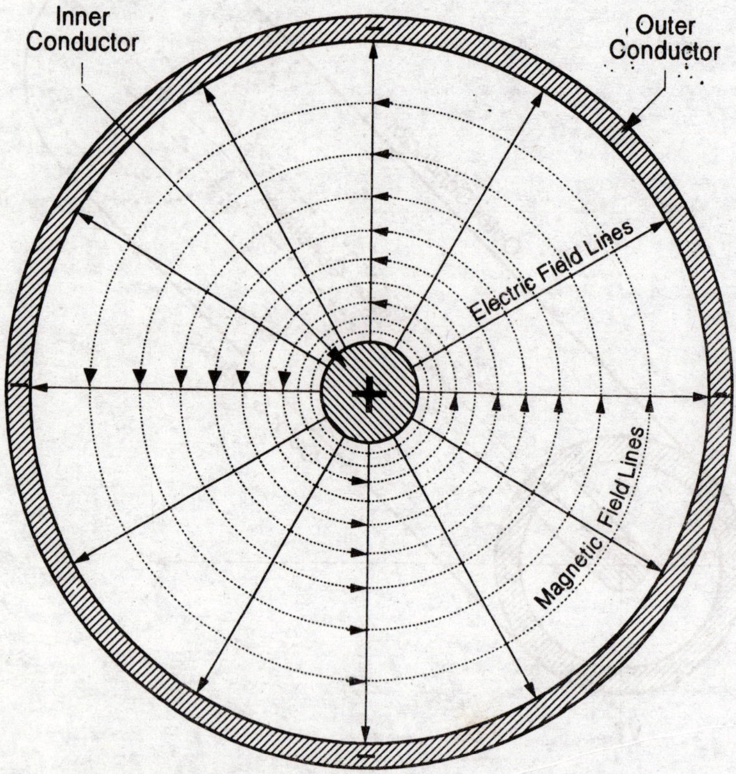

Fig. 2.4 Transverse Plane Field Configuration (Coaxial Transmission Line).

to keep the simplest of these, namely, the transmission line in mind while going through the analysis given below.

2.2.1 Circuit Representation

Assume that the geometry of the transmission line, that is, the cross-sections of, and the separation between the two conductors forming the transmission line [Ref. Fig. 2.7(a)], and the electromagnetic properties of both the conductors and the medium in which they are placed, are constant throughout the length of the transmission line. Such a transmission line is called a ***uniform transmission line***.

As the conductivity of the conductors forming the transmission line is finite, they have some resistance distributed uniformly along their lengths. Because of the skin effect, the resistance will, in general, be frequency-dependent. Further, some power may possibly be lost by the line in the form of radiation occurring uniformly throughout the length of the line. This can be accounted for by assuming a resistance distributed uniformly along the length of the line. This component of the resistance, which may be made negligibly small by an appropriate choice of the geometry of the transmission line, is also frequency-dependent.

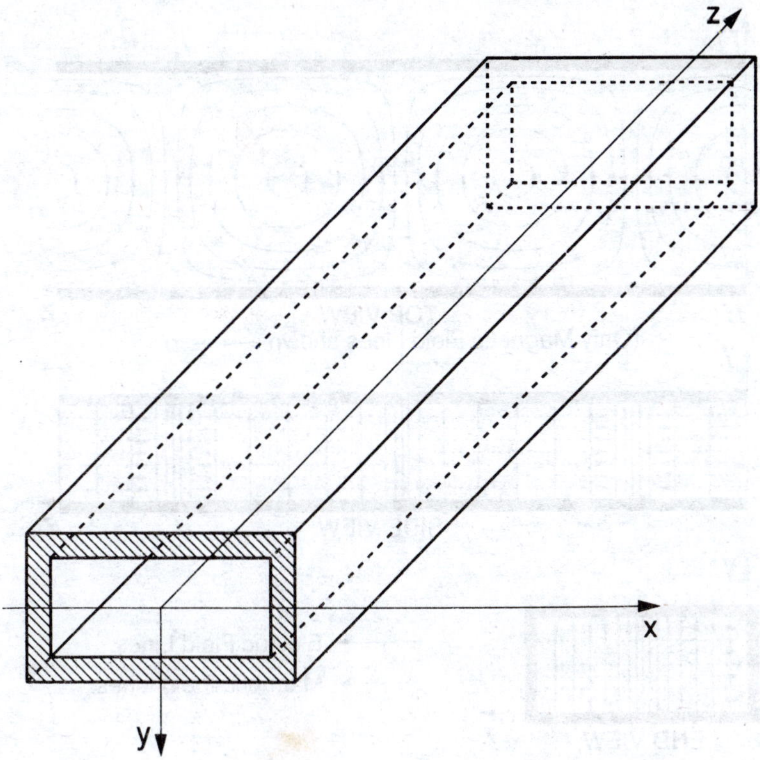

Fig. 2.5 Geometry of a Rectangular, Hollow, Metal Waveguide.

Flow of a current through the conductors establishes a magnetic field about them. linking the current. This gives rise to an inductance which is also distributed uniformly along the length of the line. The effect of this inductance is to impede the flow of current and, therefore, the inductance is effectively in series with the distributed line resistance.

The two conductors running parallel to each other in a real dielectric medium, give rise to a capacitance and a conductance between them. The capacitance and the conductance are also distributed uniformly along the length of the line.

Hence, the conductors forming the transmission line have associated with them uniformly distributed, series resistance and inductance, and also shunt capacitance and conductance between them. But as no symbols are available for representing these ***distributed*** circuit parameters, an accurate equivalent circuit of the line cannot be given. To overcome this difficulty, the line may be regarded as being composed of a large number of line sections, each of an infinitesimal length, Δz. For each of these infinitesimal length line sections, the distributed parameters may be summed up. Each of these line sections may, therefore, be represented approximately by the lumped-element network given in Fig. 2.8. In this representation, R_1, and L_1 are the

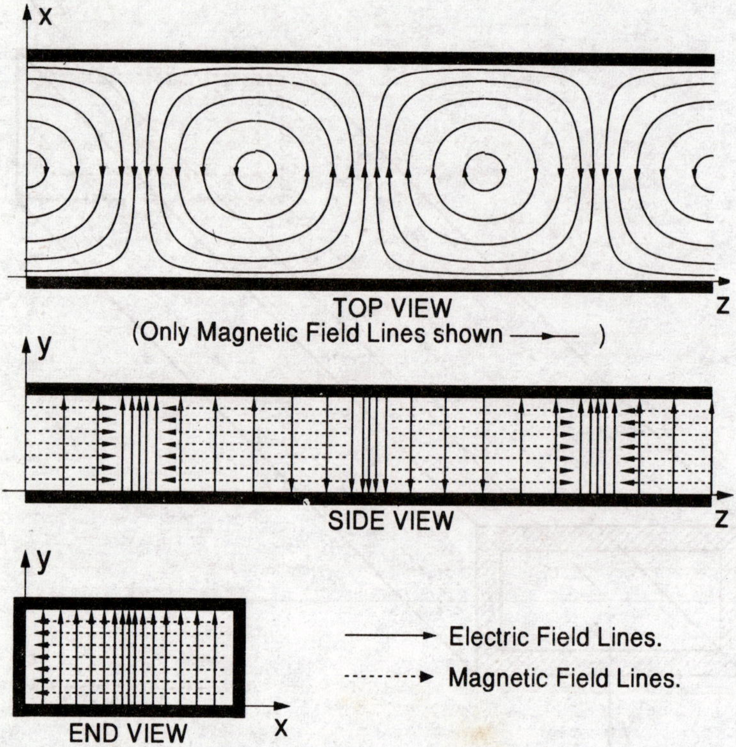

Fig. 2.6 Field Configuration in a Rectangular, Hollow, Metal Waveguide.

series resistance and inductance, respectively, per unit length of the conductor '1' and R_2 and L_2 are the corresponding values for the conductor '2'; C and G are, respectively, the capacitance and the conductance between the two conductors per unit length of the line.

2.2.2. Transmission Line Equations

Consider a semi-infinite, two-conductor uniform transmission line, at one end, $z = 0$, of which a sinusoidal generator of voltage $V_G\, e^{j\omega t}$, is connected. At any point, z, on it, the conductors carry equal and opposite current, $I(z)\, e^{j\omega t}$; as shown in Fig. 2.9. Let the voltage between the conductors at the point, z, be $V(z)\, e^{j\omega t}$. ***In the steady state***, application of Kirchhoff's voltage law[1] ***(The algebraic sum of all branch voltages around a closed loop of a network, is zero at all instants of time.)*** to the closed loop, OPQRO, formed by the infinitesimal section of the line between z and $z + \Delta z$, gives,

1 M.E. van Valkenburg, *Network Analysis*, Prentice Hall of India Pvt Ltd. 1985.

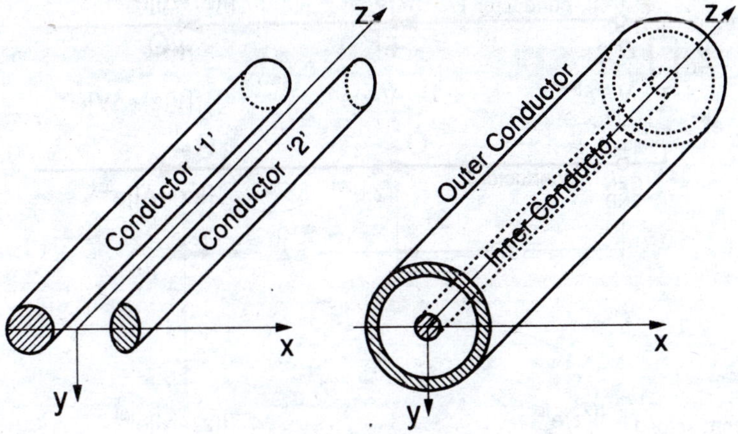

(a). Physical Representation.

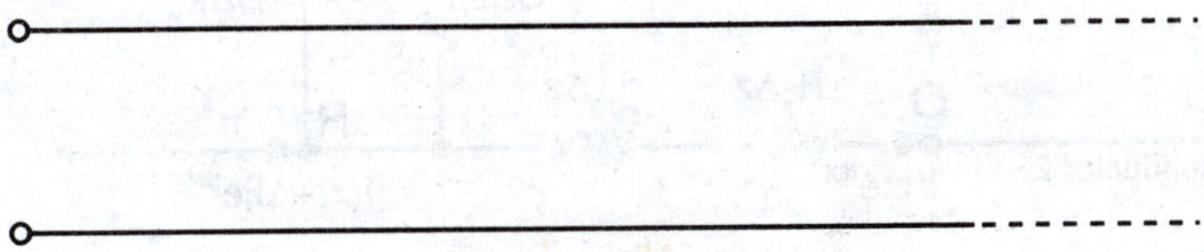

(b). Symbolic Representation.

Fig. 2.7 Two-conductor Uniform Transmission Line.

$$V(z) = (R_1 + j\,\omega L_1)\,\Delta z\, I(z) + (R_2 + j\,\omega L_2)\,\Delta z\, I(z) + [V(z) + \Delta V(z)].$$

Or, $$\Delta V(z) = -[(R_1 + R_2) + j\,\omega\,(L_1 + L_2)]\,\Delta z\, I(z)$$

Or, $$\frac{\Delta V(z)}{\Delta z} = -[(R_1 + R_2) + j\,\omega\,(L_1 + L_2)]\, I(z)$$

$$= -(R + j\,\omega\, L)\, I(z) \tag{2.1}$$

where,

$R = R_1 + R_2 =$ loop series resistance of the line per unit length (ohms/metre),

and $L = L_1 + L_2 =$ loop series inductance of the line per unit length (henrys/metre).

Similarly, in the steady state, application of Kirchhoff's current law ***(The algebraic sum of all branch currents leaving a node, is zero at all instants of time.)*** to the node, Q, of the network formed by the infinitesimal section of the line between z and $z + \Delta z$, gives

$$I(z) = I(z) + \Delta I(z) + (G + j\,\omega\, C)\,[V(z) + \Delta V(z)]$$

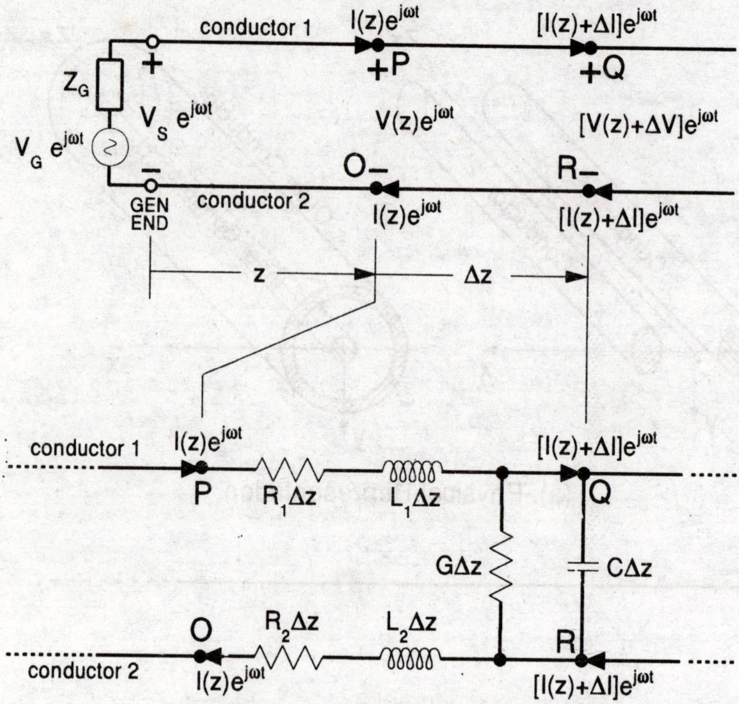

Fig. 2.8 Circuit Representation of a Line Section.

which gives, $$\Delta I(z) = -(G + j\omega C)\,\Delta z\, V(z)$$

Or, $$\frac{\Delta I(z)}{\Delta z} = -(G + j\omega C)\,[V(z) + \Delta V(z)]. \tag{2.2}$$

Now, proceed to the limit $\Delta z \to 0$. As $\Delta z \to 0$, both $\Delta V(z)$ and $\Delta I(z) \to 0$. Therefore, Eqns. (2.1) and (2.2) yield the following two simultaneous equations giving the current, $I(z)$, and the voltage, $V(z)$, at any point, z, on the line to be

$$\frac{dV(z)}{dz} = -(R + j\omega L)\, I(z), \tag{2.3}$$

and $$\frac{dI(z)}{dz} = -(G + j\omega C)\, V(z). \tag{2.4}$$

These equations are called ***Telegraphist's Equations***

Differentiating both sides of Eqn. (2.3) with respect to z, and substituting into the resulting equation the expression for $\frac{dI(z)}{dz}$ from Eqn. (2.4), give the following equation governing the distribution of the amplitude of the voltage, $V(z)$, along the line:

$$\frac{d^2 V(z)}{dz^2} = \gamma^2 V(z), \tag{2.5}$$

where,

$\gamma = \sqrt{(R + j\,\omega L)(G + j\,\omega C)}$. [γ is called the ***propagation constant*** of the line.]

$= \alpha + j\,\beta_g$. [Note that γ is, in general, a complex quantity.] (2.6)

Similarly, $V(z)$ may be eliminated from Eqns. (2.3) and (2.4) giving the following equation in $I(z)$:

$$\frac{d^2 I(z)}{dz^2} = \gamma^2 I(z). \tag{2.7}$$

Eqns. (2.5) and (2.7), which govern the distribution of the amplitudes of the voltage, $V(z)$, and the current, $I(z)$, along the transmission line, in the steady state, are called the ***Transmission Line Equations***. The most general solution of Eqn. (2.5) is

$$V(z) = V^+ e^{-\gamma z} + V^- e^{+\gamma z} \tag{2.8}$$

Substitution of Eqn. (2.8) into Eqn. (2.3) gives

$$I(z) = \frac{1}{Z_0}(V^+ e^{-\gamma z} - V^- e^{+\gamma z}), \tag{2.9}$$

where,

$Z_0 = \sqrt{\dfrac{R + j\,\omega L}{G + j\,\omega C}}$. [Z_0 is called the ***characteristic impedance*** of the line.]

$= |Z_0|\, e^{j\phi_0}$. [Note that Z_0 is, in general, a complex quantity.] (2.10)

From Eqns. (2.8) and (2.9), the input impedance[2], $Z(z)$, of the line at the point, z, looking to the right (towards the load-end) of the point, z, is given by

$$Z(z) = \frac{V(z)}{I(z)} = Z_0 \left[\frac{V^+ e^{-\gamma z} + V^- e^{+\gamma z}}{V^+ e^{-\gamma z} - V^- e^{+\gamma z}}\right] \tag{2.11}$$

2.2.3 Solution of Transmission Line Equations

It may be mentioned that Eqns.(2.8), (2.9) and (2.11) give the most general expressions, in terms of the integration constants, V^+ and V^-, for the voltage, $V(z)$, current, $I(z)$, and the input impedance, $Z(z)$, respectively, at any point z, of the line.

---

2 Note : Unless otherwise specified, the terms,
"Input impedance of the line looking to the right of the point, z,"
"Input impedance of the line at the point, z" *or simply,*
"Impedance of the line or Line impedance at the point, z," would imply $Z(z)$.

In the case of a given or practical transmission line, V^+ and V^- are to be determined by applying appropriate boundary conditions to the general solutions.

In practice, all the transmission lines used, are of finite length, having a generator or source at the one end (***generator-end*** or ***sending-end***), and a load of arbitrary impedance, Z_L, at the other (***load-end*** or ***receiving-end***), as shown in Fig. 2.9. The integration constants, V^+ and V^-, and, hence, the voltage, current and the impedance at any point on the line, may be expressed in terms of any two of the six quantities - the voltage, current and the impedance at the load-end or at the generator-end.

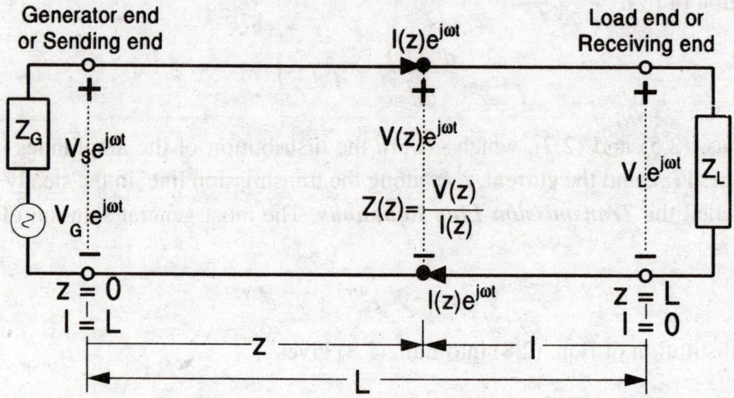

Fig. 2.9 Definitions of Line Voltages, Currents and Impedances.

Consider now a transmission line having Z_0 as its characteristic impedance and γ as the propagation constant, of a finite length 'L' and driven by a generator of open-circuit voltage, $V_G\, e^{j\omega t}$, and of an internal impedance, Z_G. For the analysis given below, reference may be made to Fig. 2.10. Let an arbitrary load impedance, Z_L, be connected across the line at the load-end, $z = L$. The line voltage, $V(z)$, the line current, $I(z)$, and the line impedance, $Z(z)$, at any point on it, at a distance of

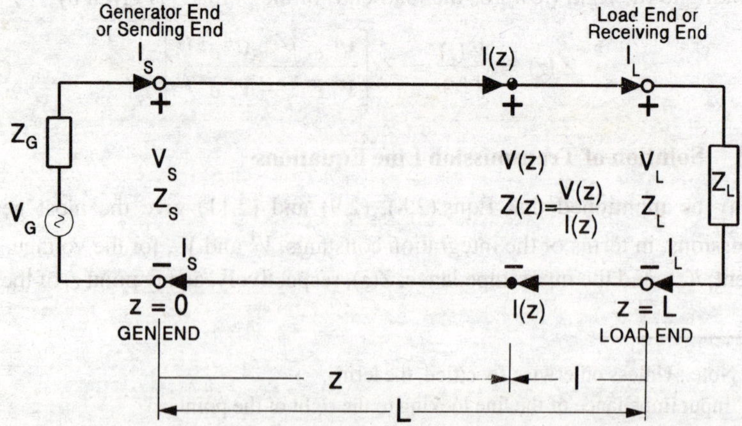

Fig. 2.10 Line Voltage and Current Phasors, and Line Impedances.

'z' from the generator-end or equivalently, at a distance of 'l' from the load-end, where, $l = L - z$, are, in general, given by

$$V(z) = V^+ e^{-\gamma z} + V^- e^{\gamma z}, \tag{2.12}$$

$$I(z) = \frac{1}{Z_0}[V^+ e^{-\gamma z} - V^- e^{\gamma z}], \tag{2.13}$$

and

$$Z(z) = \frac{V(z)}{i(z)} = Z_0\left[\frac{V^+ e^{-\gamma z} + V^- e^{+\gamma z}}{V^+ e^{-\gamma z} - V^- e^{+\gamma z}}\right] \tag{2.14}$$

(1) Solution in terms of sending end voltage, V_s, and current, I_s:

Application of boundary conditions,

$$V(0) = V_S \text{ and } I(0) = I_S, \text{ at } z = 0,$$

to Eqns. (2.12) and (2.13), gives

$$V_S = V^+ + V^-,$$

and

$$Z_0 I_S = V^+ - V^-,$$

which give,

$$V^+ = \frac{1}{2}[V_S + Z_0 I_S], \tag{2.15}$$

and

$$V^- = \frac{1}{2}[V_S - Z_0 I_S] \tag{2.16}$$

Substitution of Eqns. (2.15) and (2.16) into Eqns.(2.12) and (2.13) gives

$$V(z) = \frac{1}{2}(V_S + Z_0 I_S)\, e^{-\gamma z} + \frac{1}{2}(V_S - Z_0 I_S)\, e^{\gamma z} \tag{2.17}$$

$$= \left(\frac{Z_S + Z_0}{2 Z_S}\right) V_S\, [e^{-\gamma z} + \rho_S e^{\gamma z}] \tag{2.18}$$

$$= V_S \cosh \gamma z - Z_0 I_S \sinh \gamma z \tag{2.19}$$

$$= V_S\left[\cosh \gamma z - \frac{Z_0}{Z_S}\sinh \gamma z\right], \text{ and} \tag{2.20}$$

$$I(z) = \frac{1}{2Z_0}[(V_S + Z_0 I_S)e^{-\gamma z} - (V_S - Z_0 I_S)e^{\gamma z}] \tag{2.21}$$

$$= \left(\frac{Z_S + Z_0}{2Z_0}\right) I_S [e^{-\gamma z} - \rho_S e^{\gamma z}] \tag{2.22}$$

$$= I_S \cosh \gamma z - \frac{V_S}{Z_0} \sinh \gamma z \tag{2.23}$$

$$= I_S \left[\cosh \gamma z - \frac{Z_S}{Z_0} \sinh \gamma z\right]. \tag{2.24}$$

From Eqns. (2.20) and (2.24), the input impedance is

$$Z(z) = Z_0 \left[\frac{Z_S \cosh \gamma z - Z_0 \sinh \gamma z}{Z_0 \cosh \gamma z - Z_S \sinh \gamma z}\right] = Z_0 \left[\frac{Z_S - Z_0 \tanh \gamma z}{Z_0 - Z_S \tanh \gamma z}\right] \tag{2.25}$$

$$= Z_0 \left[\frac{e^{-\gamma z} + \rho_S e^{\gamma z}}{e^{-\gamma z} - \rho_S e^{\gamma z}}\right] = Z_0 \left[\frac{1 + \rho_S e^{2\gamma z}}{1 - \rho_S e^{2\gamma z}}\right]. \tag{2.26}$$

where, $Z_S = \frac{V_S}{I_S}$ = input impedance at the generator-end (2.27)

and, $$\rho_S = \frac{Z_S - Z_0}{Z_S + Z_0} \tag{2.28}$$

(2) Solution in terms of sending end voltage, V_S, and load-end voltage, V_L:

In this case, the boundary conditions are,

$V(0) = V_S$ at z = 0 and $V(L) = V_L$, at $z = L$ yielding

$$V(z) = \frac{1}{\sinh \gamma L}[V_S \sinh \gamma l + V_L \sinh \gamma z], \tag{2.29}$$

$$I(z) = \frac{1}{Z_0 \sinh \gamma L}[V_S \cosh \gamma l - V_L \cosh \gamma z], \tag{2.30}$$

and $$Z(z) = Z_0 \left[\frac{V_S \sinh \gamma l + V_L \sinh \gamma z}{V_S \cosh \gamma l - V_L \cosh \gamma z}\right]. \tag{2.31}$$

(3) Solution in terms of sending end voltage, V_S, and load-end current, I_L:

In this case, the boundary conditions are,

$$V(0) = V_S, \text{ at } z = 0 \text{ and } I(L) = I_L, \text{ at } z = L,$$

giving
$$V(z) = \frac{1}{\cosh \gamma L}[V_S \cosh \gamma l - Z_0 I_L \sinh \gamma z], \tag{2.32}$$

$$I(z) = \frac{1}{Z_0 \cosh \gamma L}[V_S \sinh \gamma l + Z_0 I_L \cosh \gamma z], \tag{2.33}$$

$$Z(z) = Z_o\left[\frac{V_S \cosh \gamma l - Z_0 I_L \sinh \gamma z}{V_S \sinh \gamma l + Z_0 I_L \cosh \gamma z}\right] \tag{2.34}$$

(4) Solution in terms of sending end voltage, V_S, and load impedance, Z_L:

In this case, the boundary conditions are,

$$V(0)\ V_S, \text{ at } z = 0 \text{ and } Z(L) = Z_L, \text{ at } z = L, \text{ yielding}$$

$$V(l) = V_S\left[\frac{Z_L \cosh \gamma l + Z_0 \sinh \gamma l}{Z_L \cosh \gamma L + Z_0 \sinh \gamma L}\right] = V_S\left[\frac{e^{\gamma l} + \rho_L e^{-\gamma l}}{e^{\gamma L} + \rho_L e^{-\gamma L}}\right], \tag{2.35}$$

$$I(l) = \frac{V_S}{Z_0}\left[\frac{Z_L \sinh \gamma l + Z_0 \cosh \gamma l}{Z_L \cosh \gamma L + Z_0 \sinh \gamma L}\right] = \frac{V_S}{Z_0}\left[\frac{e^{\gamma l} - \rho_L e^{-\gamma l}}{e^{\gamma L} + \rho_L e^{-\gamma L}}\right], \tag{2.36}$$

and
$$Z(l) = Z_0\left[\frac{Z_L \cosh \gamma l + Z_0 \sinh \gamma l}{Z_0 \cosh \gamma l + Z_l \sinh \gamma l}\right] = Z_0\left[\frac{Z_L + Z_0 \tanh \gamma l}{Z_0 + Z_L \tanh \gamma l}\right] \tag{2.37}$$

$$= Z_0\left[\frac{e^{\gamma l} + \rho_L e^{-\gamma l}}{e^{\gamma l} - \rho_L e^{-\gamma l}}\right] = Z_0\left[\frac{1 + \rho_L e^{-2\gamma l}}{1 - \rho_L e^{-2\gamma l}}\right] \tag{2.38}$$

where,
$$\rho_L = \frac{Z_L - Z_0}{Z_L + Z_0} \tag{2.39}$$

(5) Solution in terms of sending end current, I_S, and load-end voltage, V_L:

In this case, the boundary conditions are,

$$I(0) = I_S, \text{ at } z = 0 \text{ and } V(L) = V_L, \text{ at } z = L,$$

giving
$$V(z) = \frac{1}{\cosh \gamma L}[V_L \cosh \gamma z + Z_0 I_S \sinh \gamma l], \tag{2.40}$$

$$I(z) = \frac{1}{Z_0 \cosh \gamma L}[Z_0 I_S \cosh \gamma l - V_L \sinh \gamma z], \tag{2.41}$$

and $$Z(z) = Z_0\left[\frac{V_L \cosh\gamma z + Z_0 I_S \sinh\gamma l}{-V_L \sinh\gamma z + Z_0 I_S \cosh\gamma l}\right]. \tag{2.42}$$

(6) Solution in terms of sending end current, I_S, and load-end current, I_L:

In this case, boundary conditions are,

$$I(0) = I_S, \text{ at } z = 0 \text{ and } I(L) = I_L, \text{ at } z = L,$$

yielding $$V(z) = \frac{Z_0}{\sinh\gamma L}[I_S \cosh\gamma l - I_L \cosh\gamma z], \tag{2.43}$$

$$I(z) = \frac{1}{\sinh\gamma L}[I_S \sinh\gamma l + I_L \sinh\gamma z], \tag{2.44}$$

and $$Z(z) = Z_0\left[\frac{I_S \cosh\gamma l - I_L \cosh\gamma z}{I_S \sinh\gamma l + I_L \sinh\gamma z}\right] \tag{2.45}$$

(7) Solution in terms of sending end current, I_S, and load impedance, Z_L:

In this case, the boundary conditions are,

$$I(0) = I_S, \text{ at } z = 0 \text{ and } Z(L) = Z_L, \text{ at } z = L,$$

yielding $$V(l) = Z_0 I_S \frac{Z_L \cosh\gamma l + Z_0 \sinh\gamma l}{Z_L \sinh\gamma L + Z_0 \cosh\gamma L} = I_S Z_0\left[\frac{e^{\gamma l} + \rho_L e^{-\gamma l}}{e^{\gamma L} - \rho_L e^{-\gamma L}}\right], \tag{2.46}$$

$$I(l) = I_S \frac{Z_L \sinh\gamma l + Z_0 \cosh\gamma l}{Z_L \sinh\gamma L + Z_0 \cosh\gamma L} = I_S\left[\frac{e^{\gamma l} - \rho_L e^{-\gamma l}}{e^{\gamma L} - \rho_L e^{-\gamma L}}\right], \tag{2.47}$$

and $$Z(l) = Z_0\left[\frac{Z_L \cosh\gamma l + Z_0 \sinh\gamma l}{Z_0 \cosh\gamma l + Z_L \sinh\gamma l}\right] = Z_0\left[\frac{Z_L + Z_0 \tanh\gamma l}{Z_0 + Z_L \tanh\gamma l}\right] \tag{2.48}$$

$$= Z_0\left[\frac{e^{\gamma l} + \rho_L e^{-\gamma l}}{e^{\gamma l} - \rho_L e^{-\gamma l}}\right] = Z_0\left[\frac{1 + \rho_L e^{-2\gamma l}}{1 - \rho_L e^{-2\gamma l}}\right]. \tag{2.49}$$

(8) Solution in terms of load-end voltage, V_L, and current, I_L :

In this case, the boundary conditions are,

$$V(L) = V_L, \text{ and } I(L) = I_L, \text{ at } z = L,$$

giving $$V(l) = V_L \cosh\gamma l + Z_0 I_L \sinh\gamma l = \left(\frac{Z_L + Z_0}{2Z_L}\right) V_L\,[e^{\gamma l} + \rho_L e^{-\gamma l}], \tag{2.50}$$

$$I(l) = I_L \cosh \gamma l + \frac{V_L}{Z_0} \sinh \gamma l = \left(\frac{Z_L + Z_0}{2Z_0}\right) I_L [e^{\gamma l} - \rho_L e^{-\gamma l}], \tag{2.51}$$

and
$$Z(l) = Z_0 \left[\frac{Z_L \cosh \gamma l + Z_0 \sinh \gamma l}{Z_0 \cosh \gamma l + Z_L \sinh \gamma l}\right] = Z_0 \left[\frac{Z_L + Z_0 \tanh \gamma l}{Z_0 + Z_L \tanh \gamma l}\right] \tag{2.52}$$

$$= Z_0 \left[\frac{e^{\gamma l} + \rho_L e^{-\gamma l}}{e^{\gamma l} - \rho_L e^{-\gamma l}}\right] = Z_0 \left[\frac{1 + \rho_L e^{-2\gamma l}}{1 - \rho_L e^{-2\gamma l}}\right]. \tag{2.53}$$

(9) Solution in terms of generator voltage, V_G, and load impedance, Z_L:

In this case, the boundary conditions are,

$$Z(L) = Z_L, \text{ at } z = L,$$

and
$$V_S = \frac{Z_S}{Z_G + Z_S} V_G. \tag{2.54}$$

Taking the exponential forms of Eqns.(2.35), (2.36) and setting $l = L$ in Eqn. (2.38), gives

$$V(l) = V_S \left[\frac{e^{\gamma l} + \rho_L e^{-\gamma l}}{e^{\gamma L} + \rho_L e^{-\gamma L}}\right], \tag{2.55}$$

$$I(l) = \frac{V_S}{Z_0} \left[\frac{e^{\gamma l} - \rho_L e^{-\gamma l}}{e^{\gamma L} + \rho_L e^{-\gamma L}}\right], \tag{2.56}$$

and
$$Z(L) = Z_S = Z_0 \left[\frac{e^{\gamma L} + \rho_L e^{-\gamma L}}{e^{\gamma L} - \rho_L e^{-\gamma L}}\right] \tag{2.57}$$

Substitution of Eqn. (2.57) into Eqn. (2.54) gives

$$V_S = \left(\frac{Z_0}{Z_0 + Z_G}\right) V_G \left[\frac{e^{\gamma L} + \rho_L e^{-\gamma L}}{e^{\gamma L} - \rho_G \rho_L e^{-\gamma L}}\right] \tag{2.58}$$

Substitution of the expression for V_S from Eqn. (2.58) into Eqns.(2.55) and (2.56), gives

$$V(l) = \left(\frac{Z_0}{Z_0 + Z_G}\right) V_G \left[\frac{e^{\gamma l} + \rho_L e^{-\gamma l}}{e^{\gamma L} - \rho_G \rho_L e^{-\gamma L}}\right], \tag{2.59}$$

and
$$I(l) = \left(\frac{V_G}{Z_0 + Z_G}\right) \left[\frac{e^{\gamma l} - \rho_L e^{-\gamma l}}{e^{\gamma L} - \rho_G \rho_L e^{-\gamma L}}\right] \tag{2.60}$$

Dividing Eqn. (2.59) by Eqn. (2.60), gives the expression for the line impedance to be

$$Z(l) = Z_0 \frac{e^{\gamma l} + \rho_L e^{-\gamma l}}{e^{\gamma l} - \rho_L e^{-\gamma l}} = Z_0 \frac{1 + \rho_L e^{-2\gamma l}}{1 - \rho_L e^{-2\gamma l}}. \tag{2.61}$$

where,

$$\rho_G = \frac{Z_G - Z_0}{Z_G + Z_0}, \tag{2.62}$$

Depending on the requirement or application of the solution, the appropriate one from amongst the various alternative forms given above, may be used.

The variations in the magnitudes of the Line Voltage and Line Current phasors and that of the Line Impedance normalised with respect to the characteristic impedance of the line are given in Figs. 2.11, 2.12 and 2.13, respectively, for a practical line which does have some attenuation along it. Note that all these three quantities are oscillatory in nature – the oscillations being very prominent close to

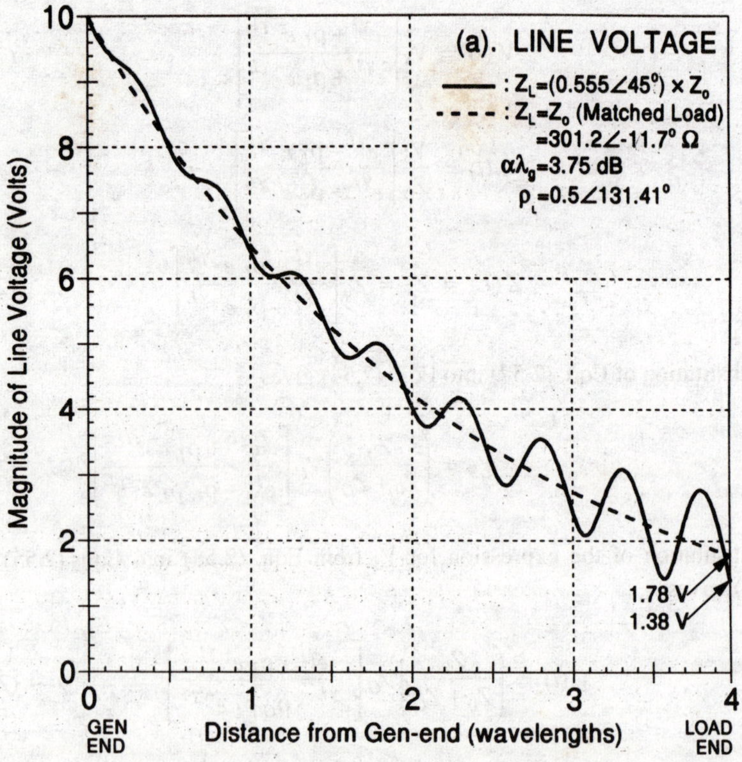

Fig. 2.11 Variation of Voltage along a Line.

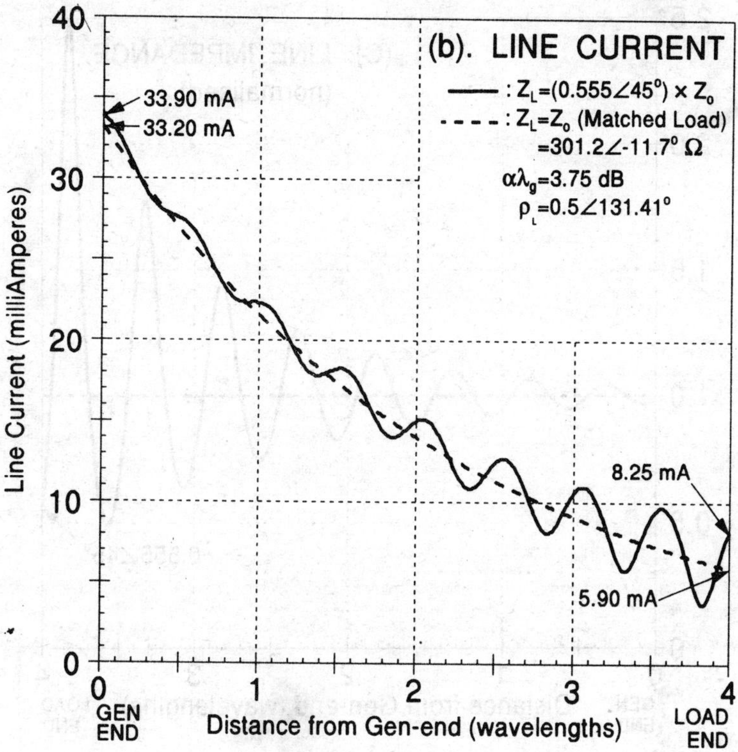

Fig. 2.12 Variation of Current along a Line.

the load-end, and becoming insignificantly small at large distances from the load-end.

2.3 LINE AS AN IMPEDANCE TRANSFORMER

From the various alternative forms of the solution of the transmission line equations, it may be observed that the input impedance, Z_{in}, of a line of a finite length, l, and terminated in an arbitrary impedance, Z_L, [Ref. Fig. 2.14], varies with the length of the line – the relation between Z_{in} and Z_L being given by [Ref. Eqn. (2.53)]

$$Z_{in} = Z(l) = Z_0 \frac{Z_L + Z_0 \tanh \gamma l}{Z_0 + Z_L \tanh \gamma l} \tag{2.63}$$

For a fixed load impedance, Z_L, the input impedance, Z_{in}, of the line at varying distance, l, from the load-end, has been plotted in the complex-Z plane in Fig. 2.15. The line may, thus, be regarded as an **impedance transformer** that transforms the impedance, Z_L, to the impedance, Z_{in}. In other words, an impedance, Z_L, when 'viewed' through a length, l, of a line of characteristic impedance, Z_0, gets

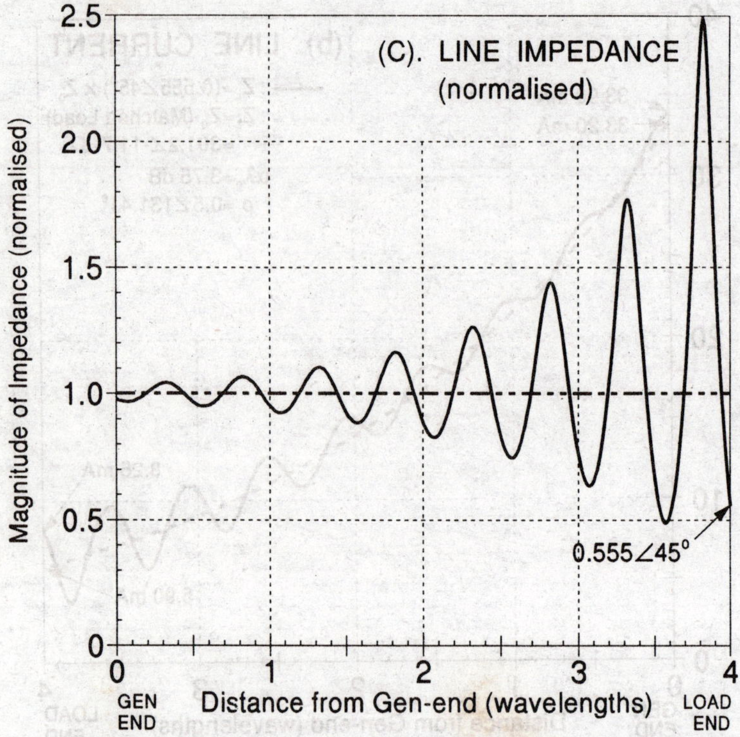

Fig. 2.13 Variation of Input Impedance along Line.

transformed to the impedance, Z_{in}, given by Eqn. (2.63). Yet another way of expressing this impedance transforming property of a line is the following:

A transmission line is a device for effecting or realising the mathematical transformation [Ref. Eqn. (2.53)] of Z_L to Z_{in}. Or, a transmission line of length, l,

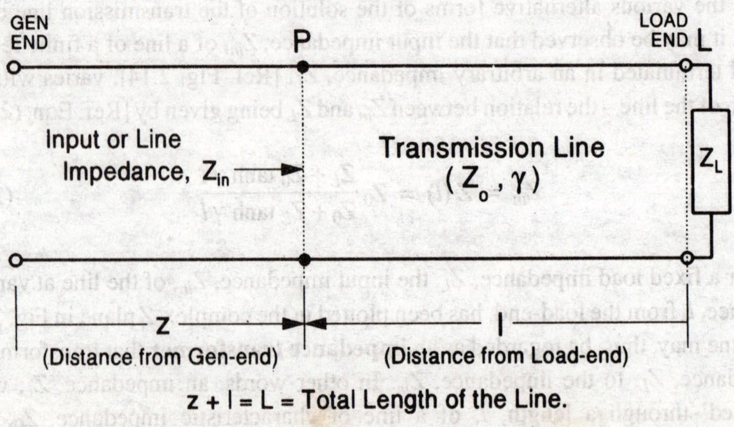

Fig. 2.14 Line as an Impedance Transformer.

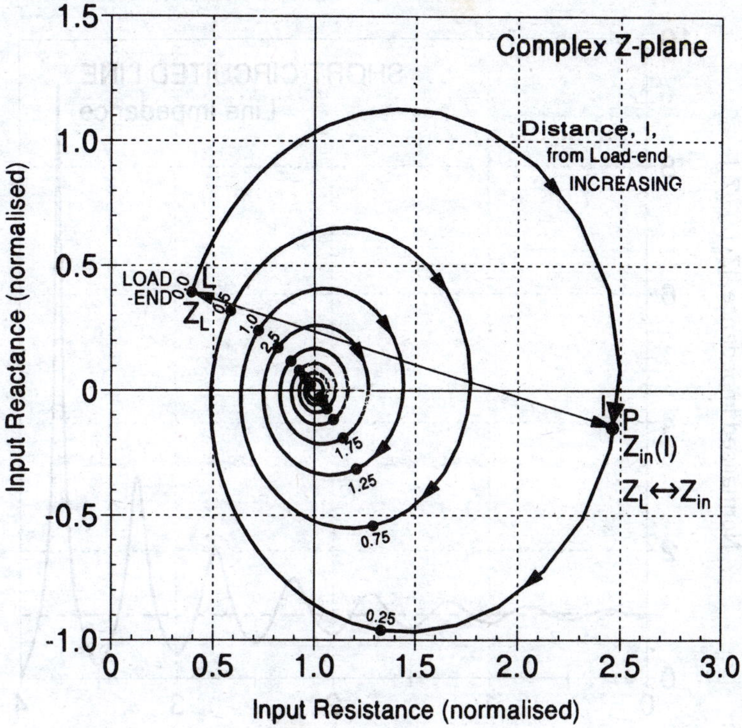

Fig. 2.15 Variation of Input Impedance along Line (Complex Z-plane representation).

maps a point, Z_L, of the complex Z_L-plane onto the point, Z_{in}, of the complex Z_{in}-plane, the transformation or the mapping (bilinear, conformal) being given by Eqn. (2.53).

The following three special cases of interest arise:

Case 1. $Z_L = 0$ (Line short-circuited at the load-end)

Substitution of this equation in Eqn. (2.63) yields,

$$Z_{SC} = Z_{in} = Z(l) = Z_0 \tanh \gamma l, \quad \text{for } Z_L = 0 \tag{2.64}$$

The variation in the magnitude of the Z_{SC}, normalised with respect to the characteristic impedance, Z_0, of the line, as the length, l, is varied, is presented in Fig. 2.16. The variations in the Line Voltage and Current magnitudes are also given in Fig. 2.17.

Case 2. $Z_L \to \infty$ (Line open-circuited at the load-end)

Substitution of this equation in Eqn. (2.63) yields,

$$Z_{OC} = Z_{in} = Z(l) = Z_0 \coth \gamma l, \quad \text{for } Z_L \to \infty \tag{2.65}$$

The variation in the magnitude of the Z_{OC}, normalised with respect to the characteristic impedance, Z_0, of the line, as the length, l, is varied, is presented in

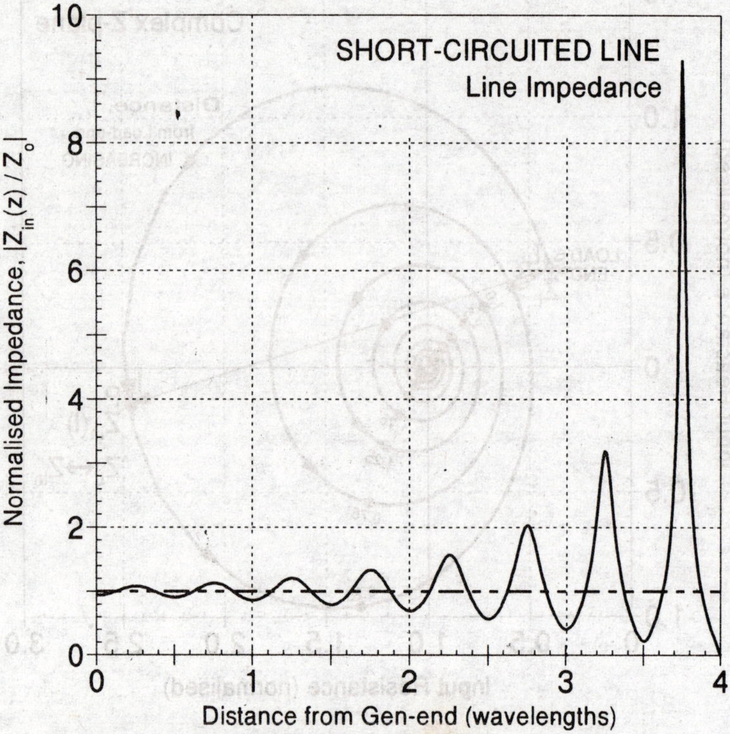

Fig. 2.16 Magnitude of Line Impedance (normalised).

Fig. 2.18. The variations in the Line Voltage and Current magnitudes are also given in Fig. 2.19.

Case 3. $Z_L = Z_0$ (Line terminated in its characteristic impedance)

Substitution of this equation in Eqn. (2.63) yields,

$$Z_{in} = Z(l) = Z_0 = \text{constant for } \textbf{\textit{ALL}} \text{ values of } l. \qquad (2.66)$$

That is, if the transmission line is terminated in its characteristic impedance, the input impedance at all points on the line equals the characteristic impedance itself. In view of this fact, the characteristic impedance of a line may be defined as **that load impedance for which the input impedance of the line equals the load impedance itself, for *ALL LENGTHS*** of the line.

SOLVED EXAMPLES

Example 2.1 (a) Derive the relation between the primary constants of a transmission line so that its characteristic impedance is a pure resistance.

(b) If the characteristic impedance of a line is a pure resistance, the line must be lossless also. Comment on this statement. Support your comments with analysis.

Solution : By definition [Ref. Eqn. (2.10)], the Characteristic Impedance, Z_0, of a transmission line is given by

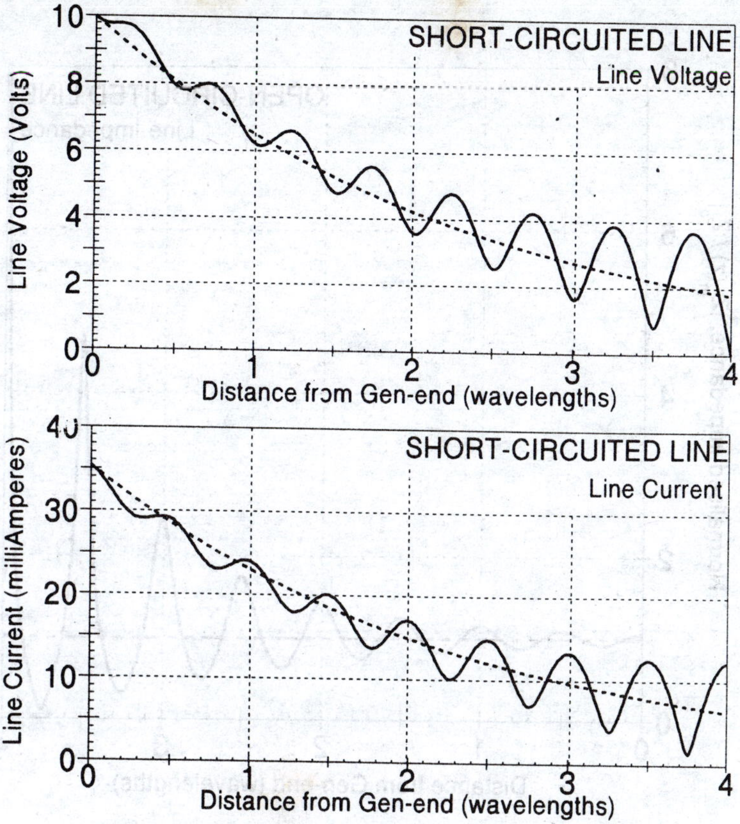

Fig. 2.17 Magnitudes of Line Voltage and Current.

$$Z_0 = \sqrt{\frac{R + j\,\omega L}{G + j\,\omega C}}$$

$$= |\,Z_0\,|\,e^{j\,\phi_0}.$$

where, $$|\,Z_0\,| = \sqrt{\frac{\sqrt{R^2 + \omega^2 L^2}}{\sqrt{G^2 + \omega^2 C^2}}}, = \sqrt[4]{\frac{R^2 + \omega^2 L^2}{G^2 + \omega^2 C^2}},$$

and, $$\phi_0 = \frac{1}{2}\left[\tan^{-1}\left(\frac{\omega L}{R}\right) - \tan^{-1}\left(\frac{\omega C}{G}\right)\right],$$

It is clear that the Characteristic Impedance would be a pure resistance, if ϕ_0 is zero, that is, if

$$\frac{\omega L}{R} = \frac{\omega C}{G} \text{ which gives, } \frac{L}{R} = \frac{C}{G}$$

or if, alternatively, $$\frac{R}{G} = \frac{L}{C}$$

The significance of this equation would become clear in the next chapter.

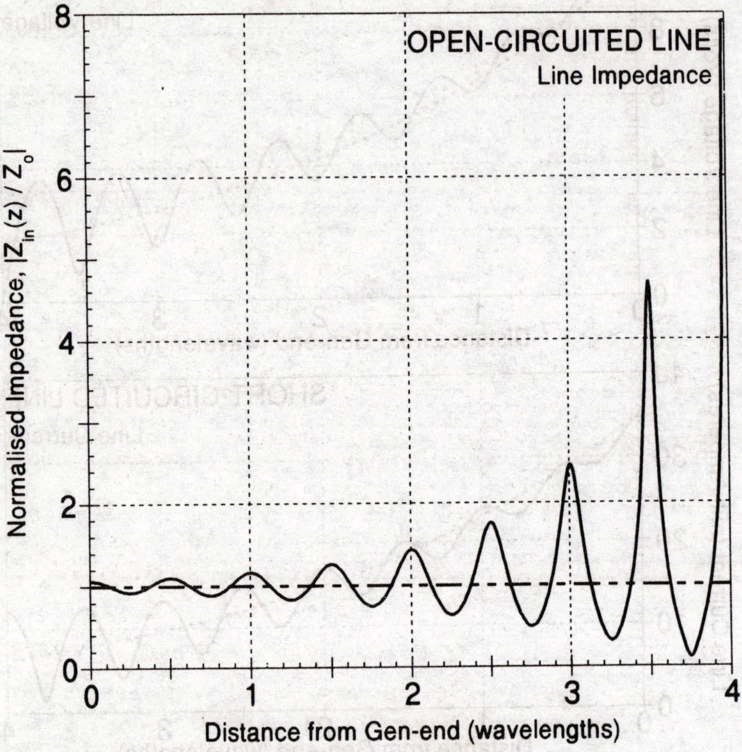

Fig. 2.18 Magnitude of Line Impedance (normalised).

Example 2.2 Input impedances at the points, P_1 and P_2, at distances of l_1 and $l_2 = l_1 + l$, respectively, from the load-end, are equal to $Z_1 = Z(l_1)$ and $Z_2 = Z(l_2)$, respectively. Then, starting from the impedance transformation formula, [Ref. Eqn. (2.63)], show that the impedance, Z_1 and Z_2 are related to each other by the relations,

$$Z_2 = Z_0 \frac{Z_1 + Z_0 \tanh \gamma l}{Z_0 + Z_1 \tanh \gamma l}, \text{ and } Z_1 = Z_0 \frac{Z_2 - Z_0 \tanh \gamma l}{Z_0 - Z_2 \tanh \gamma l}, \tag{2.67}$$

Solution : From Eqn. (2.64), the line impedance, Z_2, at the point, P_2, a distance of l_2 towards the generator from the load-end is given by

$$Z_2 = Z(l_2) = Z_0 \frac{Z_L + Z_0 \tanh \gamma l_2}{Z_0 + Z_L \tanh \gamma l_2}, \text{ where, } Z_L = \text{the load-impedance}, \tag{2.68}$$

Now,
$$\tanh \gamma l_2 = \tanh [\gamma (l_1 + l)] = \frac{\tanh \gamma l_1 + \tanh \gamma l}{1 + \tanh \gamma l_1 \tanh \gamma l} \tag{2.69}$$

Substitution of the expression for $\tanh \gamma l_2$ from Eqn. (2.69) into Eqn. (2.68) gives, on simplification,

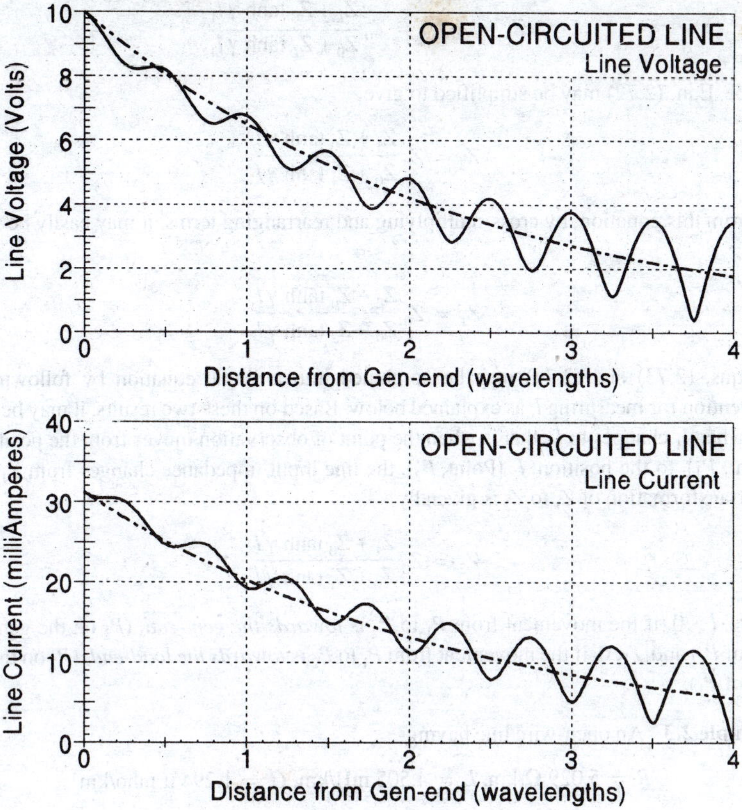

Fig. 2.19 Magnitudes of Line Voltage and Current.

$$Z_2 = Z_0 \frac{Z_L (1 + \tanh \gamma l_1 \tanh \gamma l) + Z_0 (\tanh \gamma l_1 + \tanh \gamma l)}{Z_0 (1 + \tanh \gamma l_1 \tanh \gamma l) + Z_L (\tanh \gamma l_1 + \tanh \gamma l)},$$

Regrouping the terms in the numerator and denominator of the right hand side of the above equation gives

$$Z_2 = Z_0 \frac{(Z_L + Z_0 \tanh \gamma l_1) + (Z_0 + Z_L \tanh \gamma l_1) \tanh \gamma l}{(Z_0 + Z_L \tanh \gamma l_1) + (Z_L + Z_0 \tanh \gamma l_1) \tanh \gamma l} \tag{2.70}$$

$$= Z_0 \frac{\dfrac{Z_L + Z_0 \tanh \gamma l_1}{Z_0 + Z_L \tanh \gamma l_1} + \tanh \gamma l}{1 + \dfrac{Z_L + Z_0 \tanh \gamma l_1}{Z_0 + Z_L \tanh \gamma l_1} \tanh \gamma l} \tag{2.71}$$

$$= Z_0 \frac{\dfrac{Z_1}{Z_0} + \tanh \gamma l}{1 + \dfrac{Z_1}{Z_0} \tanh \gamma l} \tag{2.72}$$

because, from Eqn. (2.68) by changing subscripts ‘2‘ to ‘1’, it may be observed that

$$Z_1 = Z(l_1) = Z_0 \frac{Z_L + Z_0 \tanh \gamma l_1}{Z_0 + Z_L \tanh \gamma l_1},$$

Hence, Eqn. (2.72) may be simplified to give

$$Z_2 = Z_0 \frac{Z_1 + Z_0 \tanh \gamma l}{Z_0 + Z_1 \tanh \gamma l} \tag{2.73}$$

From this equation, by cross-multiplying and rearranging terms, it may easily be shown that

$$Z_1 = Z_0 \frac{Z_2 - Z_0 \tanh \gamma l}{Z_0 - Z_2 \tanh \gamma l}. \tag{2.74}$$

Eqns. (2.73) and (2.74) may be combined into a single equation by following the convention for measuring l, as explained below. Based on these two results, it may be stated that when l_1 changes to l_2, that is when the point of observation moves from the position, l_1 [Point, P_1], to the position, l_2 [Point, P_2], the line input impedance changes from Z_1 to Z_2. The transformation of Z_1 to Z_2 is given by

$$Z_2 = Z_0 \frac{Z_1 + Z_0 \tanh \gamma l}{Z_0 + Z_1 \tanh \gamma l},$$

where, $l > 0$, if the movement from P_1 to P_2 is *towards the gen.-end*, (P_2 on the generator side of P_1) and, $l < 0$, if the movement from P_1 to P_2 is *towards the load-end*, (P_2 on the load side of P_1).

Example 2.3 An open-wire line having

$$R = 5.029 \ \Omega/\text{km}, L = 4.505 \ \text{mH/km}, G = 1.294 \ \mu \ \text{mho/km}$$

and $$C = 13.490 \ \text{m} \ \mu \ \text{F/km},$$

is used to convey power at a frequency of 440 Hz to a load at a distance of 247.587 km. Calculate the following:

(1) characteristic impedance, Z_0, of the line,
(2) total attenuation in dB of the line,
(3) phase constant in deg/km along the line, and
(4) the wavelength along the line.

Solution : For $f = 440$ Hz, angular frequency, $\omega = 2.7646 \times 10^3$ radians/sec.

$$R + j\omega L = (5.0290 + j\,12.4545) \ \Omega/\text{km} = 13.4315 \ \underline{/68.012^\circ} \ \Omega/\text{km},$$

$$G + j\omega C = (1.2940 + j\,37.2945) \ \mu\text{mhos/km} = 37.3169 \ \underline{/88.013^\circ} \ \mu\text{mhos/km}.$$

Substitution of these into Eqns. (2.10) and (2.6), gives, at $f = 440$ Hz.

$$Z_0 = (590.8275 - j\,104.1848) \ \Omega = 599.9430 \ \underline{/-10.001^\circ} \ \Omega,$$

and $$\gamma = (0.00465 + j\,0.02190)/\text{km} = 0.022388 \ \underline{/78.012^\circ} \ /\text{km}.$$

Hence, the attenuation and phase constants of the line at $f = 440$ Hz are

$$\alpha = 0.00465 \ \text{Nepers/km} = 0.00465 \times 8.686 \ \text{dB/km} = 0.04039 \ \text{dB/km}$$

$$\beta_g = 0.02190 \ \text{radian/km} = 0.02190 \times \frac{180}{\pi} \ \text{degrees/km} = 1.25477 \ \text{degrees/km}.$$

From these, the total attenuation of the line (end-to-end) and the wavelength along the line λ_g, are obtained as

Total line attenuation $= \alpha\, l = 0.0404 \times 247.587$ dB $= 10.000$ dB,

and, the wavelength, $\lambda_g = \dfrac{2\,\pi}{\beta_g} = \dfrac{2\,\pi}{0.02190}$ km $= 286.906$ km.

EXERCISES

1. If a quantity, $\rho\,(l)$, at a distance of l from the load-end is defined by the equation,

$$\rho\,(l) = \frac{Z\,(l) - Z_0}{Z\,(l) + Z_0}$$

then, prove that $\rho_1 = \rho(l_1)$ and $\rho_2 = \rho(l_2)$ are related to each other by the relations, $\rho_2 = \rho_1\, e^{-2\gamma l}$ and $\rho_1 = \rho_2\, e^{+2\gamma l}$, where $l_2 = l_1 + l$.
Hence, it may be concluded that if ρ is known to be equal to ρ_1 at a point, $P_1\,(l_1)$, then its value, ρ_2 at a point, $P_2(l_2)$, at a distance, l, from P_1, may be obtained by multiplying ρ_1 by the factor, $e^{-2\gamma l}$, l being regarded positive, if the movement is towards the generator-end, that is, if the point, P_2, is on the generator side of the point, P_1.

2. A 10.383 km long open-wire line has the following parameters:

$R = 26.160\ \Omega$/km, $L = 0.621$ mH/km, $G = 0.932\ \mu$ mho/km

and $C = 38.526$ m μ F/km.

Calculate the characteristic impedance, attenuation constant in dB/km and the phase constant is deg/km, at a frequency of 1 kHz. What is the velocity of propagation of electromagnetic waves along this line?

[Ans. $Z_0 = 330.552/\!\underline{-40.648°}$, $\alpha = 0.455$ dB/km,
$\beta_g = 3.467°$/km and $v_p = 10.384 \times 10^4$ km/sec.]

3. If a load of impedance, $(295.414 - j\,52.090)\ \Omega$, is connected at the load-end of the line of Example 2.3, calculate the input impedances of the line at the following distances from the load-end:

(1) 247.587 km (the generator-end or the input of the line),
(2) 143.453 km, and
(3) 71.727 km. **[Ans.** (1) $Z_{in} = (588.567 - j\,144.297)\ \Omega = 605.998\,/\!\underline{-13.775°}\ \Omega$,
(2) $Z_{in} = (495.456 - j\,87.367)\ \Omega = 503.100\,/\!\underline{-10.000°}\ \Omega$,
and, (3) $Z_{in} = (834.690 - j\,147.196)\ \Omega = 847.570\,/\!\underline{-10.001°}\ \Omega$.]

4. A 62.440 km long telephone line is found to have the following characteristics at a frequency of 1575 Hz:

$Z_0 = 880.028\,/\!\underline{-24.216°}\ \Omega$, $\alpha = 0.103$ dB/km,

$\beta_g = 1.441°$/km, $v_p = 2.4976 \times 10^5$ km/sec.

If the input impedance of this line at the generator-end is measured to be $(933.302 - j\,419.737)\ \Omega$, calculate the impedance of the load connected to the line at its load-end. **[Ans.** $(401.295 - j\,180.484)\ \Omega = 440.014\,/\!\underline{-24.216°}\ \Omega$].

Suggestion : It may be more convenient to work with exponential than with hyperbolic functions. Therefore, first calculate ρ_S [Eqn. (2.28)], then ρ_L [Ref. Ex. 1] and finally, R_L using Eqn. (2.39).

5. A 25.250 km long telephone line is found to have the following characteristics at a frequency of 1 kHz:

$$Z_0 = 184.789 \underline{/-38.976^\circ}\ \Omega,\ \alpha = 0.396 \text{ dB/km},$$

$$\beta_g = 3.204^\circ/\text{km},\ v_p = 1.1237 \times 10^5 \text{ km/sec}.$$

If the input impedance of this line at the generator-end is measured to be $(132.535 - j\,111.885)\ \Omega$, calculate the impedance of the load connected to the line at its load-end. [**Ans.** $(287.314 - j\,232.463)\ \Omega = 369.579 \underline{/-38.976^\circ}\ \Omega$].

6. An 8.381 km long telephone line is found to have the following characteristics at a frequency of 1 kHz:

$$Z_0 = 261.984 \underline{/-41.964^\circ}\ \Omega,\ \alpha = 0.597 \text{ dB/km},$$

$$\beta_g = 4.343^\circ/\text{km},\ v_p = 0.82896 \times 10^5 \text{ km/sec}.$$

If the input impedance of this line at the generator-end is measured to be $(165.862 - j\,223.938)\ \Omega$, calculate the impedance of the load connected to the line at its load-end. [**Ans.** $(389.605 - j\,350.359)\ \Omega = 523.969 \underline{/-41.964^\circ}\ \Omega$].

7. An open-wire line has

$$R = 3.604\ \Omega/\text{km},\ L = 2.200 \text{ mH/km},\ G = 0.621\ \mu \text{ mho/km}$$

and $C = 5.530$ mμF/km,

At a frequency of 1 kHz, calculate the following:

(1) characteristic impedance, Z_0, of the line,

(2) attenuation constant α in dB/km of the line, and the

(3) velocity of propagation along the line.

Hypothetically, if the inductance of the line is increased to $L = 32.094$ mH/km without altering the other three line constants, re-calculate the above-noted characteristics of the line and compare them with their earlier values.

[**Ans.** $Z_0 = 641.079 \underline{/-6.795^\circ}$, $\alpha = 0.026$ dB/km, and $v_p = 2.847 \times 10^5$ km/sec.

For the modified (but hypothetical) line,

$Z_0 = 2409.07 \underline{/-0.000^\circ}$, $\alpha = 0.013$ dB/km, and $v_p = 0.751 \times 10^5$ km/sec.]

Chapter 3

TRANSMISSION LINE PARAMETERS

3.1 INTRODUCTION

It may be recalled that, for a given ***uniform*** transmission line, the per-unit-length-quantities, namely, *R, L, G* and *C* appearing in the transmission line analysis presented in Chapter 2, are constant along the length of the line. As their values are dependent on, and, in principle, may be calculated from, (1) the geometry of the line [Ref. Fig. 3.1], (2) electromagnetic properties of the materials of the two line conductors and the medium in which they are placed, and (3) the frequency of operation, they are called the ***Primary Characteristics*** or ***Primary Constants*** of the line, or, simply the ***Line Parameters.*** Although the concept of line parameters and the theory of transmission lines have been developed, in Chapter 2, for the two-wire or open-wire line yet they are applicable to all transmission lines and, in fact, to all waveguiding structures — the latter, although under restricted conditions. All transmission structures possess all the four parameters though their calculation may be involved and the magnitudes entirely different from structure to structure.

Derived from these four line parameters, are the two ***Secondary Characteristics*** or ***Secondary Constants,*** of the line, Z_0 and γ, given by Eqns.(2.10) and (2.6), respectively. At a given frequency, Z_0 and γ are constant for a uniform transmission line. These are of great practical importance in that they govern the behaviour of a transmission line. That is, Z_0 and γ characterise the transmission line. These are called the ***Characteristic Impedance*** and the ***(Complex) Propagation Constant,*** respectively, of the line. Both the characteristic impedance, Z_0, and propagation constant, γ, are, in general, complex quantities and involve the frequency of

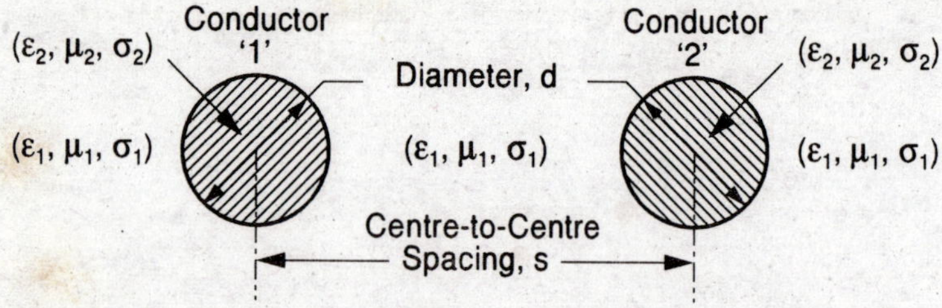

(a) Open-Wire, Two-Wire or Two-Conductor Line.

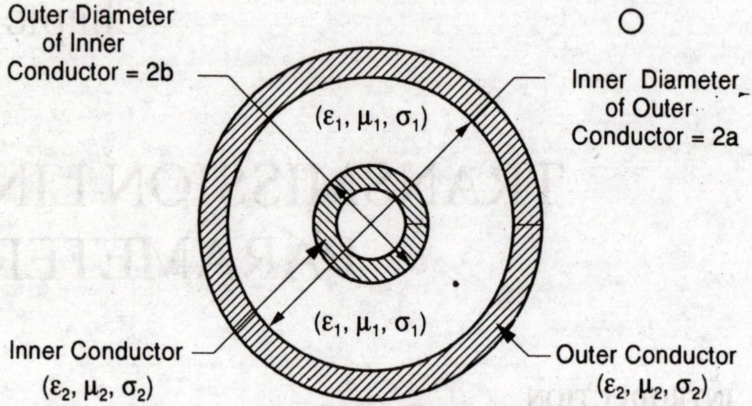

(b) Coaxial Line.

Fig. 3.1 Structure of Open-wire and Coaxial Lines.

operation both explicitly and also implicitly — through the frequency dependence of the line parameters themselves. The explicit variations of these characteristics with the frequency of operation, are discussed later in this chapter.

3.2 LINE PARAMETERS — *R, L, G* AND *C*

As stated above, expressions (available in any text book[1] on electromagnetics) for the primary line parameter, *R, L, G,* and *C* may be obtained in terms of the line geometry and electromagnetic properties of the line and the medium. For ready reference, expressions for the line parameters of the open-wire line and the coaxial line are given in Table 3.1.

---

1 For example,

1. S. Ramo, J.R. Whinnery and T. Van Duzer, *Fields and Waves in Communication Electronics*, John Wiley and Sons, Inc.,

2. J.D. Kraus *Electromagnetics*, Third Ed., McGraw-Hill Book Co.,

3. E.C. Jordan and K.G. Balmain, *Electromagnetic Waves and Radiating Systems*, Prentice-Hall of India Pvt. Ltd.

Table 3.1 Expression for Line Parameters at high frequencies. [Ref. Fig. 3.1]

Parameter	*Open-wire or Two-wire or Parallel-wire line*	*Coaxial line*
Resistance, R (ohms/metre)	$\dfrac{2\sqrt{\dfrac{f\mu_2}{\pi\sigma_2}}}{d\sqrt{1-\left(\dfrac{d}{s}\right)^2}}$	$\dfrac{1}{2}\sqrt{\dfrac{f\mu_2}{\pi\sigma_2}}\cdot\left(\dfrac{1}{a}+\dfrac{1}{b}\right)$
External Inductance, L (henrys/metre)	$\dfrac{\mu_1}{\pi}\cdot\ln\left\{\left(\dfrac{s}{d}\right)+\sqrt{\left(\dfrac{s}{d}\right)^2-1}\right\}$	$\dfrac{\mu_1}{2\pi}\cdot\ln\left(\dfrac{a}{b}\right)$
Conductance, G (mhos/metre)	$\dfrac{\pi\sigma_1}{\ln\left\{\left(\dfrac{s}{d}\right)+\sqrt{\left(\dfrac{s}{d}\right)^2-1}\right\}}$	$\dfrac{2\pi\sigma_1}{\ln\left(\dfrac{a}{b}\right)}$
Capacitance, C (farads/metre)	$\dfrac{\pi\varepsilon_1}{\ln\left\{\left(\dfrac{s}{d}\right)+\sqrt{\left(\dfrac{s}{d}\right)^2-1}\right\}}$	$\dfrac{2\pi\varepsilon_1}{\ln\left(\dfrac{a}{b}\right)}$
NOTES: $\varepsilon_1, \varepsilon_2$ = Permittivities, μ_1, μ_2 = Permeabilities, σ_1, σ_2 = Conductivities, of the Media.	Line having identical conductors having circular cross-sections of Diameters = d, and with the Centre-to-Centre Spacing = s between the two conductors.	a = Inner radius of the outer conductor, b = Outer radius of the inner conductor of the Coaxial line.
Subscript '1' refers to the dielectric medium between the two conductors; and Subscript '2' refers to the material of the two conductors, assumed to be the same for both.		

Note : The formulae for R given in the Table are valid at frequencies, $f >> \dfrac{4}{\pi\,\mu_2\,\sigma_2\,d^2}$, (frequencies at which the *skin depth*, $\delta = \dfrac{1}{\sqrt{\pi\,\mu_2\,\sigma_2 f}} << \dfrac{d}{2}$ or b, as the case may be).

At very low frequencies, R may be calculated using the dc approximation (the bulk resistance of the conductors).

3.3 CHARACTERISTIC IMPEDANCE

The expression for the *characteristic impedance* is [Ref. Eqn.(2.10)]

$$Z_0 = \sqrt{\frac{R + j\,\omega L}{G + j\,\omega C}}, \tag{3.1}$$

so that,
$$|Z_0| = \sqrt[4]{\frac{R^2 + \omega^2 L^2}{G^2 + \omega^2 C^2}}, \tag{3.2}$$

and, $$\phi_0 = \arg(Z_0) = \frac{1}{2}\left[\tan^{-1}\left(\frac{\omega L}{R}\right) - \tan^{-1}\left(\frac{\omega C}{G}\right)\right],$$

$$= \frac{1}{2}\tan^{-1}\left[\frac{\omega(GL - RC)}{RG + \omega^2 LC}\right],$$

$$= \frac{1}{2}\tan^{-1}\left[\frac{\left(\dfrac{Z_{0\infty}}{Z_{00}} - \dfrac{Z_{0,0}}{Z_{0\infty}}\right)}{\left(\dfrac{\omega}{\omega_0} + \dfrac{\omega_0}{\omega}\right)}\right], \text{ where, } \omega_0 = \sqrt{\frac{RG}{LC}}. \tag{3.3}$$

For a typical telephone line, the variation of the characteristic impedance, Z_0, in the complex-Z_0 plane, with frequency, assuming the line parameters, R, L, G and C, to be frequency-independent, is given in Fig. 3.2. Discussed in the following sections are the variations of the magnitude and angle of the characteristic impedance with frequency, assuming the line parameters to be frequency-independent.

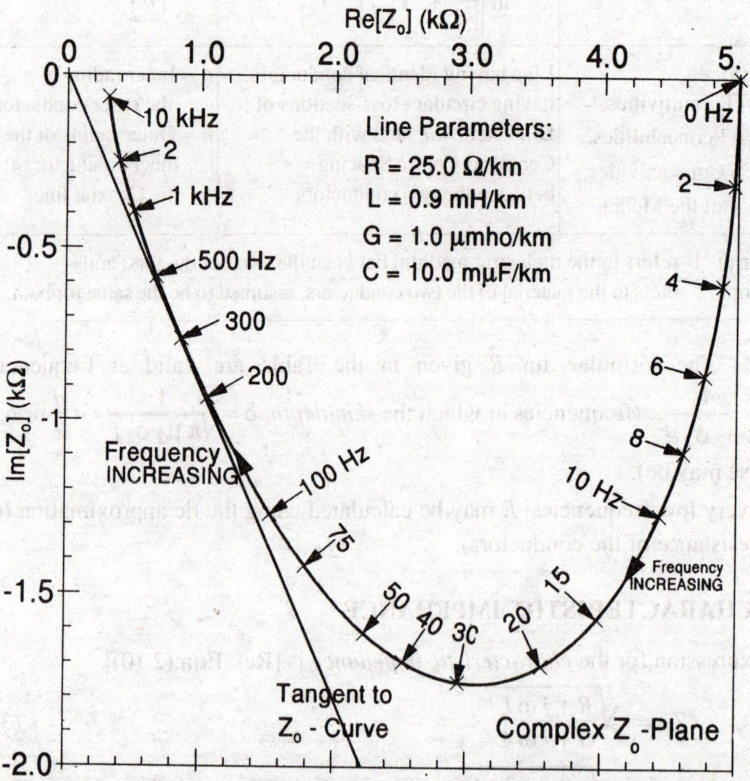

Fig. 3.2 Variation of Z_0 (in the Complex Z_0-plane) with Frequency.

3.3.1 Magnitude of Characteristic Impedance

From, Eqn. (3.2) it follows that

$$\therefore \quad |Z_0|^4 = \frac{R^2 + \omega^2 L^2}{G^2 + \omega^2 C^2}. \tag{3.4}$$

Therefore,
$$\frac{d|Z_0|^4}{d\omega} = \frac{2\omega (G^2 L^2 - R^2 C^2)}{(G^2 + \omega^2 C^2)^2} \tag{3.5}$$

and,
$$\frac{d^2|Z_0|^4}{d\omega^2} = \frac{2(G^2 L^2 - R^2 C^2)(G^2 - \omega^2 C^2)}{(G^2 + \omega^2 C^2)^3} \tag{3.6}$$

From Eqns. (3.4), (3.5) and (3.6), it may be observed that

(i) $|Z_0| = Z_{00} = \sqrt{\frac{R}{G}}$ at $\omega = 0$ and $|Z_0| = Z_{0\infty} = \sqrt{\frac{L}{C}}$ as $\omega \to \infty$.

(ii) $\frac{d|Z_0|^4}{d\omega} = 0$ at $\omega = 0$ and as $\omega \to \infty$

(iii) at $\omega = 0$, $\frac{d^2|Z_0|^4}{d\omega^2} \begin{matrix} < \\ = \\ > \end{matrix} 0$ acc. as $Z_{00} \begin{matrix} > \\ = \\ < \end{matrix} Z_{0\infty}$, i.e. $|Z_0| = \begin{cases}\text{Maximum.} \\ \text{Constant.} \\ \text{Minimum}\end{cases}$

(iv) as $\omega \to \infty$, $\frac{d^2|Z_0|^4}{d\omega^2} \begin{matrix} < \\ = \\ > \end{matrix} 0$ acc. as $Z_{00} \begin{matrix} < \\ = \\ > \end{matrix} Z_{0\infty}$, i.e. $|Z_0| = \begin{cases}\text{Maximum.} \\ \text{Constant.} \\ \text{Minimum}\end{cases}$

(v) For all values of ω, $\frac{d|Z_0|^4}{d\omega} \begin{matrix} < \\ = \\ > \end{matrix}$ acc. as $Z_{00} \begin{matrix} > \\ = \\ < \end{matrix} Z_{0\infty}$.

In view of the observations (i) to (v) made above, the variation of the magnitude of the characteristic impedance, $[Z_0]$, with frequency may be described thus:

Case 1. $Z_{00} > Z_{0\infty}$: [Ref. Fig. 3.3. (Practical lines)]

At $\omega = 0$ (dc), $|Z_0|$ is maximum $= Z_{00} = \sqrt{\frac{R}{G}}$. As the frequency of operation increases, $|Z_0|$ decreases monotonically and asymptotically assumes the value,

$$Z_{0\infty} = \sqrt{\frac{L}{C}}, \text{ when } \omega \to \infty.$$

Case 2. $Z_{00} < Z_{0\infty}$: [Ref. Fig. 3.4 (Fictitious line)]

At $\omega = 0$ (dc), $|Z_0|$ is minimum $= Z_{00} = \sqrt{\frac{R}{G}}$. As the frequency of operation increases, $|Z_0|$ increases monotonically and asymptotically assumes the value,

$$Z_{0\infty} = \sqrt{\frac{L}{C}}, \text{ when } \omega \to \infty.$$

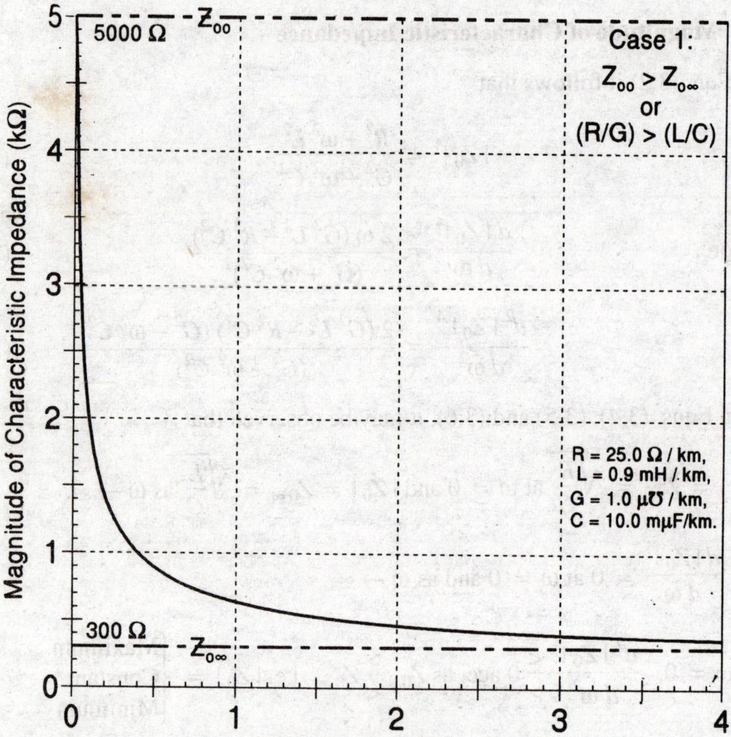

Fig. 3.3 | Z_0 | versus Frequency.
Case 1: $Z_{00} > Z_{0\infty}$ or $(R/G) > (L/C)$.

Case 3. $Z_{0\,0} = Z_{0\,\infty}$:

In this case, $|Z_0|$ remains constant $= Z_{0\,0} = \sqrt{\dfrac{R}{G}} = Z_{0\,\infty} = \sqrt{\dfrac{L}{C}}$, at all frequencies.

3.3.2 Phase Angle of Characteristic Impedance

Turning now to the phase angle, ϕ_0, of the characteristic impedance, the expression for ϕ_0 is: [Ref. Eqn. (3.3)]

$$\phi_0 = \frac{1}{2}\tan^{-1}\left[\frac{\left(\dfrac{Z_{0\,\infty}}{Z_{0\,0}} - \dfrac{Z_{0\,0}}{Z_{0\,\infty}}\right)}{\left(\dfrac{\omega}{\omega_0} + \dfrac{\omega_0}{\omega}\right)}\right] \tag{3.7}$$

which gives,

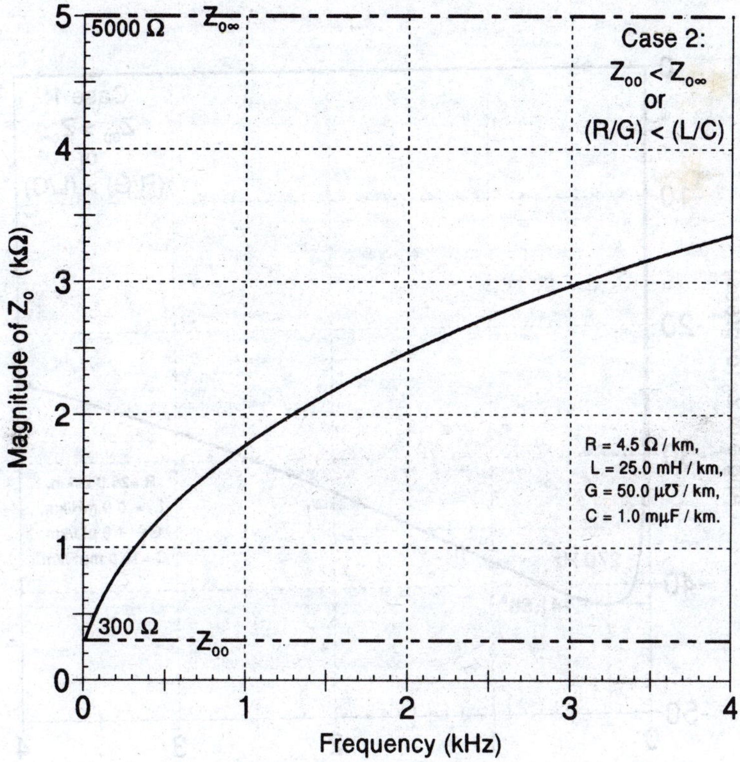

Fig. 3.4 $|Z_0|$ versus Frequency
Case 2: $Z_{00} < Z_{0\infty}$ or $(R/G) < (L/C)$.

$$\frac{d\phi_0}{d\omega} = \frac{\dfrac{1}{2\omega}\left(\dfrac{\omega}{\omega_0} - \dfrac{\omega_0}{\omega}\right)\left(\dfrac{Z_{00}}{Z_{0\infty}} - \dfrac{Z_{0\infty}}{Z_{00}}\right)}{\left(\dfrac{\omega}{\omega_0} + \dfrac{\omega_0}{\omega}\right)^2 + \left(\dfrac{Z_{00}}{Z_{0\infty}} - \dfrac{Z_{0\infty}}{Z_{00}}\right)^2} \tag{3.8}$$

Note that:

(i) $\phi_0 = 0$ at $\omega = 0$ and also as $\omega \to \infty$

(ii) $\dfrac{d\phi_0}{d\omega} = 0$ at $\omega = \omega_0 = \sqrt{\dfrac{RG}{LC}} = \dfrac{G}{C}\dfrac{Z_{00}}{Z_{0\infty}}$, and also as $\omega \to \infty$

(iii) In the range, $0 \le \omega < \omega_0$, $\dfrac{d\phi_0}{d\omega} \lesseqgtr 0$, according as $Z_{00} \gtreqless Z_{0\infty}$,

and in the range, $\omega_0 < \omega < \infty$, $\dfrac{d\phi_0}{d\omega} \gtreqless 0$, according as $Z_{00} \gtreqless Z_{0\infty}$.

From the observations (i), (ii) and (iii) made above, it may be inferred that at $\omega = 0$ (dc), $\phi_0 = 0$ and as ω increases [Ref. Fig. 3.5 for Case 1 and Fig. 3.6 for

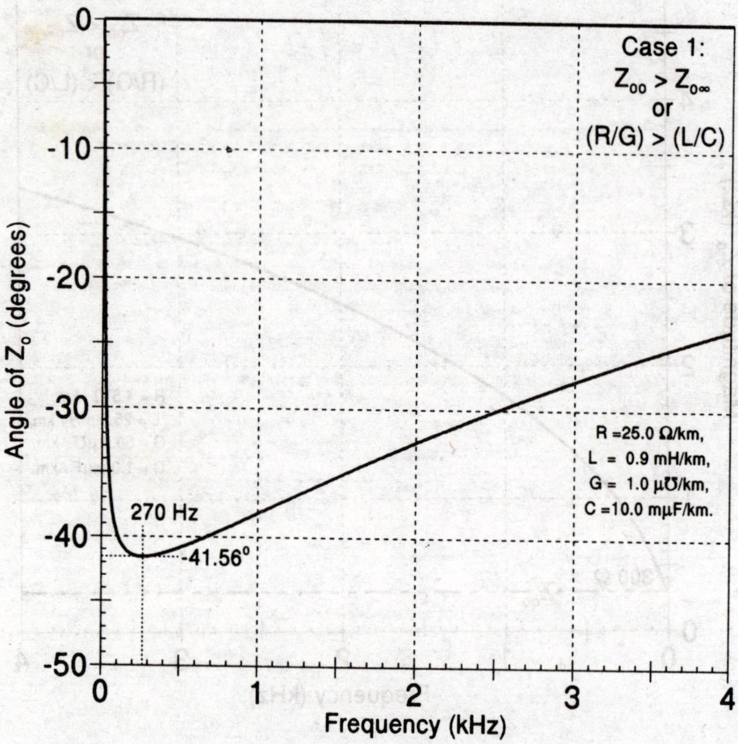

Fig. 3.5 Angle of Characteristic Impedance versus Frequency.
Case 1: $Z_{00} > Z_{0\infty}$ or $(R/G) > (L/C)$.

Case 2], ϕ_0 decreases (or increases) until it reaches its minimum (or maximum) value, $\phi_{0_{extr}}$ given by

$$\phi_{0_{extr}} = \frac{1}{2}\tan^{-1}\left[\frac{1}{2}\left(\sqrt{\frac{GL}{RC}} - \sqrt{\frac{RC}{GL}}\right)\right]$$

$$= \frac{1}{2}\tan^{-1}\left[\frac{1}{2}\left(\frac{Z_{0\,\infty}}{Z_{0\,0}} - \frac{Z_{0\,0}}{Z_{0\,\infty}}\right)\right], \tag{3.9}$$

at $$\omega = \omega_0 = \sqrt{\frac{RG}{LC}} = \frac{G}{C}\frac{Z_{0\,0}}{Z_{0\,\infty}}. \text{ [Ref. Eqn. 3.3)]} \tag{3.10}$$

As ω increases, further, ϕ_0 increases (or decreases) and approaches the value zero asymptotically as $\omega \to \infty$.

All practical lines of interest are low loss lines, so that $Z_{0\,0} > Z_{0\,\infty}$ and hence the variation of $|Z_0|$ and of ϕ_0 with ω is similar to that shown in Fig. 3.7, wherein frequency is plotted on a linear scale, and also in Fig. 3.8 wherein logarithmic scale has been used for frequency.

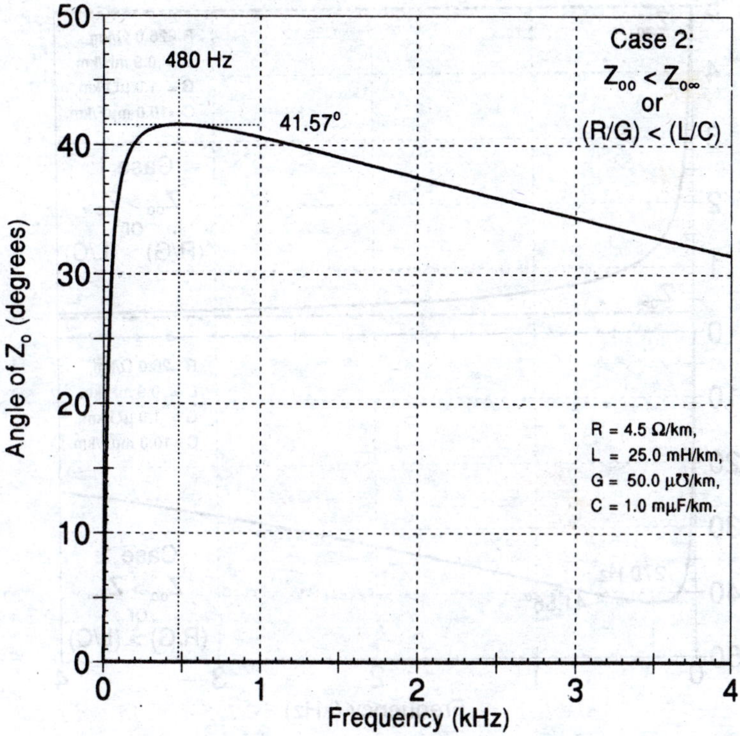

Fig. 3.6 Angle of Characteristic Impedance versus Frequency.
Case 2: $Z_{00} < Z_{0\infty}$ or $(R/G) < (L/C)$.

3.4 PROPAGATION CONSTANT

The other secondary characteristic of a transmission line that is of practical importance, is its ***propagation constant***. The expression for the propagation constant is [Ref. Eqn. (2.6)]

$$\gamma = \sqrt{(R + j\,\omega\,L)\,(G + j\,\omega\,C)}\,. \tag{3.11}$$

In general, γ is a complex number and may be expressed as $\gamma = \alpha + j\,\beta_g$, where α and β_g are real quantities called the ***attenuation constant*** and the ***phase constant***, respectively, of the line.

Now, $$\gamma = \alpha + j\,\beta_g = \sqrt{(R + j\,\omega\,L)\,(G + j\,\omega\,C)}$$

$$= \sqrt{(RG - \omega^2\,LC) + j\,\omega(RC + LG)} \tag{3.12}$$

so that

$$\alpha = \frac{1}{\sqrt{2}}\sqrt{\sqrt{(RG - \omega^2\,LC)^2 + \omega^2\,(RC + LG)^2} + (RG - \omega^2\,LC)}\,, \tag{3.13}$$

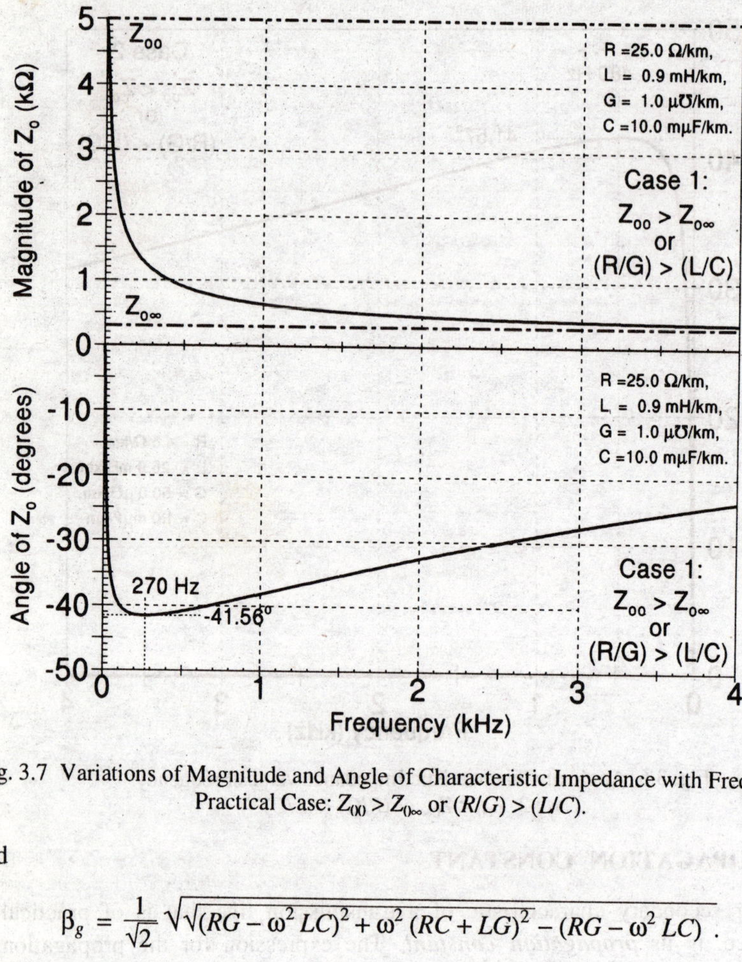

Fig. 3.7 Variations of Magnitude and Angle of Characteristic Impedance with Frequency Practical Case: $Z_{00} > Z_{0\infty}$ or $(R/G) > (L/C)$.

and

$$\beta_g = \frac{1}{\sqrt{2}} \sqrt{\sqrt{(RG - \omega^2 LC)^2 + \omega^2 (RC + LG)^2} - (RG - \omega^2 LC)}\ . \quad (3.14)$$

For a typical telephone line, the variation of the propagation constant, γ, in the complex γ-plane, with frequency, assuming the line parameters, R, L, G and C, to be frequency-independent, is given in Fig. 3.9. Variations in the quantities, $|\gamma|$, $\underline{/\gamma}$, α and β_g, with the frequency of operation, are presented, Fig. 3.10. Discussed in the following sections are the variations of the real and imaginary parts of the propagation constant with frequency, assuming the line parameters to be frequency-independent.

3.4.1 Attenuation Constant

From Eqn. (3.13), the attenuation constant, α, of the line is given by

$$\alpha = \frac{1}{\sqrt{2}} \sqrt{\sqrt{(RG - \omega^2 LC)^2 + \omega^2 (RC + LG)^2} + (RG - \omega^2 LC)}\ . \quad (3.15)$$

At very low frequencies, $\omega << \omega_l$, defined below, it may be shown that the variation of the attenuation constant, α, with the frequency of operation, is given by

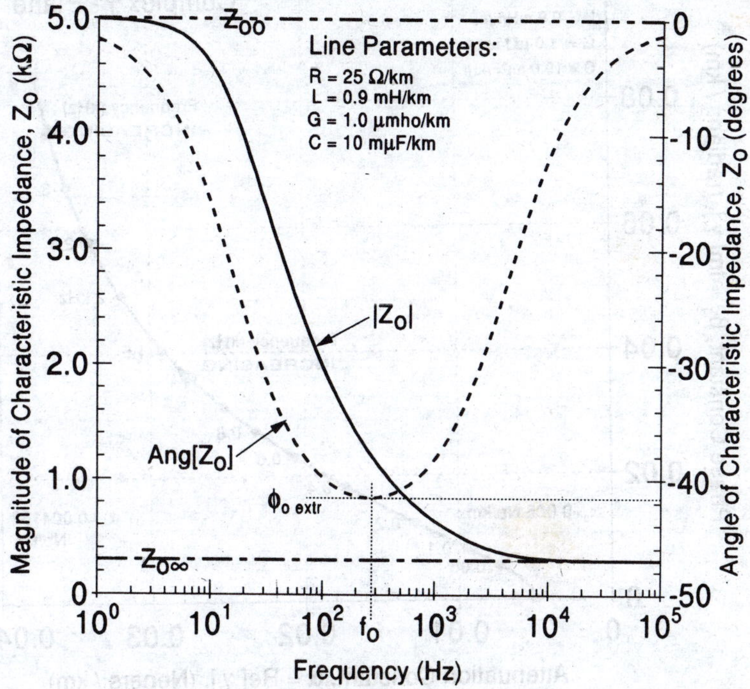

Fig. 3.8 Variations of the Magnitude and Angle of the Characteristic Impedance of Practical Transmission Line with Frequency.

$$\alpha = \sqrt{RG}\left[1 + \frac{1}{8}\left(\frac{Z_{00}}{Z_{0\infty}} - \frac{Z_{0\infty}}{Z_{00}}\right)^2 \left(\frac{\omega}{\omega_0}\right)^2 + \ldots\right], \text{ for } \omega << \omega_l, \tag{3.16}$$

where,

$$\omega_l = \frac{2\sqrt{2}\,\omega_0}{\left|\dfrac{Z_{00}}{Z_{0\infty}} - \dfrac{Z_{0\infty}}{Z_{00}}\right|}, \tag{3.17}$$

and,

$$\omega_0 = \sqrt{\frac{RG}{LC}}. \tag{3.18}$$

And, at frequencies, $\omega >> \omega_h$, defined below, the variation of the attenuation constant, α, with the frequency of operation may be shown to be given by

$$\alpha = \frac{1}{2}\left(\frac{R}{Z_{0\infty}} + GZ_{0\infty}\right)\left[1 - \frac{1}{8}\left(\frac{Z_{00}}{Z_{0\infty}} - \frac{Z_{0\infty}}{Z_{00}}\right)^2 \left(\frac{\omega_0}{\omega}\right)^2 + \ldots\right] \text{ for } \omega >> \omega_h, \tag{3.19}$$

where,

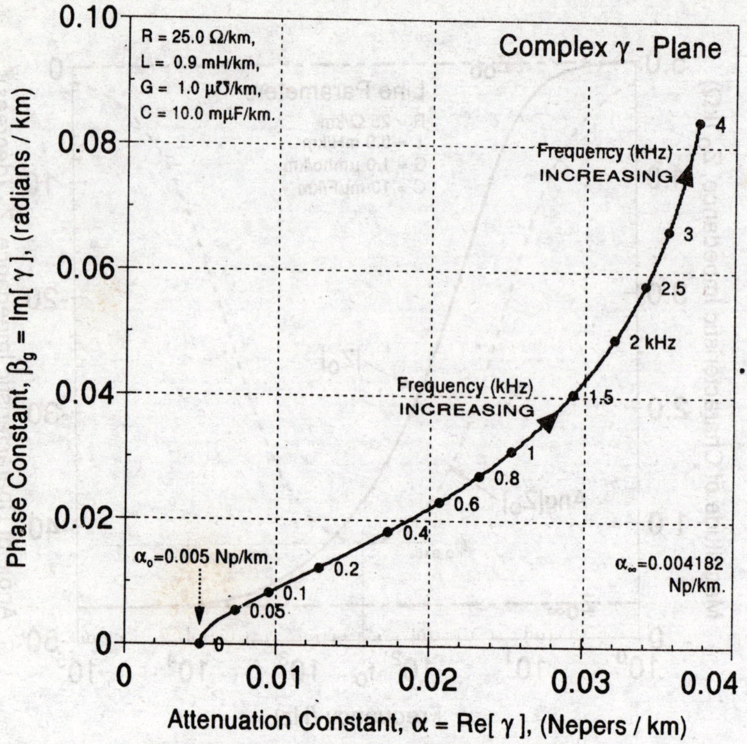

Fig. 3.9 Propagation Constant vs Frequency.
(in the Complex γ-plane).

$$\omega_h = \frac{\omega_0}{2\sqrt{2}}\left|\frac{Z_{00}}{Z_{0\infty}} - \frac{Z_{0\infty}}{Z_{00}}\right| \tag{3.20}$$

From Eqn. (3.16), it may be noted that

at $$\omega = 0, \alpha = \alpha_0 = \sqrt{RG}. \tag{3.21}$$

Also, from Eqn. (3.19), it is clear that

as $$\omega \to \infty, \alpha = \alpha_\infty = \frac{1}{2}\left(\frac{R}{Z_{0\infty}} + GZ_{0\infty}\right) \geq \alpha_0, \tag{3.22}$$

where,

$$Z_{00} = \sqrt{\frac{R}{G}}. \tag{3.23}$$

= the characteristic impednace of the line at $\omega = 0$

$$Z_{0\infty} = \sqrt{\frac{L}{C}} \tag{3.24}$$

= the characteristic impedance of the line at frequencies, $\omega \to \infty$

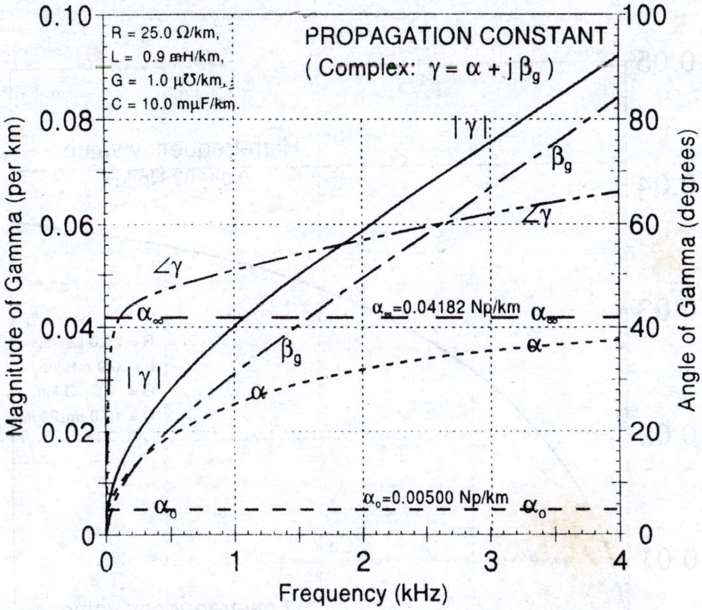

Fig. 3.10 Complex Propagation Constant vs Frequency

Based on the discussion given above, variation of α with frequency may be described thus: Starting from the value $\alpha_0 = \sqrt{RG}$ at $\omega = 0$, the attenuation constant, α, increases rapidly, initially, as the frequency is increased; and later the rate of increase diminishes, although remaining positive always, as the attenuation constant asymptotically approaches the value, $\alpha_\infty = \frac{1}{2}\left[\frac{R}{Z_{0\infty}} + GZ_{0\infty}\right]$. Fig. 3.11 depicts the variation of the attenuation constant, α, with frequency.

3.4.2 Phase Constant

From Eqn. (3.14), the phase constant, β_g, of the line is given by

$$\beta_g = \frac{1}{\sqrt{2}} \sqrt{\sqrt{(RG - \omega^2 LC)^2 + \omega^2 (RC + LG)^2} + (RG - \omega^2 LC)} \,. \quad (3.25)$$

At very low frequencies, $\omega << \omega_l$ [Ref. Eqn. (3.17)], it may be shown that the variation of the phase constant, β_g, with the frequency of operation, is given by

$$\beta_g = \frac{1}{2}\left(\frac{Z_{00}}{Z_{0\infty}} + \frac{Z_{0\infty}}{Z_{00}}\right) \omega \sqrt{LC} \left[1 - \frac{1}{8}\left(\frac{Z_{00}}{Z_{0\infty}} - \frac{Z_{0\infty}}{Z_{00}}\right)^2 \left(\frac{\omega_0}{\omega}\right)^2 + \ldots\right],$$

$$\text{for } \omega << \omega_l. \quad (3.26)$$

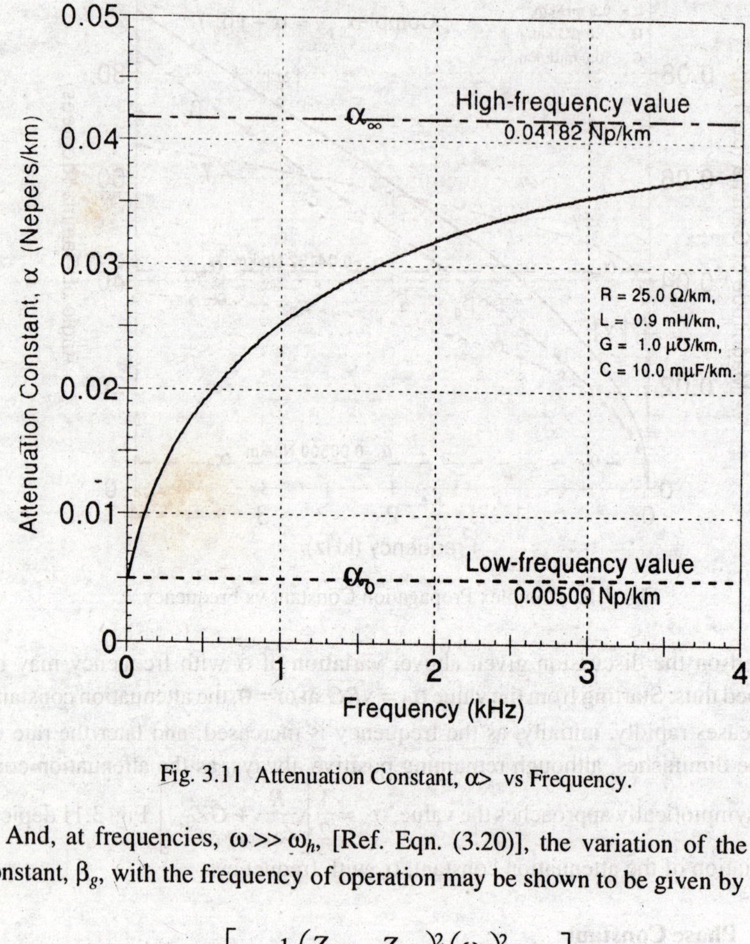

Fig. 3.11 Attenuation Constant, α vs Frequency.

And, at frequencies, $\omega >> \omega_h$, [Ref. Eqn. (3.20)], the variation of the phase constant, β_g, with the frequency of operation may be shown to be given by

$$\beta_g = \omega\sqrt{LC}\left[1+\frac{1}{8}\left(\frac{Z_{00}}{Z_{0\infty}}-\frac{Z_{0\infty}}{Z_{00}}\right)^2\left(\frac{\omega_0}{\omega}\right)^2+\ldots\right] \text{ for } \omega >> \omega_h. \quad (3.27)$$

From Eqn. (3.26), it may be noted that

at $$\omega = 0,\ \beta_g = \beta_{g0} = 0 \quad (3.28)$$

In the immediate vicinity of $\omega = 0$ [Ref. Eqn. (3.26)], β_g increases linearly with ω as

$$\beta_g = \frac{1}{2}\left(\frac{Z_{00}}{Z_{0\infty}}+\frac{Z_{0\infty}}{Z_{00}}\right)\omega\sqrt{LC}.$$

Also, from Eqn. (3.27), it is clear that

as $$\omega \to \infty, \beta_g = \beta_{g\infty} \to \infty. \tag{3.29}$$

However, at high frequencies, $\omega >> \omega_h$, [Ref. Eqn. (3.27)], β_g increases linearly with ω as

$$\beta_g = \omega\sqrt{LC} \tag{3.30}$$

Based on the above discussion variation of β_g with frequency may be described thus: Starting from the value $\beta_{g0} = 0$ at $\omega = 0$, the phase constant, β_g, increases initially rapidly — the slope of the $(\beta_g - \omega)$ curve being equal to $\frac{1}{2}\left(\frac{Z_{00}}{Z_{0\infty}} + \frac{Z_{0\infty}}{Z_{00}}\right)\sqrt{LC} > \sqrt{LC}$. As the frequency is increased, the slope decreases, and very soon, when $\omega >> \omega_l$, it attains the value of $\sqrt{LC}$. Figure 3.12 depicts the variation of the phase-constant, β_g, with frequency.

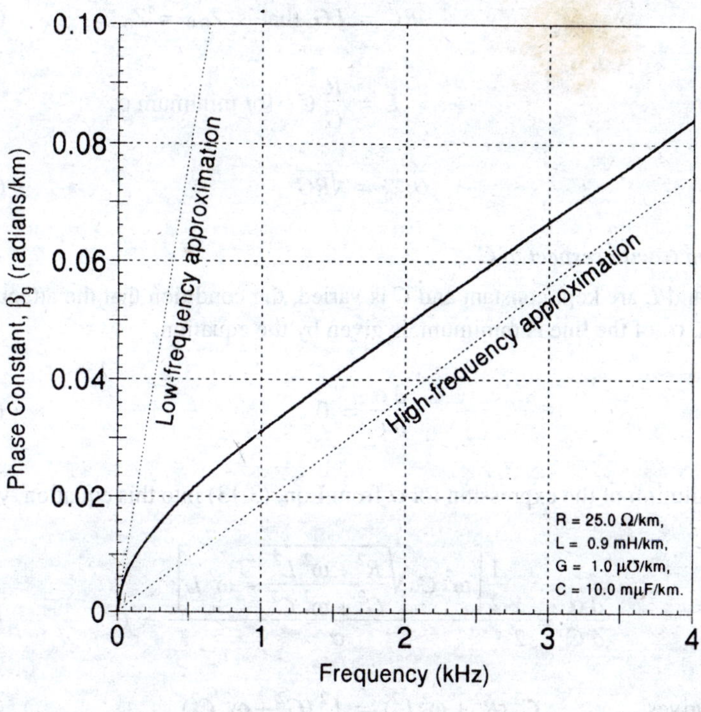

Fig. 3.12 Phase Constant, β_g, vs Frequency.

3.5 CONDITIONS FOR MINIMUM ATTENUATION

It may be interesting to determine the condition that the line parameters must satisfy in order that the line attenuation is minimum at any given frequency and under various constraints.

Minimum α with respect to L

If R, G and C are held constant and L is varied, the condition that the attenuation constant, α, of the line is minimum, is given by the equation,

$$\frac{\partial \alpha}{\partial L} = 0 \tag{3.31}$$

Substitution of the expression for α from Eqn. (3.13) into this equation, yields

$$\frac{\partial \alpha}{\partial L} = \frac{1}{2}\,\frac{\frac{1}{2}\left[\omega^2 L \sqrt{\dfrac{G^2+\omega^2 C^2}{R^2+\omega^2 L^2}} - \omega^2 C\right]}{\alpha} = 0 \tag{3.32}$$

This gives, $$C^2 (R^2 + \omega^2 L^2) = L^2 (G^2 + \omega^2 C^2)$$

Or, $$RC = LG, \text{ that is, } Z_{0\,0} = Z_{0\,\infty} \tag{3.33}$$

Therefore, $$L = \frac{R}{G} C \quad \text{for minimum } \alpha.$$

and, $$\alpha_{min} = \sqrt{RG} \tag{3.34}$$

Minimum α with respect to C

If R, G and L are kept constant and C is varied, the condition that the attenuation constant, α, of the line is minimum, is given by the equation,

$$\frac{\partial \alpha}{\partial C} = 0 \tag{3.35}$$

Substitution of the expression for α from Eqn. (3.13) into this equation, yields

$$\frac{\partial \alpha}{\partial C} = \frac{1}{2}\,\frac{\frac{1}{2}\left[\omega^2 C \sqrt{\dfrac{R^2+\omega^2 L^2}{G^2+\omega^2 C^2}} - \omega^2 L\right]}{\alpha} = 0 \tag{3.36}$$

This gives, $$C^2 (R^2 + \omega^2 L^2) = L^2 (G^2 + \omega^2 C^2)$$

Or, $$RC = LG, \text{ that is, } Z_{0\,0} = Z_{0\,\infty} \tag{3.37}$$

Therefore, $$C = \frac{G}{R} L \quad \text{for minimum } \alpha,$$

and, $$\alpha_{min} = \sqrt{RG} \tag{3.38}$$

Minimum α with respect to R and G

α would be minimum with respect to R and G, if $\frac{\partial \alpha}{\partial R}$ and $\frac{\partial \alpha}{\partial G}$ are zero, respectively. Using the expression for α from Eqn. (3.13), these two conditions give,

$$\frac{\partial \alpha}{\partial R} = \frac{1}{2} \frac{\frac{1}{2}\left[R\sqrt{\frac{G^2+\omega^2 C^2}{R^2+\omega^2 L^2}}+G\right]}{\alpha} = 0 \tag{3.39}$$

and,

$$\frac{\partial \alpha}{\partial G} = \frac{1}{2} \frac{\frac{1}{2}\left[G\sqrt{\frac{R^2+\omega^2 L^2}{G^2+\omega^2 C^2}}+R\right]}{\alpha} = 0 \tag{3.40}$$

Note that in the numerators of $\frac{\partial \alpha}{\partial R}$ and $\frac{\partial \alpha}{\partial G}$, both terms are always positive because both R and G are non-negative quantities (physically realisable). Therefore, neither $\frac{\partial \alpha}{\partial R}$ nor $\frac{\partial \alpha}{\partial G}$ can become zero **unless both** R and G are zero.

Hence, only if **both** R and G are zero, α has a minimum and this minimum value of $\alpha = \alpha_{min} = 0$

However, if $G \neq 0$, then it may be noted that as R is decreased, α decreases and attains its least (not minimum) value at $R = 0$. The least value of α for $R = 0$ is given by

$$\alpha_{least} = \sqrt{\frac{\omega L}{2}}\sqrt{\sqrt{G^2+\omega^2 C^2}-\omega C}$$

And, the least value of α for $G = 0$ is given by

$$\alpha_{least} = \sqrt{\frac{\omega C}{2}}\sqrt{\sqrt{R^2+\omega^2 L^2}-\omega L},$$

These two cases, particularly that of $R = 0$, are of academic interest only as these would never be found in practical two-conductor transmission lines.

In view of the above discussion, it may be concluded that a transmission line has minimum attenuation at a given frequency, if the line parameters are related by the following equation:

$$RC = LG \text{ or } Z_{00} = Z_{0\infty}.$$

The minimum value of the attenuation constant is given by

$$\alpha_{min} = \sqrt{RG} \tag{3.41}$$

CAUTION : The above formula may lead one to believe that $\alpha_{min} = 0$ even if only one of the two parameters, R and G, that is, even if $R = 0$ but $G \neq 0$, or $G = 0$ but $R \neq 0$. But, this cannot be true because based on purely physical or philosophical reasoning, one cannot expect attenuation to be zero, if there exists in the system, even one dissipative element like R or G! Indeed, it is not true mathematically also because, in both of these cases, the condition,

$$RC = LG, \text{ that is, } \frac{R}{G} = \frac{L}{C},$$

or $$Z_{00} = Z_{0\infty}, \text{ [Ref. Eqns. (3.33) and (3.37)]} \tag{3.42}$$

which has been derived from Eqn. (3.32) and (3.36) and which yields the formula, Eqn. (3.34), for α_{min}, cannot itself be satisfied by non-trivial values of L and C. If both R and G are zero, then, of course, the above condition is satisfied and the formula, Eqn. (3.34), holds good, giving $\alpha_{min} = 0$, as already discussed above.

3.6 EXPERIMENTAL DETERMINATION

As mentioned earlier Z_0 and γ are of great practical importance because these govern the properties of the transmission line. Both are, in general, complex quantities, explicitly dependent on the frequency of operation. For a given line, at a given frequency, these may be determined experimentally by measuring the input impedance of the line for two different values of the load impedance, say, zero (short-circuit) and infinity (open-circuit), as explained below:

Consider a uniform transmission line of known length, l. Let the measured values of its input impedance by Z_{SC} when the load-end is short-circuited and let it be Z_{OC} when the load-end is open-circuited. [Ref. Fig. 3.13)

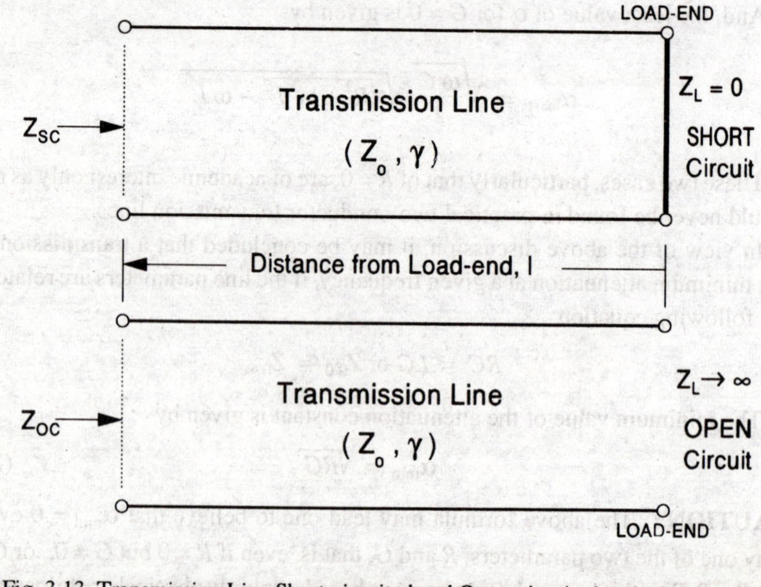

Fig. 3.13 Transmission Line Short-circuited and Open-circuited at the Load-end.

NOTE: In practice, whereas a near-perfect short-circuit may be obtained quite easily, obtaining an open-circuit ($Z_L \to \infty$) poses a problem. If the load-end terminals are left *open* or *unconnected*, the resulting load impedance may although be very large yet not infinity. Perhaps, better results may be obtained by simulating an *open-circuit* by employing a short-circuited quarterwave line. A short-circuited quarterwave low-loss line, presents at its input terminals an impedance of

$$\frac{Z_0}{\alpha \dfrac{\lambda_g}{4}}$$

This is certainly the method that must be used for lossless ($\alpha = 0$) lines. Notwithstanding the difficulty in the realisation of an open circuit, the theory of the method for experimental determination of line parameters is discussed below:

From Eqn. (2.63), it is clear that [Ref. Eqns. (2.64) and 2.65)]

$$Z_{SC} = Z_0 \tanh \gamma l, \tag{3.43}$$

and,

$$Z_{OC} = Z_0 \coth \gamma l. \tag{3.44}$$

From Eqns. (3.43) and (3.44), it is clear that

$$Z_0 = \sqrt{Z_{OC} Z_{SC}}, \tag{3.45}$$

and,

$$\tanh \gamma l = \sqrt{\frac{Z_{SC}}{Z_{OC}}} \tag{3.46}$$

Therefore, if Z_{OC} and Z_{SC} are determined or measured for a given line experimentally, at a given frequency, from Eqns. (3.45) and (3.46) the characteristic impedance Z_0 and $\tanh \gamma l$ are readily obtained.

Remembering that $\gamma = \alpha + j \beta_g$, consider

$$\tanh (\gamma l + \gamma^* l) = \frac{\tanh \gamma l + \tanh \gamma^* l}{1 + \tanh \gamma l \,.\, \tanh \gamma^* l} \tag{3.47}$$

This gives,

$$\tanh 2 \alpha l = \frac{2 \,\text{Re}\, [\tanh \gamma l]}{1 + |\tanh \gamma l|^2} = a \,(say) \tag{3.48}$$

Now consider,

$$\tanh (\gamma l - \gamma^* l) = \frac{\tanh \gamma l - \tanh \gamma^* l}{1 - \tanh \gamma l \,.\, \tanh \gamma^* l} \tag{3.49}$$

which gives,

$$\tan 2 \beta_g l = \frac{2 \,\text{Im}\, [\tanh \gamma l]}{1 - |\tanh \gamma l|^2} = b \,(say) \tag{3.50}$$

From Eqn. (3.48)

$$\alpha = \frac{1}{2\,l}\tanh^{-1}(a) = \frac{1}{4\,l}\ln\left[\frac{1+a}{1-a}\right] \tag{3.51}$$

Hence, the attenuation constant, α can be obtained.

The most general solution of Eqn. (3.50) is

$$2\,\beta_g\,l = k\,\pi + \tan^{-1}(b),\ \text{where } k = 0, 1, 2, \ldots \tag{3.52}$$

As '*k*' may be given any integral value, Eqn. (3.52) does not given a unique value for the phase constant, β_g. Therefore some more information is required to remove the ambiguity or to determine '*k*' and , hence, the exact value of β_g. This additional information may be a knowledge of anyone of the following:

(i) An approximate value of the phase constant, β_g or

(ii) the measured value of the wavelength along the line, or

(iii) the ***roundtrip time*** of the wave for the given line which is the time the waves take to go to the far-end of the line and come back. This may be determined by measuring the time delay (***echo time***) after which the echo of a pulse sent along the line from the generator-end, arrives at the generator-end.

From the Item (ii) or (iii) an approximate value of β_g may be computed. Hence, from the additional information provided in the Items (i), (ii) or (iii) above, an approximate value of $2\,\beta_g\,l = \theta_{\text{approx}}$ may be calculated. Then an appropriate value of '*k*' is chosen such that when this value of '*k*' is substituted in Eqn. (3.52), the value of $2\beta_g\,l = \theta_{exact}$ obtained approximates the value of θ_{approx}. Hence, the exact value of $\beta_g = \dfrac{\theta_{exact}}{2\,l}$ is obtained. In this way, $\gamma = \alpha + j\,\beta_g$ is computed and then noting that

$$\gamma\,Z_0 = R + j\,\omega\,L, \tag{3.53}$$

and

$$\frac{\gamma}{Z_0} = G + j\,\omega\,C, \tag{3.54}$$

the primary constants of the line, *R, L, G* and *C*, may be obtained from the real and imaginary parts of $\gamma\,Z_0$ and $\dfrac{\gamma}{Z_0}$, as given below:

$$\left.\begin{aligned} R &= \text{Re}\,[\gamma\,Z_0] \\ L &= \frac{1}{\omega}\,\text{Im}\,[\gamma\,Z_0] \\ G &= \text{Re}\left[\frac{\gamma}{Z_0}\right] \\ C &= \frac{1}{\omega}\,\text{Im}\left[\frac{\gamma}{Z_0}\right] \end{aligned}\right\} \tag{3.55}$$

SOLVED EXAMPLES

Example 3.1 At a frequency of 1 kHz, the open-circuited and short-circuited impedance of a 63.054 km length of a line are measured to be

$$Z_{0C} = 258.560 \angle -34.240^\circ\ \Omega \text{ and } Z_{SC} = 1931.498 \angle -8.979^\circ\ \Omega, \text{ respectively.}$$

Calculate the value of the characteristic impedance, Z_0, attenuation constant, α in dB/km and other primary constants of the line, given that the phase velocity of waves along the line is 27.9×10^4 km/sec approximately.

Solution : The calculation of the characteristic impedance is quite straightforward. It is given by

$$Z_0 = \sqrt{Z_{0C} Z_{SC}} = \sqrt{258.560\, e^{-j\,34.240^\circ} \,.\, 1931.498\, e^{j\,8.979^\circ}}\ \Omega = 706.688 \angle -12.630^\circ\ \Omega$$

The calculation of the propagation constant is illustrated below:

$$\tanh(\gamma l) = \sqrt{\frac{Z_{SC}}{Z_{0C}}} = \sqrt{\frac{1931.498\, e^{j8.979^\circ}}{258.560\, e^{-j\,34.240^\circ}}}\ /\text{km}$$

$$= 2.7332 \angle 21.610^\circ/\text{km} = (2.5411 + j1.0066)/\text{km}.$$

From Eqn. (3.48),

$$\tanh(2\alpha l) = \frac{2 \operatorname{Re}[\tanh \gamma l]}{1 + |\tanh \gamma l|^2} = \frac{2 \times 2.5411}{1 + (2.7332)^2} = 0.6000 = a, \text{ say}$$

Now, from Eqn. (3.51), α may be calculated easily as

$$\alpha = \frac{1}{2l} \tanh^{-1}(a) = \frac{1}{4l} \ln\left[\frac{1+a}{1-a}\right] = \frac{1}{4 \times 63.054} \ln\left[\frac{1+0.6000}{1-0.6000}\right] \qquad (3.56)$$

$$(3.57)$$

$$= 0.005496 \text{ Nepers/km.} = 0.005496 \times 8.686 \text{ dB/km.} \qquad (3.58)$$

$$= 0.047742 \text{ dB/km.} \qquad (3.59)$$

For the calculation of β_g, use of Eqn. (3.50) gives,

$$\tanh 2\beta_g l = \frac{2 \operatorname{Im}[\tanh \gamma l]}{1 - |\tanh \gamma l|^2} = \frac{2 \times 1.0066}{1 - (2.7332)^2} = -0.311 = b, \text{ say} \qquad (3.60)$$

$$(3.61)$$

Therefore, $2\beta_g l = \tan^{-1} b = -17.2830^\circ + 180^\circ k$, where, $k = 0, 1, 2, \ldots$ (3.62)

To remove the ambiguity in the determination of the value of β_g, additional information is required. In the present example, approximate value of the velocity of propagation is given. From such additional bit of information (in whatever form it may have been given), ***for accurate results, approximate value of*** **2** β_g l **(NOT** β_g**),** ***must be calculated*** and the value of k chosen such that the value of $2\beta_g l$ obtained above matches its approximate value calculated now. This is illustrated below:

It is given that the velocity of propagation, $v_p = 2.79 \times 10^5$ km/sec approximately.

Therefore, approximate value of

$$2\,\beta_g\,l \;=\; 2\,\frac{2\,\pi f}{v_p}\,l \;=\; 2\times\frac{360\times 1000}{2.79\times 10^5}\times 63.054 \text{ deg.} \;=\; 162.720 \text{ deg. approximately.}$$

Comparing it with the calculated value of $2\,\beta_g\,l$ above, it is noted that if $k = 1$, then the calculated value becomes,

$$2\,\beta_g\,l \;=\; -\,17.2830 + 180 = 162.717 \text{ degrees.}$$

This compares well with the approximate value supplied. Hence, the **exact value** (which is the calculated value agreeing well with the approximate value supplied), is

Exact value of $\quad 2\,\beta_g\,l \;=\; 162.7170$ degrees.

This gives $\quad \beta_g \;=\; 1.2903$ degrees/ km $\;=\; \dfrac{\pi}{180}\times 1.2903$ radian/km.

$$= 0.022520 \text{ radian/km}$$

Hence, $\quad \gamma \;=\; (0.005496 + j0.022520)\text{/km} \;=\; 0.023181\;\underline{/76.284^\circ}\text{ /km}$

Having determined Z_0 and γ, the calculation of the primary constants is straight forward as given by Eqn. (3.55), giving, in the present case,

$$R \;=\; 7.270\ \Omega\text{/km},\ L \;=\; 2.336\text{ mH/km},\ G = 0.622\ \mu\text{ mho/km and } C \;=\; 5.220\text{ m}\,\mu\text{ F/km},$$

EXERCISES

1. Two No.8 (3.264 mm diameter) copper wires, stretched parallel to each other in air with a centre-to-centre spacing of 15.240 cm, are used to form a parallel-wire transmission line.

Calculate the primary constants of the line at very low frequencies as well as at high frequencies.

[Take conductivity of copper to be $\sigma_{Cu} = 5.75\times 10^7\ \Omega$/metre]

2. A telephone cable is made by twisting together two No. 16 (1.295 mm diameter) paper-insulated copper wires (paper thickness = 0.15 mm and relative permittivity of 3.5).

Calculate the inductance and capacitance of the cable per km length of the cable.

3. (a) Design an open-wire transmission line for operation at 100 MHz employing two 0.5 mm diameter wires, embedded in a dielectric of relative permittivity of 2.1. The characteristic impedance is required to be $Z_{0\,\infty} = 292.0\ \Omega$.

(b) If the dielectric is now replaced by air, what would be the characteristic impedance of the line?

4. A coaxial line for use at UHF employs a copper wire of 6.189 mm diameter as its inner conductor and a copper tube of 22.225 mm inner diameter and 1.588 mm wall thickness, as its outer conductor. Calculate the inductance and capacitance per unit length of the line and also its characteristic impedance when the dielectric filling up the annular space between the inner and the outer conductor is air and also when it is polyethylene (a dielectric of relative permittivity of 2.3).

5. (a) Design a coaxial transmission line for operation in the UHF band of frequencies, having a characteristic impedance of $Z_{0\,\infty} = 73.43\ \Omega$. Use a hollow copper tube of 1 mm wall thickness and an inner diameter of 8.50 mm.

(b) Redesign the line, if the annular space between the inner and the other conductors is required to be filled up with polyethylene of relative permittivity of 2.3.

(c) Calculate the characteristic impedances of the lines designed above, if the line of (a) above has polyethylene as the dielectric in place of air; and the line of (b) has air as the dielectric in place of polyethylene.

6. The angle of the characteristic impedance of a transmission line varies with the frequency of operation. When the angle of the characteristic impedance attains its extremum with respect to the frequency of operation, assuming the line constants to be independent of the frequency of operation, what can you say about the angle of its propagation constant?

7. A transmission line has a given load impedance connected at its load-end. Determine the distance from the load-end (towards the generator) at which
(1) the magnitude of the angle of the input line impedance is maximum.
(2) the magnitude of the input line reactance is maximum.
(3) the angle of the input line impedance is 45°
(4) the angle of the input line impedance is zero.

8. For a lossless ($\alpha = 0$) line, $\frac{\lambda_g}{8}$ long, and terminated in an arbitrary impedance, $Z_L = (R_L + jX_L)$, determine the condition that R_L and X_L must satisfy in order that the input impedance of the line is a pure resistance.

9. Carry out an analysis to prove that the magnitude of the input impedance, Z_{OC}, of any given line (α and β_g arbitrarily prescribed) when its load-end is open-circuited, equals the magnitude of its input impedance, Z_{SC}, when its load-end is short-circuited, if the length of the line is one eighth of the wavelength. Further, show that the magnitude of the normalised input impedance is unity.

10. Results of measurements made on a 38.821 km long line at a frequency of 1 kHz, are. (a) $Z_{OC} = 816.200 \underline{/-54.113^\circ}\ \Omega$, (*b*) $Z_{SC} = 1453.543 \underline{/-7.77^\circ}\ \Omega$ and (c) the approximate phase velocity along the line is 22.100×10^4 km/sec.
Calculate all the constants/parameters of the line.

[**Ans.** $Z_0 = 1089.212 \underline{/-30.941^\circ}\ \Omega$, $\gamma = 0.033602 \underline{/57.903^\circ}$ per km, $\alpha = 0.699$ dB/km, $\beta_g = 4.940°$/km, $R = 32.622\ \Omega$/km, $L = 2.641$ mH/km, $G = 0.622\ \mu$ mho/km, $C = 4.909$ m μ F/km.]

11. Measurements made at 1 kHz at the input of a 8.932 km length of a transmission line yield

$$Z_{0C} = (160.995 - j\,461.848)\ \Omega \text{ and } Z_{SC} = (436.517 - j\,115.406)\ \Omega$$

Further, the round-trip time for a pulse to travel along the line to the far-end and return to the sending-end is measured to be 240 μ sec approximately.
Calculate the characteristic impedance, attenuation constant in dB/km, the phase constant in degrees/km and the primary constants of the line. Also calculate the quantities, $\sqrt{\frac{R}{G}}$ and $\sqrt{\frac{L}{C}}$, for the line.

[**Ans.** $Z_0 = 469.934 \underline{/-42.796^\circ}\ \Omega$, $\gamma = 0.113753 \underline{/46.984^\circ}$ per km, $\alpha = 0.674$ dB/km, $\beta_g = 4.765°$/km, $R = 53.314\ \Omega$/km, $L = 0.621$ mH/km, $G = 0.933\ \mu$ mho/km, $C = 38.525\ m\,\mu$ F/km, $Z_{00} = 7559.269\ \Omega$, $Z_{0\infty} = 162.962\ \Omega$.]

12. The open-circuited and short-circuited impedance of a 161.25 km line are observed to be

1208.000 $\underline{/-47.294^\circ}\ \Omega$ and 310.579 $\underline{/-36.599^\circ}\ \Omega$, respectively, at $f = 1$ kHz.

Calculate the value of its characteristic impedances at $f = 1$ kHz and also at extremely high frequencies. What is its attenuation in dB/km?

[**Ans.** $Z_0 = 612.519 \angle -5.348^\circ\ \Omega$, $\gamma = 0.021882 \angle 84.363^\circ$ per km, $\alpha = 0.0187$ dB/km, $Z_{0\infty} = 606.885\ \Omega$.]

13. Measurements made on a 7.949 km length of a transmission line (D8 cable, twisted, dry, unloaded) at 1000 Hz yield the following:

$$Z_{0C} = 508.468 \angle -41.261^\circ\ \Omega \text{ and } Z_{SC} = 689.436 \angle -26.379^\circ\ \Omega$$

Calculate the values of the characteristic impedance, attenuation constant in dB/km and the phase constant, given that the approximate value of the phase constant is 8.75°/km.
[**Ans.** $Z_0 = 592.078 \angle -33.820^\circ\ \Omega$, $\alpha = 1.258$ dB/km, $\beta_g = 8.789$°/km]

14. At a frequency of 1 kHz, the input impedance of a 73.385°/km long line when its far-end is open-circuited, is found to be 233.061 $\angle -4.081^\circ\ \Omega$, and equal to 2050.847 $\angle -19.418^\circ\ \Omega$, when the far-end is short-circuited.
Calculate the characteristic impedance of the line. Is it a lossy line? If it is so, determine the length of the line whose total loss is 10 dB.

[**Ans.** $Z_0 = 691.356 \angle -11.750^\circ\ \Omega$, YES, $\alpha = 0.041$dB/km, $l_{10\,dB} = 243.778$ km.]

15. A transmission line has the following primary constants:

$$R = 25.0\ \Omega/\text{km},\ L = 0.9\ \text{mH/km},\ G = 1.0\ \mu\text{mho/km},\ C = 10.0\ \text{m}\mu\text{F/km}.$$

For this line calculate the following quantities:

(1) $Z_{0\,0}$; Z_0 at $f = 83.882$ Hz, $f = 265.258$ Hz, $f = 838.820$ Hz; and $Z_{0\infty}$.

(2) the frequency, f_0, at which the angle, $\phi_{0_{extr}}$, of the characteristic impedance, attains its extremum. Value of (1) $\phi_{0_{extr}}$, and (2) $|Z_0|$.

(3) upper limit f_l, of the low frequency approximation region ($f << f_l$) [Ref. Eqn. (3.18)].

(4) lower limit, f_h, of high frequency approximation region ($f >> f_h$)[Ref .Eqn .(3.20)].

[**Ans.**(1) $Z_{0\,0} = 5000\ \Omega$;

$Z_0 = 2158.961 \angle -39.085^\circ\ \Omega$, $1224.745 \angle -41.566^\circ\ \Omega$, $694.779 \angle -39.085^\circ\ \Omega$;

$Z_{0\infty} = 300\ \Omega$;

(2) $f_0 = 265.258\ Hz$. (1)$\phi_{0_{extr}} = -41.566^\circ$, and (2) $|Z_0| = 1224.745\ \Omega$;

(3) $f_l = 45.178$ Hz.

(4) $f_h = 1.557$ kHz.]

Chapter 4

TRANSMISSION LINE DISTORTION

4.1 CAUSES OF DISTORTION

A communication signal is nothing but a complicated voltage (or current) waveform which is equivalent to a linear combination of a large number of sinusoidal waveforms of different frequencies, amplitudes and phases. One of the basic requirements of a communication system is that the signal be transmitted without distortion; that is, the signal be transmitted without any change in its waveform. This, in turn, requires that the relative amplitudes of the various frequency components of the signal be maintained during transmission; and the propagation time, that is, the time taken to arrive at the load-end, for various frequency components of the signal be the same. The relative amplitudes of the various frequency components of the signal will be unaltered, if the attenuation suffered by the various frequency components is the same; or, in other words, if α is independent of ω at least over the band of frequencies involved, which, in general, is not the case for a transmission line and, therefore, ***frequency-distortion*** is introduced. Secondly, the propagation time of the frequency components would be the same, if the group velocity of the wave on the line is independent of frequency, at least over the band of frequencies involved. This requires β_g to be a linear function of ω at least over the band of frequencies involved which, again, is in general not true and ***phase-distortion or delay-distortion*** results. Plotted in Fig. 4.1 are the attenuation and phase constants of a typical telephone line, as functions of the frequency of operation. It may be observed from the figure that neither is α constant with respect to frequency nor does β_g vary linearly with the frequency.

Hence, in general, a transmission line introduces distortion. It may be noted that both the frequency-distortion and the phase-distortion are 'linear distortions' because these are not due to any non-linearity of the system.

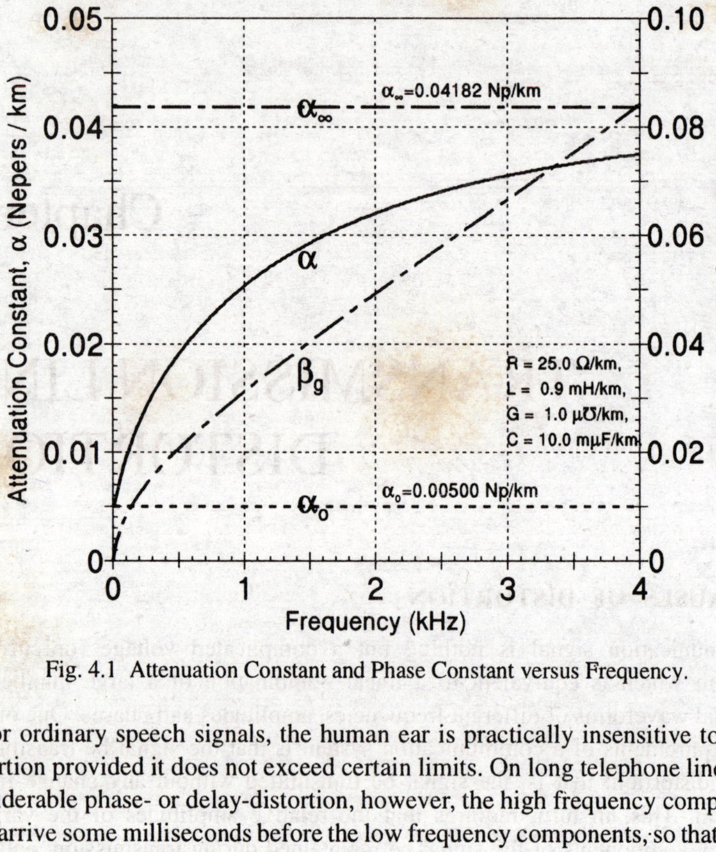

Fig. 4.1 Attenuation Constant and Phase Constant versus Frequency.

For ordinary speech signals, the human ear is practically insensitive to phase distortion provided it does not exceed certain limits. On long telephone lines with considerable phase- or delay-distortion, however, the high frequency components may arrive some milliseconds before the low frequency components, so that words are preceded by a distinct high pitch whistle.

The frequency-distortion introduced by the transmission line may be corrected by employing a two-port network, called ***Equaliser***, at the receiving end. The Equaliser is designed to have a transmission loss complementary to that of the particular transmission line used at least over the bandwidth of interest.

In the case of telegraph and TV signals, phase distortion is quite harmful although a limited amount of frequency distortion does not seriously affect the signals. Phase distortion results in the displacement of the zero-crossings, and if it is excessive, it may cause errors in the telegraph signals, and in the case of TV signals, the intensity at the crest of the wave is highly distorted.

Phase-distortion or delay-distortion or veocity-distortion is measured in terms of the ***delay-distortion coefficient***, also called ***differential delay***, which is the difference between the maximum and minimum group delays for frequencies within a specified band. It is usually expressed in microseconds. Two-port networks, called ***Delay Equalisers***, may be designed to correct phase distortion to acceptably small limits over the specified band of frequencies. It may be mentioned that for phase distortion not to occur, it is not sufficient for the transmission lines to have a phase characteristic ($\beta_g - \omega$ characteristic) which is just linear at least over the bandwidth of interest but the $\beta_g - \omega$ straight line over the bandwidth of interest must also have an intercept of $m\,\pi$, $m = 0, \pm 1, \pm 2, \ldots$, on the β_g-axis. If the latter condition is not satisfied, a distortion, called ***intercept distortion***, would occur.

4.2 CONDITIONS FOR NO DISTORTION

As discussed above, the distortion arises from the facts that α is frequency-dependent and β_g does not vary linearly with frequency. The general expressions for the parameters, α and β_g, are: [Ref. Eqns. (3.13) and (3.14)]

$$\alpha = \frac{1}{\sqrt{2}}\sqrt{\sqrt{(RG - \omega^2 LC)^2 + \omega^2 (RC + LG)^2} + (RG - \omega^2 LC)} \tag{4.1}$$

and

$$\beta_g = \frac{1}{\sqrt{2}}\sqrt{\sqrt{(RG - \omega^2 LC)^2 + \omega^2 (RC + LG)^2} + (RG - \omega^2 LC)} \tag{4.2}$$

An inspection of these expressions shows that the distortion is absent, if $R = 0$ and $G = 0$, then $\alpha = 0$ and $\beta_g = \omega\sqrt{LC}$, for all ω. That is, the $\beta_g - \omega$ characteristic of the line is a straight line with zero intercept on the β_g-axis. Though obtaining $G = 0$, in practice, is not impossible yet achieving $R = 0$ is impracticable. Therefore, these conditions are of academic interest only.

Alternatively, the distortion would be absent, if

$$\sqrt{(RG - \omega^2 LC)^2 + \omega^2 (RC + LG)^2} \equiv k + \omega^2 LC, \text{ for all } \omega, \tag{4.3}$$

where k is a constant with respect to frequency, k may be determined by insisting that the condition (4.3) holds.

Squaring both sides of the identity (4.3) in ω and equating the coefficients of like powers of ω on the two sides of the resulting equation, gives

$$k^2 = R^2 G^2, \tag{4.4}$$

and,

$$2\,LCk = R^2 C^2 + L^2 G^2. \tag{4.5}$$

Eqn. (4.4) gives, $k = \pm RG$.

Note that as the constants, R, L, G and C are real, the RHS of Eqn. (4.5) cannot be negative. Further, as the constants, L and C are positive (so are the constants, R and G), k cannot be negative. Hence, from Eqn. (4.4) and (4.5), it follows that

$$\left.\begin{aligned} k &= +RG, \\ \text{and,}\quad RC - LG &= 0 \text{ or, } \frac{R}{G} = \frac{L}{C} \text{ or, } Z_{00} = Z_{0\infty}. \end{aligned}\right\} \tag{4.6}$$

That is, the identity (4.3) holds, if Eqns. (4.6) are satisfied.

Hence, if the line parameters are related to each other as given in Eqn. (4.6), the line will not distort the signals, or, the line would be distortionless. Substitution of Eqns. (4.6) into Eqns. (4.1) and (4.2) yields that for such a line,

$$\alpha = \sqrt{RG} = \text{independent of the frequency of operation,} \tag{4.7}$$

and, $$\beta_g = \omega\sqrt{LC} = \text{linearly dependent on the frequency of operation,} \tag{4.8}$$

provided, the parameters, R, L, G and C, do not themselves vary with the frequency of operation.

It may be recalled that [Ref. Sec. 3.5., Eqn. (3.33)] that the condition (4.6) is the same as that for the line to have minimum attenuation. Therefore, if the line parameters satisfy Eqn. (4.6), then the line will not only have no distortion but will also have the minimum attenuation.

4.3 LOADING

The condition that a line is distortionless, namely, $\frac{R}{G} = \frac{L}{C}$, is seldom encountered in practice as the ratio, $\frac{R}{G}$, is usually very large compared to the ratio, $\frac{L}{C}$. For instance, for a telephone cable,

$$\frac{R}{G} = \frac{26.31}{0.940 \times 10^{-6}} = 2.779 \times 10^{7}\ \Omega^2,$$

and, $$\frac{L}{C} = \frac{0.63 \times 10^{-3}}{38.75 \times 10^{-9}} = 1.626 \times 10^{4}\ \Omega^2,$$

so that the ratio, $\frac{R}{G}$, is more than 1000 times the ratio, $\frac{L}{C}$.

Theoretically, at least, the following four methods are possible to modify the line parameters so as to make the line distortionless:

1. Reduce R. This calls for the use of much larger conductors resulting in an appreciable increase in cable size and cost. Reduction in R also decreases $|Z_0|$.
2. Increase G. This may be done by using low quality insulation between the conductors but it would result in increasing line loss. Increase in G also decreases $|Z_0|$.
3. Reduce C. This requires the spacing between the two conductors to be increased, which again results in an increase in the cable size and cost. Decreasing C raises $|Z_0|$ and lowers the attenuation constant. α.

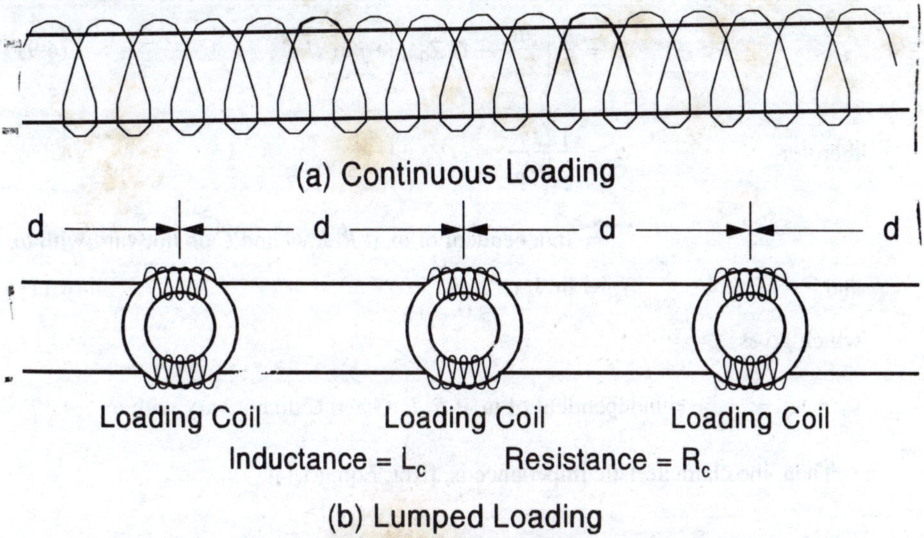

Fig. 4.2 Types of Loading.

4. Increase L. This results in an increase of $|Z_0|$ and a reduction in α, and is, therefore, the best method of improving the cable response. Increasing the inductance of the line is called ***Loading***.

Loading may be applied to a line [Ref. Fig. 4.2] in the two ways discussed in what follows.

4.3.1 Continuous Loading

The inductance of the line is increased uniformly along the length of the line. But addition of distributed inductance upto the value given by $L = \frac{R}{G}C$ is rarely practicable or even desirable. This is so because an increase in L results in a decrease of the phase velocity along the line thereby causing large transmission delays which may not be acceptable for certain applications like telephony. However, beneficial results of loading are obtained by loading a cable to only a small fraction of the theoretical value corresponding to the no-distortion condition, if the increase in inductance is such that $\omega L >> R$ at the mid-band frequency (1 kHz).

For cables loaded in this manner, $\frac{R}{\omega L} << 1$ and $\frac{G}{\omega C} << 1$. Therefore, the propagation constant for the cable is [Ref. Eqn. (2.6)]

$$\gamma = \alpha + j\beta_g = \sqrt{j\omega L\left(1 - j\frac{R}{\omega L}\right) j\omega C\left(1 - j\frac{G}{\omega C}\right)}$$

$$\approx j\omega\sqrt{LC}\left[1 - j\left(\frac{R}{2\omega L} + \frac{G}{2\omega C}\right)\right]$$

$$= \frac{1}{2}\left(\frac{R}{Z_{0\infty}} + G\,Z_{0\infty}\right) + j\,\omega\,\sqrt{LC} \tag{4.9}$$

therefore,

$$\alpha = \frac{1}{2}\left(\frac{R}{Z_{0\infty}} + G\,Z_{0\infty}\right) \tag{4.10}$$

= Independent of ω, if R, L, G and C do not vary with ω.

and,

$$\beta_g = \omega\,\sqrt{LC} \tag{4.11}$$

which gives,

$$v_p = v_g = \frac{1}{\sqrt{LC}} = \text{independent of } \omega, \text{ if } R, L, G \text{ and } C \text{ do not vary with } \omega. \tag{4.12}$$

Then, the characteristic impedance is, [Ref. Eqn. (3.1)].

$$Z_0 = \sqrt{\frac{R + j\,\omega\,L}{G + j\,\omega\,C}}$$

$$= \sqrt{\frac{j\,\omega\,L\left(1 - j\dfrac{R}{\omega\,C}\right)}{j\,\omega\,C\left(1 - j\dfrac{G}{\omega\,C}\right)}}$$

$$\approx \sqrt{\frac{L}{C}}\left[1 + j\left(-\frac{R}{2\,\omega\,L} + \frac{G}{2\,\omega\,C}\right)\right]$$

$$\approx Z_{0\infty}\Big/\frac{G}{2\,\omega\,C} - \frac{R}{2\,\omega\,L} \tag{4.13}$$

Ideally, loading must not result in increasing R, G or C. Or, even if they are increased, at least their effect on the attenuation must be negligible compared with the improvement obtained by increasing the inductance. The best method of achieving continuous loading is to wind a fine high-permeability ferromagnetic wire or tape round the copper conductors. This ferromagnetic layer tends to increase C and R — the latter increasing rapidly with frequency. Usually, the latter effect is so serious that this method of loading does not show any net improvement above a few kilohertz. Continuous loading is, therefore, useful for very low-frequency lines such as the long-distance submarine d.c. telegraph cables for which the loading tapes or wires also form a protective layer for the cable.

4.3.2 Lumped Loading

In this type of loading, which is quite common in telephone practice, inductance coils usually wound on magnetic dust cores of toroidal form, are connected, periodically, in series with the line. To maintain the balance in the case of balanced transmission lines, the coils are wound on the toroidal core so as to put an equal inductance into each conductor. If L_c is the total inductance of coils on one core and d is the spacing between two successive loading coils, the effective equivalent smoothed inductance added to the line is $\frac{L_c}{d}$ per unit length of the line. This simplification applies only when 4 or more coils are there in one operating wavelength. At a frequency corresponding to π coils per wavelength, the attenuation rises steeply owing to the low-pass characteristic exhibited by the combination of the line shunt capacitance and the series lumped inductance added by the coils — the cut-off frequency being given by

$$\omega_c = \frac{2}{\sqrt{(Ld + L_c)\,Cd}} \approx \frac{2}{\sqrt{L_c\,Cd}}, \tag{4.14}$$

as $L_c >> Ld$ for the cables to which this type of loading is applied.

It may be pointed out that loading is normally not applied to open-wire lines which have an appreciable inductance of their own and the advantage of loading is not so great. The distortion on these lines is much less than that in cables.

Because of the low-pass filter action of the lumped loading, cables with carrier systems are not loaded. It may be noted that in these cases the bandwidth of interest is a small fraction of the carrier frequency and, therefore, over this bandwidth, the attenuation characteristics of the cable is more or less constant and the phase characteristic sensibly linear rendering line distortions insignificant in so far as the modulated carrier is concerned.

The cut-off frequency of the loaded cable may be raised by decreasing the inductance per coil, or by decreasing the spacing between the coils, or both. Note that decreasing the inductance of the loading coils will increase the attenuation in the transmission band. Decreasing the spacing between coils increases the cost. Typical data for loading coils is:

Coils with inductance of 88 mH and 7.3 Ω resistance introduced after every 1.816 km in cables having R = 53.625 Ω/km, L = 0.625 mH/km, G = 0.9375 μ mho/km and = 0.03875 μF/km with an attenuation constant, α = 0.6781 dB/km at 1 kHz. ($f_c \approx$ 4.045 kHz).

4.4 LOW-DISTORTION LINES

It has already been pointed out that loading of a physical transmission line to the extent required by the condition that the line is distortionless, is neither feasible nor is it desirable. It will now be demonstrated that the line distortion is practically negligible, if the following conditions are satisfied:

$$\omega L >> R \text{ and } \omega C >> G \text{ that is, } \frac{R}{\omega L} << 1 \text{ and } \frac{G}{\omega C} << 1. \tag{4.15}$$

For such a line, the characteristic impedance, Z_0, is [Ref. Eqn. (2.10) or (3.1)]

$$Z_0 = \sqrt{\frac{R + j\omega L}{G + j\omega C}}$$

$$= \sqrt{\frac{L}{C}\left[\frac{1 - j\dfrac{R}{\omega L}}{1 - j\dfrac{R}{\omega C}}\right]} \tag{4.16}$$

Under the conditions, (4.15), the expression for the characteristic impedance may be simplified to give

$$Z_0 \approx \sqrt{\frac{L}{C}}\left[1 - j\frac{1}{2}\left(\frac{R}{\omega L} - \frac{G}{\omega C}\right)\right]. \tag{4.17}$$

$$= Z_{0\infty} \underline{/ -\frac{1}{2}\left(\frac{R}{\omega L} - \frac{G}{\omega C}\right)}, \tag{4.18}$$

Also, the propagation constant, γ, is [Ref. Eqn. (2.6)]

$$\gamma = \sqrt{(R + j\omega L)(G + j\omega C)}$$

$$= j\omega\sqrt{LC\left[\left(1 - j\frac{R}{\omega L}\right)\left(1 - j\frac{G}{\omega C}\right)\right]} \tag{4.19}$$

Again, under the conditions, (4.15), the expression for the propagation constant may be simplified to give

$$\gamma = \alpha + j\beta_g \approx j\omega\sqrt{LC}\left[1 - j\frac{1}{2\omega}\left(\frac{R}{L} + \frac{G}{C}\right)\right]$$

$$= \frac{1}{2}\left(\frac{R}{Z_{0\infty} + G Z_{0\infty}}\right) + j\omega\sqrt{LC}. \tag{4.20}$$

Therefore,

$$\alpha = \frac{1}{2}\left(\frac{R}{Z_{0\infty}} + G Z_{0\infty}\right) \tag{4.21}$$

and, $$\beta_g = \omega\sqrt{LC}, \tag{4.22}$$

giving, $$v_p = v_g = \frac{1}{\sqrt{LC}} \tag{4.23}$$

It may be noted that under these approximations, α is independent of the frequency, and β_g is linearly dependent on the frequency, if the parameters R, L, G and C, do not themselves change with frequency. The characteristic impedance is also nearly a pure resistance which is also independent of frequency. Hence, the line is practically distortionless.

SOLVED EXAMPLES

Example 4.1 A 37.381 km long 19 AWG line has the following primary constants:

$R = 53.315\ \Omega/\text{km},\ L = 0.621\ \text{mH/km},\ G = 0.932\ \mu\ \text{mho/km}$ and $C = 38.526\ \text{m}\mu\text{F/km}$.

It is lightly loaded using coils, each having an inductance of 31 mH and a resistance of 2.7 Ω, at intervals of 1.829 km.

Assuming the loading to be uniformly distributed, calculate Z_0, total attenuation ($= \alpha l$) and travel time τ, over the given length of the line at frequencies of 300 Hz, 1.0 kHz and 3.4 kHz. Hence, calculate the delay distortion or the differential delay, Δ τ, introduced by the line into the standard telephone channel (300 Hz to 3400 Hz). Also calculate the cut-off frequency, f_{co} of the line.

Solution : Assuming the loading to be uniformly distributed, the transmission line may be treated as a uniform transmission line having an effective per unit length inductance, L_{eff}, and resistance, R_{eff}, given by

$$L_{eff} = L + \frac{L_c}{d} = 0.621 + \frac{31.0}{1.829}\ \text{mHz/km} = 17.5702\ \text{mH/km}$$

$$R_{eff} = R + \frac{R_c}{d} = 53.315 + \frac{2.7}{1.829}\ \text{mHz/km} + 54.7912\ \Omega/\text{km}$$

$$R_{eff} + j\,\omega\,L_{eff} = (54.791 + j\,2\,\pi \times 1000 \times 17.570)\ \Omega/\text{km}$$

$$= (54.791 + j\,110.3965)\ \Omega/\text{km}$$

$$= 123.2456\ \underline{/63.604^\circ}\ \Omega/\text{km}$$

$$G + j\,\omega\,C = (0.932 + j\,2\,\pi \times 1000 \times 0.038526)\ \mu\ \text{mhos/km}$$

$$= (0.9320 + j\,242.0660)\ \mu\ \text{mho/km}$$

$$= 242.0678\ \underline{/89.779^\circ}\ \mu\ \text{mhos/km}$$

Therefore, $$Z_0 = \sqrt{\frac{123.2456\ \underline{/63.604^\circ}}{242.0678 \times 10^{-6}\ \underline{/89.779^\circ}}} = 713.538\ \underline{/-13.088^\circ}\ \Omega$$

and, $$\gamma = \sqrt{(123.2456\ \underline{/63.604^\circ}) \times (242.0678 \times 10^{-6}\ \underline{/89.779^\circ})}$$

$$= 0.172725\ \underline{/76.692^\circ}/\text{km}$$

$$= (0.03976 + j\,0.16809)/\text{km}$$

From this, $$\alpha = 0.03976\ \text{Neper/km} = 0.03976 \times 8.686\ \text{dB/km} = 0.345\ \text{dB/km}$$

Hence,

Total Attenuation $= \alpha l = 0.34534 \times 37.381\ \text{dB} = 12.909\ \text{dB}$.

Also, $$\beta_g = 0.16809\ \text{radian/km} = 0.16809 \times \frac{180}{\pi}\ \text{deg/km} = 9.63063\ \text{deg/km}$$

which gives,

Phase Velocity $= v_p = 3.738 \times 10^4$ km/sec

Therefore,

Total Travel Time $\tau = \dfrac{37.381}{3.738 \times 10^4}$ sec. = 1.000 millisec.

Proceeding in the same way as given above, the characteristics of the line may be obtained at the other two frequencies, namely, 300 Hz and 3400 Hz. The following results are obtained:

[**Ans.** $Z_0 = 938.907 \underline{/-29.057^\circ}\ \Omega$, $\alpha l = 11.000$ dB, $\tau = 1.174$ millisec at 0.3 kHz;

$Z_0 = 713.538\underline{/-13.088^\circ}\ \Omega$, $\alpha l = 12.909$ dB, $\tau = 1.000$ millisec at 1.0 kHz;

$Z_0 = 678.891 \underline{/-04.120^\circ}\ \Omega$, $\alpha l = 13.233$ dB, $\tau = 0.975$ millisec at 3.4 kHz;

$f_{C0} = 6.689$ kHz; and Differential Delay, $\Delta\tau = (1.174 - 0.975)$ msec $= 199$ μsec.]

EXERCISES

1. Study the distortion introduced by the following line of length, l, into a standard telephone channel (300 Hz to 3400 Hz frequency band) by calculating their Z_0, αl and τ = time to travel the given length of the line, at frequencies of 300 Hz, 1.0 kHz and 3.4 kHz.

$R = 5.468\ \Omega$/km, $L = 2.274$ mH/km, $G = 0.621$ μ mho/km

and $C = 5.344$ m μ F/km. $l = 282.390$ km.

[**Ans.** $Z_0 = 829.714 \underline{/-24.190^\circ}\ \Omega$, $\alpha l = 3.553$ dB, $\tau = 1.111$ millisec at 0.3 kHz,

$Z_0 = 674.939 \underline{/-9.941^\circ}\ \Omega$, $\alpha l = 10.655$ dB, $\tau = 1.000$ millisec at 1.0 kHz,

$Z_0 = 654.374 \underline{/-3.055^\circ}\ \Omega$, $\alpha l = 10.751$ dB, $\tau = 0.986$ millisec at 3.4 kHz]

2. Study the distortion introduced by the following line of length, l into a standard telephone channel (300 Hz to 3400 Hz frequency band) by calculating their Z_0, αl and τ = time to travel the given length of the line, at frequencies of 300 Hz, 1.0 kHz and 3.4 kHz.

$R = 8.699\ \Omega$/km, $L = 2.336$ mH/km, $G = 0.621$ μ mho/km

and $C = 5.520$ mμF/km. $l = 276.096$km

[**Ans.** $Z_0 = 994.451 \underline{/-29.771^\circ}\ \Omega$, $\alpha l = 12.957$ dB, $\tau = 1.195$ millisec at 0.3 kHz,

$Z_0 = 721.257 \underline{/-14.783^\circ}\ \Omega$, $\alpha l = 15.513$ dB, $\tau = 1.000$ millisec at 1.0 kHz,

$Z_0 = 673.981 \underline{/-4.785^\circ}\ \Omega$, $\alpha l = 16.052$ dB, $\tau = 0.968$ millisec at 3.4 kHz.]

3. Study the distortion introduced by the following line of length, l, into a standard telephone channel (300 Hz to 3400 Hz frequency band) by calculating their Z_0, αl and τ = time to travel the given length of the line, at frequencies of 300 Hz, 1.0 kHz and 3.4 kHz.

$R = 27.341\ \Omega$/km, $L = 0.621$ mH/km, $G = 0.621$ μ mho/km

and $C = 40.390$ mμF/km. $l = 99.452$ km

[**Ans.** $Z_0 = 599.531 \underline{/-43.541^\circ}\ \Omega$, $\alpha l = 27.354$ dB, $\tau = 1.733$ millisec at 0.3 kHz,

$Z_0 = 329.892 \underline{/-40.866^\circ}\ \Omega$, $\alpha l = 37.452$ dB, $\tau = 1.000$ millisec at 1.0 kHz,

$Z_0 = 187.670 \underline{/-32.038^\circ}\ \Omega$, $\alpha l = 74.268$ dB, $\tau = 0.637$ millisec at 3.4 kHz.]

4. Study the distortion introduced by the following line of length, l, into a standard telephone channel (300 Hz to 3400 Hz frequency band) by calculating their Z_0, αl and τ = time to travel the given length of the line, at frequencies of 300 Hz, 1.0 kHz and 3.4 kHz.

$R = 54.682\ \Omega$/km, $L = 0.621$ mH/km, $G = 0.621$ μ mho/km

and $C = 40.390$ mμF/km. $l = 72.870$ km

[**Ans.** $Z_0 = 847.573 \underline{/-44.153^\circ}\ \Omega$, $\alpha l = 28.656$ dB, $\tau = 1.776$ millisec at 0.3 kHz,

$Z_0 = 464.779 \underline{/-42.888^\circ}\ \Omega$, $\alpha l = 50.941$ dB, $\tau = 1.000$ millisec at 1.0 kHz,

$Z_0 = 255.367 \underline{/-38.161^\circ}\ \Omega$, $\alpha l = 86.346$ dB, $\tau = 0.597$ millisec at 3.4 kHz.]

5. Examine the effect of loading the line of the Example 4.1 moderately as described below:
Loading coils' data: Resistance, R_c = 7.3 Ω/coil, Inductance, L_c = 88 mH/coil, and Inter-coil Spacing, d_c = 1.829 km.
[**Ans.** Z_0 = 1220.990 $\underline{/-15.611^\circ}$ Ω, αl = 8.103 dB, τ = 1.687 millisec at 0.3 kHz,
Z_0 = 1134.432 $\underline{/-5.190^\circ}$ Ω, αl = 8.407 dB, τ = 1.626 millisec at 1.0 kHz,
Z_0 = 1125.565 $\underline{/-1.543^\circ}$ Ω, αl = 8.439 dB, τ = 1.620 millisec at 3.4 kHz.
f_{c0} = 4.016 kHz; and Differential Delay, Δ τ = (1.687 – 1.620) msec = 67 μ sec.]

6. Examine the effect of loading the line of Example 4.1 heavily as described below:
Loading coils' data: Resistance, R_c = 7.3 Ω/coil, Inductance, L_c = 88 mH/coil, and Inter-coil Spacing, d_c = 0.914 km.
[**Ans.** Z_0 = 1628.704 $\underline{/-8.909^\circ}$ Ω, αl = 6.434 dB, τ = 2.313 millisec at 0.3 kHz,
Z_0 = 1589.941 $\underline{/-2.764^\circ}$ Ω, αl = 6.507 dB, τ = 2.287 millisec at 1.0 kHz,
Z_0 = 1586.289 $\underline{/-0.816^\circ}$ Ω, αl = 6.514 dB, τ = 2.284 millisec at 3.4 kHz.
f_{c0} = 5.700 kHz; and Differential Delay, Δ τ = (2.313 – 2.284) msec = 29 μ sec.]

7. A 38.152 km long 16 AWG line has the following primary constants:

R = 26.160 Ω/km, L = 0.621 mH/km, G = 0.932 μ mho/km

and C = 38.526 mμF/km.

It is lightly loaded using coils, each having an inductance of 31 mH and a resistance of 2.7 Ω, at intervals of 1.829 km.
Assuming the loading to be uniformly distributed, calculate Z_0, total attenuation (= αl) and travel time, τ, over the given length of the line at frequencies of 300 Hz, 1.0 kHz and 3.4 kHz. Hence, calculate the delay distortion or the differential delay, Δ τ introduced by the line into the standard telephone channel (300 Hz to 3400 Hz).
Also calculate the cutoff frequency, f_{co}, of the line.
[**Ans.** Z_0 = 770.672 $\underline{/-19.554^\circ}$ Ω, αl = 6.431 dB, τ = 1.063 millisec at 0.3 kHz,
Z_0 = 685.660 $\underline{/-6.971^\circ}$ Ω, αl = 6.834 dB, τ = 1.000 millisec at 1.0 kHz,
Z_0 = 676.235 $\underline{/-2.073^\circ}$ Ω, αl = 6.880 dB, τ = 0.993 millisec at 3.4 kHz.
f_{c0} = 6.689 kHz ; and Differential Delay, Δτ = (1.063 – 0.993) msec = 70 μsec.]

8. Examine the effect of loading the line of Example 4.1 moderately as described below:

Loading coils' data: Resistance, R_c = 7.3 Ω/coil, Inductance, L_c = 88 mH/coil, and Inter-coil Spacing, d_c = 1.829 km.
[**Ans.** Z_0 = 1153.806 $\underline{/-8.718^\circ}$ Ω, αl = 4.561 dB, τ = 1.673 millisec at 0.3 kHz,
Z_0 = 1127.427 $\underline{/-2.701^\circ}$ Ω, αl = 4.610 dB, τ = 1.655 millisec at 1.0 kHz,
Z_0 = 1124.950 $\underline{/-0.797^\circ}$ Ω, αl = 4.615 dB, τ = 1.653 millisec at 3.4 kHz.
f_{c0} = 4.016 kHz ; and Differential Delay, Δτ = (1.687 – 1.620) msec = 67 μsec.]

9. Examine the effect of loading the line of Example 4.1 heavily as described below:
Loading coils' data: Resistance, R_c = 7.3 Ω/coil, Inductance, L_c = 88 mH/coil, and Inter-coil Spacing, d_c = 0.914 km.
[**Ans.** Z_0 = 1599.555 $\underline{/-4.927^\circ}$ Ω, αl = 3.798 dB, τ = 2.340 millisec at 0.3 kHz,
Z_0 = 1587.182 $\underline{/-1.495^\circ}$ Ω, αl = 3.811 dB, τ = 2.287 millisec at 1.0 kHz,
Z_0 = 1586.050 $\underline{/-0.440^\circ}$ Ω, αl = 3.812 dB, τ = 2.331 millisec at 3.4 kHz.
f_{c0} = 5.700 kHz; and Differential Delay, Δτ = (2.340 – 2.331) msec = 9 μsec.]

10. A 22.514 km long 40 lb PCQT underground cable has the following primary constants:

$$R = 27.341\ \Omega/\text{km},\ L = 0.621\ \text{mH/km},\ G = 0.621\ \mu\ \text{mho/km}$$
$$C = 40.390\ \text{m}\mu\text{F/km}.$$

It is loaded using coils, each having an inductance of 88 mH and a resistance of 4.3 Ω, at intervals of 1.829 km.

Assuming the loading to be uniformly distributed, calculate Z_0, total attenuation = αl and travel time, τ, over the given length of the cable at frequencies of 300 Hz, 1.0 kHz and 3.4 kHz. Hence, calculate the delay distortion or the differential delay, Δ τ, introduced by the cable into the standard telephone channel (300 Hz to 3400 Hz). Also calculate the cut-off frequency, f_{co} of the cable.

[**Ans.** $Z_0 = 1126.066 \underline{/-8.7}$ [illegible] l = 2.678 dB, τ = 1.011 millisec at 0.3 kHz,
$Z_0 = 1101.027 \underline{/-2}$ [illegible] = 2.707 dB, τ = 1.000 millisec at 1.0 kHz,
$Z_0 = 1098.678 \underline{/-}$ [illegible] = 2.709 dB, τ = 0.999 millisec at 3.4 kHz.
f_{c0} = 3.923 kHz ; [illegible]l Delay, Δτ = (1.011 – 0.999) msec = 12 μsec.]

11. A 22.446 km long 20 lb [illegible]nd cable has the following primary constants:

$R = 54.682\ \Omega$ [illegible] mH/km, $G = 0.621\ \mu$ mho/km
and $C = 40.390$ m[illegible]

It is loaded using coils, [illegible] inductance of 88 mH and a resistance of 4.3 Ω, at intervals of 1.829 k[illegible]

Assuming the loading t[illegible]y distributed, calculate Z_0, total attenuation (= $\alpha\ l$) and travel time, τ, [illegible] length of the cable at frequencies of 300 Hz, 1.0 kHz and 3.4 kHz. Hen[illegible] the delay distortion or the differential delay, Δ τ, introduced by the cable [illegible]tandard telephone channel (300 Hz to 3400 Hz). Also calculate the cut-off fre[illegible], f_{co}, of the cable.

[**Ans.** $Z_0 = 1191.717 \underline{/-15.6}$[illegible] Ω, αl = 4.921 dB, τ = 1.038 millisec at 0.3 kHz,
$Z_0 = 1107.859 \underline{/-5.205^\circ}\ \Omega$, αl = 5.106 dB, τ = 1.000 millisec at 1.0 kHz,
$Z_0 = 1099.278 \underline{/-1.547^\circ}\ \Omega$, αl = 5.126 dB, τ = 0.996 millisec at 3.4 kHz.
f_{c0} = 3.923 kHz; and Differential Delay, Δ τ = (1.038 – 0.996) msec = 42 μsec.]

Chapter 5

REFLECTION ON TRANSMISSION LINES

5.1 WAVES ON TRANSMISSION LINES

It may be recalled that if a sinusoidal voltage, $V_G e^{j\omega t}$, generator is connected at one end, $z = 0$, [Ref. Fig. 5.1] of a two-conductor, uniform transmission line, characterised by Z_0 and γ, then the amplitude, $V(z)$, of the voltage and the amplitude, $I(z)$, of the current at a point, z, on the line are given by [Ref. Eqns.(2.12) and (2.13)]

$$V(z) = V^+ e^{-\gamma z} + V^- e^{\gamma z} \tag{5.1}$$

and,

$$I(z) = \frac{1}{Z_0} [V^+ e^{-\gamma z} - V^- e^{\gamma z}] \tag{5.2}$$

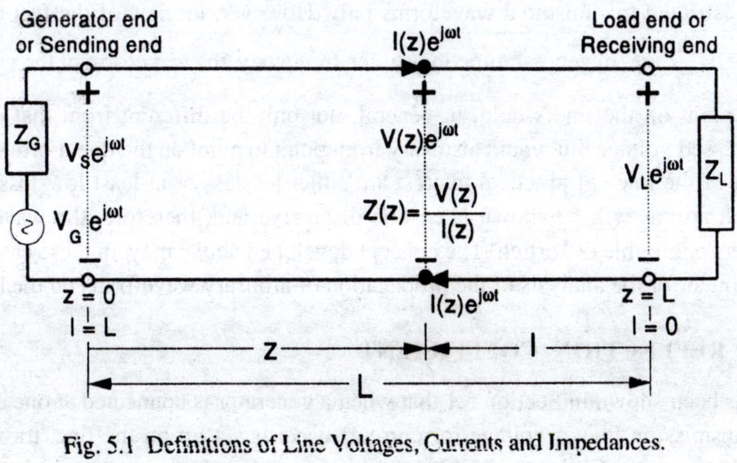

Fig. 5.1 Definitions of Line Voltages, Currents and Impedances.

Therefore, the voltage and current on the line as functions of the distance, z and time, t, are given by

$$v(z, t) = V(z)\, e^{j\omega t} = V^{+} e^{-\alpha z} e^{j(\omega t - \beta_g z)} + V^{-} e^{\alpha z} e^{j(\omega t + \beta_g z)} \quad (5.3)$$

and,

$$i(z, t) = I(z)\, e^{j\omega t} = \frac{1}{Z_0} [V^{+} e^{-\alpha z} e^{j(\omega t - \beta_g z)} - V^{-} e^{\alpha z} e^{j(\omega t + \beta_g z)}] \quad (5.4)$$

In passing, it may be observed that if $\alpha \neq 0$, then for the case of an infinite line, V^{-} must be set equal to zero to ensure that the physical quantities $v(z, t)$ and $i(z, t)$ are finite everywhere on the line. In this case, then $\frac{v(z, t)}{i(z, t)} = Z_0$ = constant everywhere on the line. This gives a physical significance of the characteristic impedance, Z_0, of a transmission line. That is, ***the characteristic impedance, Z_0, is the input impedance of an infinitely long line.***

Inspection of these expressions [Ref. Eqns. (5.3) and (5.4)] for the voltage and the current on the line, shows that the first terms represent sinusoidal travelling waves along the line in the $+z$-direction with the phase velocity of $v_p = \frac{\omega}{\beta_g}$, and the second terms represent sinusoidal travelling waves along the line in the $-z$-direction with the phase velocity v_p. The factors, $e^{\pm \alpha z}$ in these expressions show that the amplitude of each individual wave decreases as it progresses. Hence, the voltage (or current) at any point on the line at any instant is the result of superposition of a positively travelling ***voltage (or current) wave*** and a negatively travelling ***voltage (or current) wave*** at that point and at that instant of time.

It may be pointed out here that as the differential equations governing the voltage and currents on the line are linear, the principle of superposition is applicable. Further, as any arbitrary impressed voltage waveform may be regarded as a linear combination of sinusoidal voltage waveforms of various frequencies, the observations made above are applicable to any arbitrary impressed voltage waveform and not restricted to sinusoidal waveforms only. However, in view of the fact that α and $v_p = \frac{\omega}{\beta_g}$ are, in general, functions of the frequency, the waveform of the voltage at a point on the line would, in general, not only be different from that of the impressed voltage but would also vary from point to point on the line. Fortunately, most of the lines of practical interest are either lossless or at least low-loss lines which are, as will be shown later, non-dispersive and, therefore, the waveform suffers negligible distortion. The concept developed above may, in these cases, be exploited for the analysis of the propagation of arbitrary waveforms on the line.

5.2 REFLECTION COEFFICIENT

It has been shown in Section 5.1 that when a generator is connected at one end of a transmission line, a voltage (or current) wave is set up on the line, travelling towards the load-end (positively travelling wave). Also present on the line, in the steady state, is a ***reflected wave***, that is, voltage (or current) wave travelling towards the generator. The voltage (or current) at any point on the line at any instant is the result of the superposition of these two voltage (or current) waves at that point and at that instant of time, as given by the pairs of Eqns. (5.3) and (5.4) or (2.18) and (2.22) or (2.50) and (2.51). The ratio of the instantaneous amplitude of the reflected voltage (or current) wave to that of the forward voltage (or current) wave at any point on the line, is defined as the ***voltage (or current) reflection coefficient*** at that point.

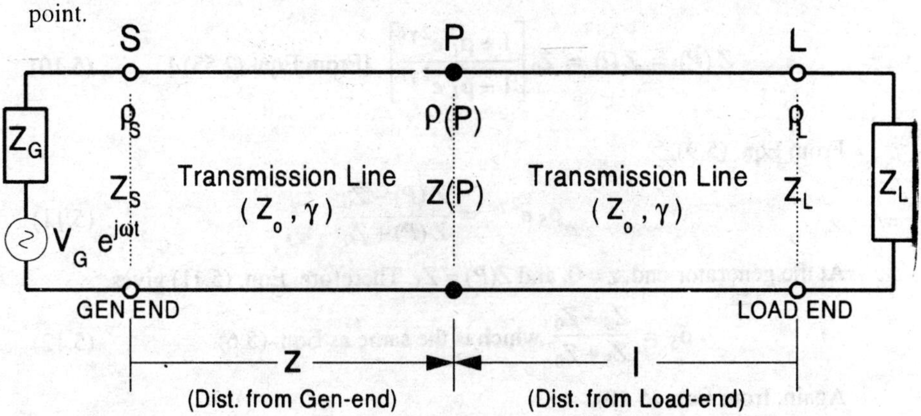

z + l = L = Total Length of the Line.

Fig. 5.2 Voltage Reflection Coefficients and Impedances on a Transmission Line.

Consider a point, P, on the line [Ref. Fig. 5.2] which is at a distance of z from the generator-end towards the load, or equivalently, at a distance of l from the load-end towards the generator. Following the definition of the voltage reflection coefficient, given above, the voltage reflection coefficient, ρ (P), at the point, P, may be obtained from any of the expressions for the line voltage given in Chapter 2 [Ref., for example, Eqns.(2.18) or (2.50)]. Therefore, from Eqn. (2.20), the voltage reflection coefficient ρ (P), which is, in general, a complex quantity, is given by

$$\rho(P) = \rho(z) = |\rho(z)|\, e^{j\phi(z)} = \rho_S\, e^{2\gamma z} \qquad (5.5)$$

where, ρ_s is the voltage reflection coefficient at the generator-end or at the point, $z = 0$ and is given by [Ref. Eqn. (2.28)]

$$\rho_S = \rho(z)|_{z=0} = |\rho_S|\, e^{j\phi_S} = \frac{Z_S - Z_0}{Z_S + Z_0} \qquad (5.6)$$

And, if Eqn. (2.50) is used, the voltage reflection coefficient, ρ (P), is given by

$$\rho(P) = \rho(l) = |\rho(l)|\, e^{j\phi(l)} = \rho_L\, e^{-2\gamma l} \qquad (5.7)$$

where, ρ_L is the voltage reflection coefficient at the load-end, and is given by [Ref. Eqn. (2.39)]

$$\rho_L = \rho(l)|_{l=0} = |\rho_L| e^{j\phi_L} = \frac{Z_L - Z_0}{Z_L + Z_0} \tag{5.8}$$

Now, the line impedance at the point, P, may be obtained either from Eqn. (2.26) or from Eqn. (2.53), that is,

$$Z(P) = Z(z) = Z_0\left[\frac{1 + \rho_S e^{2\gamma z}}{1 - \rho_S e^{2\gamma z}}\right]. \text{ [From Eqn. (2.26).]} \tag{5.9}$$

$$Z(P) = Z(l) = Z_0\left[\frac{1 + \rho_L e^{2\gamma l}}{1 - \rho_L e^{2\gamma l}}\right]. \text{ [From Eqn. (2.53).]} \tag{5.10}$$

From Eqn. (5.9),

$$\rho_S e^{2\gamma z} = \frac{Z(P) - Z_0}{Z(P) + Z_0} \tag{5.11}$$

At the generator-end, $z = 0$, and $Z(P) = Z_S$. Therefore, Eqn. (5.11) gives,

$$\rho_S = \frac{Z_S - Z_0}{Z_S + Z_0}, \text{ which is the same as Eqn. (5.6).} \tag{5.12}$$

Again, from Eqn. (5.10),

$$\rho_L e^{-2\gamma l} = \frac{Z(P) - Z_0}{Z(P) + Z_0} \tag{5.13}$$

At the load-end, $l = 0$ and $Z(P) = Z_L$. Therefore, Eqn. (5.13) gives,

$$\rho_L = \frac{Z_L - Z_0}{Z_L + Z_0}, \text{ which is the same as Eqn. (5.8)} \tag{5.14}$$

Now, combining Eqns. (5.5) and (5.11) with Eqns.(5.7) and (5.13), gives the equation

$$\rho(P) = \rho_S e^{-2\gamma z} = \rho_L e^{-2\gamma l} = \frac{Z(P) - Z_0}{Z(P) + Z_0}. \tag{5.15}$$

This equation is an important one because it gives not only a method of calculating the voltage reflection coefficient at any point on the line in terms of impedances but also the transformation of the voltage reflection coefficient effected by the line. Both of these are discussed below and in the next section in detail.

Eqn. (5.15) spells out the following method to determine the voltage reflection coefficient at any given point like P on the line:

If the line impedance at a point, P, is $Z(P)$, then the voltage reflection coefficient, $\rho(P)$, at that point is given by

$$\rho(P) = \frac{Z(P) - Z_0}{Z(P) + Z_0} \tag{5.16}$$

The following points are borne out by Eqn. (5.16):

1. It follows from Eqn. (5.16) that the voltage reflection coefficient is non-zero, that is, the reflected wave is present, if and only if the line impedance, $Z(P) \neq Z_0$. From this, it may be inferred that the reflected wave is present at a point only if the line impedance at that point differs from the characteristic impedance of the line.
2. Therefore, in the case of a practical line [of finite length, terminated in a load impedance, Z_L], the reflected wave originates at the load-end, and the voltage reflection coefficient at the load-end is given by Eqn. (5.8).
3. As a corollary to the above, if the line is of infinite length, its load-end can never be reached and, therefore, there can never be a reflected wave. Everywhere on such an infinite line, the input impedance of the line is Z_o, [Ref. Sec. 5.1], and, therefore, the reflection coefficient is zero everywhere on the line. The fact that the reflected wave is absent if the line is infinite in length, also follows from the requirement that the line voltage be finite everywhere on the line and, therefore, if the line is infinite in length, the integration constant V^- in the line solutions (2.12) and (2.13) must be equal to zero, which physically means that there is no reflected wave on the line. [Ref. Section 5.1]
4. From the point 1, noted above, it is clear that the voltage reflection coefficient is zero for the special case when the load-impedance, Z_L, is equal to the characteristic impedance, Z_0. [Also Ref. Eqn. (5.8)].

Other special cases of the load impedance are those of the short-circuit, $Z_L = 0$, and of the open-circuit, $Z_L \rightarrow \infty$.

1. If the line is short-circuited at the load-end, that is, if $Z_L = 0$, the voltage reflection coefficient at the load terminals, $\rho_L = -1 = e^{-j\pi}$; [Ref. Eqn. (5.8)], i.e., magnitude and angle are: $|\rho_L| = 1$ and $\phi_L = \pi$ radians.
2. If the line is open-circuited at the load-end, that is, if $Z_L \rightarrow \infty$, the voltage reflection coefficient at the load terminals, $\rho_L = +1 = e^{+j0}$; [Ref. Eqn. (5.8)], i.e., magnitude and angle are: $|\rho_L| = 1$ and $\phi_L = 0$ radians.

Lastly, note from Eqns. (5.4) or (2.22) or (2.51) that the reflection coefficient for the current wave, that is, the ***current reflection coefficient*** differs from the voltage reflection coefficient only in its sign: it is the negative of the voltage reflection coefficient. Therefore, only one of these, that is, either the voltage reflection coefficient or the current reflection coefficient, need be discussed. Hence, throughout this Volume only the ***voltage*** reflection coefficient, denoted by ρ, has been considered.

5.2.1 Variation of Reflection Coefficient along the Line

Eqn. (5.15) also describes the manner, explained below, in which the voltage reflection coefficient is transformed by the transmission line.

1. If the voltage reflection coefficient, ρ_L, at a point, L (the load-end, for example) [Ref. Fig. 5.2], is known, then the voltage reflection coefficient, $\rho(P)$, at a point, P, which is at a distance of l from the point, L, ***towards the Generator***, is obtained from ρ_L by multiplying it by the factor, $e^{-2\gamma l}$, that is,

$$\rho(P) = \rho_L e^{-2\gamma l} \tag{5.17}$$

2. And, if the voltage reflection coefficient, ρ_S, at a point, S (the generator-end), is known, then the voltage reflection coefficient, $\rho(P)$, at a point, P, which is at a distance of z from the point, S, ***towards the Load,*** is obtained from ρ_S by multiplying it by the factor, $e^{-2\gamma z}$, that is,

$$\rho(P) = \rho_S e^{2\gamma z} \tag{5.18}$$

3. Hence, if ***movements towards the generator are taken to be positive***, then single Eqn. (5.17) suffices to describe the manner in which the voltage reflection coefficient gets transformed by a movement along the line. [Eqn. (5.18) is contained in Eqn. (5.17) in view of the aforestated sign convention for movements along the line.]

In general, the voltage reflection coefficient at a distance, l, from the load-end towards the generator is given by Eqn. (5.15) which may be rewritten as,

$$\rho(l) = |\rho(l)|\, e^{j\phi(l)} = u(l) + j\,v(l) = \frac{Z(l) - Z_0}{Z(l) + Z_0} \tag{5.19}$$

Notation : In the parlance of transmission line theory, it is convenient to deal with ***normalised*** quantities: All impedances are normalised with respect to the characteristic impedance of the line and represented by the corresponding lower case letters. For example, normalised line impedance is given by

$$z(l) = \frac{Z(l)}{Z_0} = \frac{R(l) + j\,X(l)}{Z_0} = r(l) + j\,x(l). \tag{5.20}$$

All lengths are expressed in terms of the wavelength of the waves along the line or the guide-wavelength, λ_g. Although not necessarily yet quite often it may be convenient to give the position of a point on the line by its distance, l, from the load-end: ***increasing or positive-direction of l being a movement towards the generator.*** Further, for brevity, let the argument, "(l)", be dropped.

Hence, in general, the voltage reflection coefficient at a distance of l from the load-end towards the generator, where the line impedance is, Z, is given by [Ref. Eqn. (5.19)].

$$\rho = u + j\,v = \frac{Z - Z_0}{Z + Z_0} = \frac{\dfrac{Z}{Z_0} - 1}{\dfrac{Z}{Z_0} + 1}$$

$$= \frac{z - 1}{z + 1} \tag{5.21}$$

where,
$$z = \frac{Z}{Z_0} = r + j\,x \tag{5.22}$$

Equating the real and imaginary parts on the two sides of Eqn. (5.21), gives

$$\text{Re}\,[\,\rho\,] = u = \frac{|\,z\,|^2 - 1}{|\,z\,|^2 + 2r + 1}, \tag{5.23}$$

$$\text{Im}\,[\,\rho\,] = v = \frac{2\,x}{|\,z\,|^2 + 2r + 1}. \tag{5.24}$$

Also,
$$|\,\rho\,|^2 = u^2 + v^2 = 1 - \frac{4\,r}{|\,z\,|^2 + 2\,r + 1}, \tag{5.25}$$

and,
$$\phi = \tan^{-1}\left[\frac{2\,x}{|\,z\,|^2 - 1}\right] \tag{5.26}$$

From Eqn. (5.25), it is clear that for all practically realisable impedances, that is, for $r \geq 0$, x = arbitrary, the magnitude of the reflection coefficient, $|\,\rho\,| \leq 1$.

Further, from Eqns. (5.23), (5.24), and (5.26), it is observed that

$$\left.\begin{array}{lll}
(i) & 0 \leq \phi \leq \dfrac{\pi}{2}, & \text{if } x \geq 0 \text{ and } \;|\,z\,| > 1 \\
(ii) & \dfrac{\pi}{2} \leq \phi \leq \pi, & \text{if } x \geq 0 \text{ and } \;|\,z\,| < 1 \\
(iii) & 0 \geq \phi \geq -\dfrac{\pi}{2}, & \text{if } x \leq 0 \text{ and } \;|\,z\,| > 1 \\
(iv) & -\dfrac{\pi}{2} \geq \phi \geq -\pi, & \text{if } x \leq 0 \text{ and } \;|\,z\,| < 1 \\
\text{and } (v) & \phi = 0, & \text{if } x = 0 \text{ and } \;r > 1 \\
 & \;\;= \pi & \text{if } x = 0 \text{ and } \;r < 1
\end{array}\right\} \tag{5.27}$$

The voltage reflection coefficient, $\rho(l)$, at a distance, l, from the load-end towards the generator, in the general case, $\alpha \neq 0$, is given by

Therefore,

$$\left.\begin{aligned} \rho &= |\rho| e^{j\phi} = |\rho_L| e^{-2\alpha l} e^{j(\phi_L - 2\beta_g l)} \\ |\rho(l)| &= |\rho_L| e^{-2\alpha l} \\ \phi(l) &= \phi_L - 2\beta_g l \end{aligned}\right\} \quad (5.28)$$

and

As the distance, l, from the load-end towards the generator is increased from zero (load-end), the magnitude, $|\rho(l)|$, of the reflection coefficient decreases exponentially as $|\rho_L| e^{-2\alpha l}$ from its value, $|\rho_L|$, at the load-end; and the angle, $\phi(l)$, of the reflection coefficient also decreases linearly as $\phi_L - 2\beta_g l$ from its value, ϕ_L, at the load-end. Therefore, if the reflection coefficient is plotted in the complex (u,v)-plane (ρ-plane), then, as the distance, l, from the load-end towards the generator is increased from zero (load-end), the point representing the reflection coefficient starts from its initial position, $\rho_L = u_L + j\,v_L$, at the load-end, and moves in the clockwise direction on a contracting spiral. Therefore, at far off distances from the load-end, the reflection coefficient becomes practically zero [Ref. Fig. 5.3] because the amplitude of the reflected voltage wave becomes negligibly small compared to that of the incident voltage wave, as may be seen from Fig. 5.4. The magnitude of the normalised line impedance which oscillates wildly at distances close to the load-end, also settles down to unity at far off distances. This behaviour of the line impedance is illustrated in Fig. 5.5.

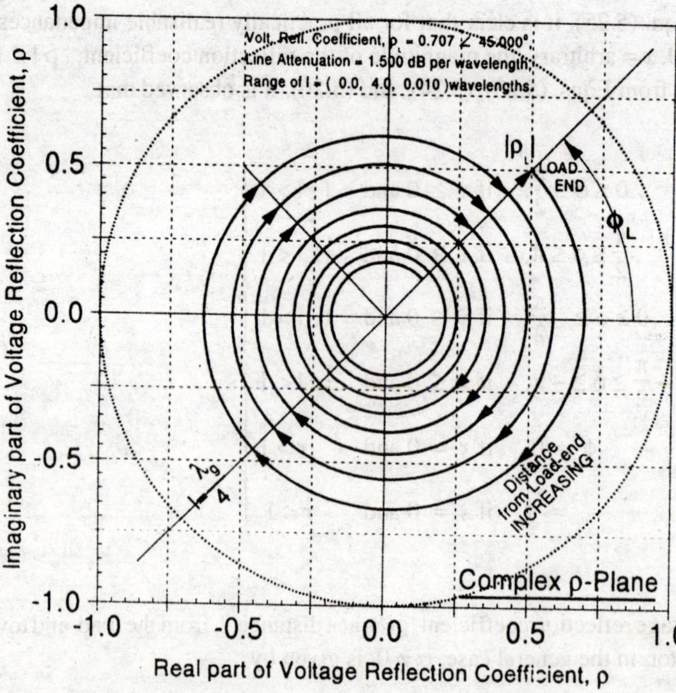

Fig. 5.3 Variation of Voltage Reflection Coefficient with Distance from the Load-end.

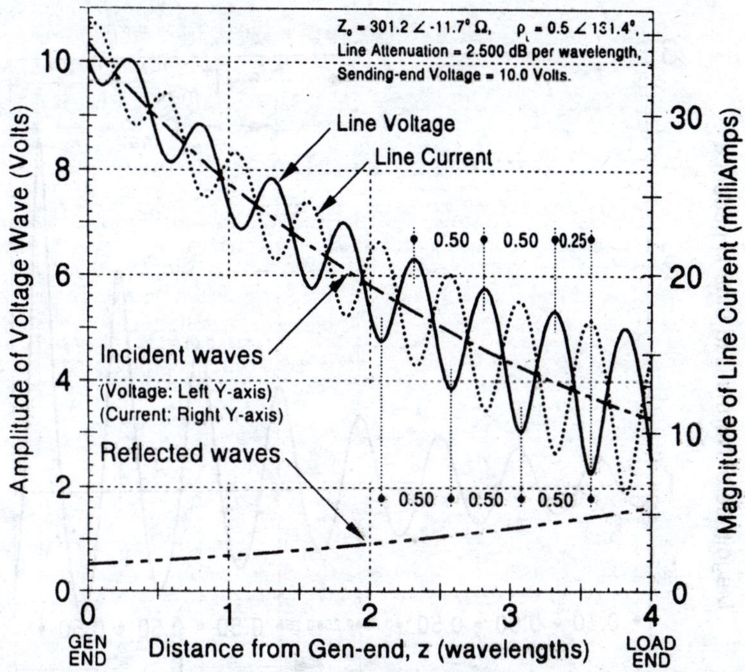

Fig. 5.4 Amplitudes of Waves on a Line.

5.3 STANDING WAVE RATIO

Based on the above discussion [Ref. Secs. 5.1 and 5.2], it may be stated that if a voltage generator of voltage, V_G, is connected at one end ($z = 0$) of a two-conductor, uniform (Z_0, γ) transmission line of length, L, terminated in a load impedance, $Z_L \neq Z_0$, then, in the steady state, the following two voltage waves are present everywhere on the line and at all times: [Ref. Fig. 5.4]

1. One wave is travelling from the generator-end to the load-end. This may be called the ***incident wave*** and represented mathematically by

$$V_{inc} = V^+ e^{-\gamma z} = V^+ e^{-\alpha z} e^{-j\beta_g z} \tag{5.29}$$

The incident wave has an amplitude of V^+ and a phase of, say, 0 radian at the generator-end. As the wave travels towards the load-end, its amplitude goes on decreasing exponentially and its phase goes on decreasing linearly with the distance, z, from the generator-end. The wave, therefore, arrives at the load-end with a reduced amplitude of $V^+ e^{-\alpha L}$ and a phase of $-\beta_g L$ radians.

2. The other wave present on the line is travelling from the load-end to the generator-end and may, therefore, be called the ***reflected wave.*** It arises out of the incident wave and is produced by the difference between the load impedance and the characteristic impedance. The amplitude and phase of

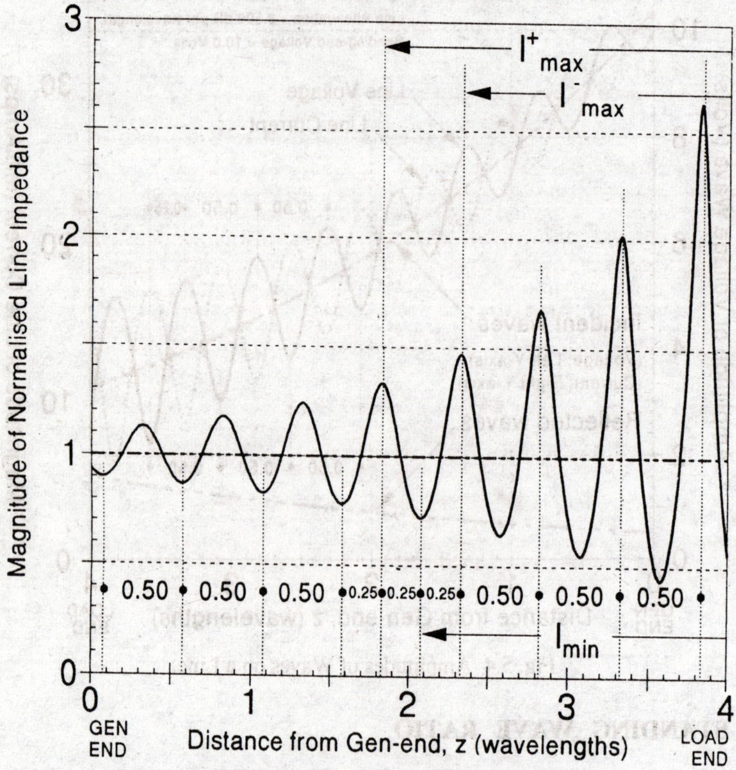

Fig. 5.5 Variation of Impedance along a Line.

the reflected wave at the load-end are $|\rho_L| V^+ e^{-\alpha L}$ and $(-\beta_g L + \phi_L)$, respectively. As the reflected wave travels towards the generator-end, its amplitude goes on decreasing exponentially and its phase goes on decreasing linearly with the distance, l, from the load-end. The wave, therefore, arrives at the generator-end with a reduced amplitude of $V^- = [V^+ e^{-\alpha L}] e^{-\alpha L} = V^+ e^{-2\alpha L}$ and a phase of $(-2\beta_g L)$.

At any point on the line, at a distance of l from the load-end, as the phase difference between these two waves is a function of l only and not of time, a stable stationary or standing wave would be formed on the line. At some points, l_{max}, the two waves would be in-phase with each other so that their amplitudes add. These points are called the antinodes. At some other points, l_{min}, the two waves would be opposite in phase so that the net amplitude is the difference of their amplitudes. These points are called the nodes.

It may be pointed out that, in the general case, $\alpha \neq 0$, the amplitudes of the two waves are nowhere equal except at the load-end and that too for the special cases of the load being either a short-circuit or an open-circuit. Therefore, the amplitude at the nodes will never be zero and that at antinodes will never be twice the amplitude of the incident wave at that point. For the special cases mentioned, $|\rho_L| = 1$ ($z_L = 0$ or ∞), the voltage at the load-end is zero for the short-circuit and is twice that of the incident wave for the case of the open-circuit. The ratio of the amplitude of the voltage standing wave at an antinode to its value at a node is defined as the ***voltage standing wave ratio (VSWR)***. Similarly, the ***current standing wave ratio*** may be defined.

Defined in this way, therefore, the voltage standing wave ratio, denoted by S, is a real number greater than or equal to unity. As discussed in the previous Section, if $Z_L = Z_0$, $\rho_L = 0$, and there is no reflected wave on the line. Therefore, no nodes or antinodes are formed and there exists only a forward travelling wave. Voltage standing wave ratio for this case would, therefore, be unity. By definition, the voltage standing wave ratio, S, is given by

$$S = \frac{|V(l)|_{\text{max at } l = l_{max}}}{|V(l)|_{\text{min at } l = l_{min}}} \geq 1 \tag{5.30}$$

It is clear from this equation defining the voltage standing wave ratio (VSWR), that, in the general case, $\alpha \neq 0$, the VSWR at any point on the line, say, a distance l from the load-end, is a function of l. Therefore, to determine the VSWR at the point, l, values of the voltage at its antinode and node are required to be obtained. Now, the variation of the line voltage, $V(l)$, with the distance, l, from the load-end towards the generator, is given by [Ref. Eqn. (2.35)]

$$V(l) = A[e^{\gamma l} + \rho_L e^{-\gamma l}], \tag{5.31}$$

where, $$A = \frac{V_S}{e^{\gamma L} + \rho_L e^{-\gamma L}} = \text{a constant with respect to } l.$$

That is, $$V(l) = Ae^{\gamma l}[1 + \rho(l)] \text{ from Eqn. (5.7)}$$

$$= Ae^{(\alpha l + j\beta_g l)}[1 + |\rho_L| e^{-2\alpha l} e^{j(\phi_L - 2\beta_g l)}] \tag{5.32}$$

Similarly, $$I(l) = \frac{A}{Z_0} e^{(\alpha l + j\beta_g l)}[1 - |\rho_L| e^{-2\alpha l} e^{j(\phi_L - 2\beta_g l)}] \tag{5.33}$$

Taking the magnitudes of both sides of the above equation, gives

$$\left.\begin{aligned} |V(l)| &= |A| e^{\alpha l}\sqrt{1 + 2|\rho_L| e^{-2\alpha l}\cos(\phi_L - 2\beta_g l) + |\rho_L|^2 e^{-4\alpha l}} \\ \text{and,} \\ |I(l)| &= \frac{|A|}{|Z_0|} e^{\alpha l}\sqrt{1 - 2|\rho_L| e^{-2\alpha l}\cos(\phi_L - 2\beta_g l) + |\rho_L|^2 e^{-4\alpha l}} \end{aligned}\right\} \tag{5.34}$$

This shows that, with respect to l, the line voltage is oscillatory in nature with respect to l, and has its maxima and minima (antinodes and nodes) located at the distances of l_{max} and l_{min}, respectively, from the load-end towards the generator. Variations in the amplitudes of the voltage and current waves as well as the variations in the line voltage and current, are illustrated in Fig. 5.4. Locations, l_{max}, [Ref. Fig. 5.5] of the voltage maxima or antinodes on the line, are given by

$$\cos(\phi_L - 2\,\beta_g\, l_{max}) = 1,$$

that is,

$$\phi_L - 2\,\beta_g\, l_{max} = 0, -2\,\pi, -4\,\pi, \ldots$$
$$= -2\,(k-1)\,\pi, k = 1, 2, 3, \ldots$$

Therefore, $$l_{max} = \frac{\phi_L}{4\,\pi}\,\lambda_g + (k-1)\,\frac{\lambda_g}{2}, k = 1, 2, 3, \ldots \qquad (5.35)$$

Similarly, locations, l_{min}, of the voltage minima or nodes on the line, are given by

$$\cos(\phi_L - 2\,\beta_g\, l_{min}) = -1,$$

that is,

$$\phi_L - 2\,\beta_g\, l_{min} = \mp\pi, -3\,\pi, -5\,\pi, \ldots$$
$$= -2\,(k-1)\,\pi \mp \pi, k = 1, 2, 3, \ldots$$

Therefore, $$l_{min} = \frac{\phi_L}{4\,\pi}\,\lambda_g + (k-1)\,\frac{\lambda_g}{2} \pm \frac{\lambda_g}{4}, k = 1, 2, 3, \ldots \qquad (5.36)$$

From Eqns. (5.35) and (5.36), it is observed that the maxima and minima repeat on the line with the periodicity of $\frac{\lambda_g}{2}$ and the distance between a maximum and the neighbouring minima is $\frac{\lambda_g}{4}$ as shown in Figs. 5.4 and 5.5. That is,

$$l_{max} = l_{min} \pm \frac{\lambda_g}{4}. \qquad (5.37)$$

Note from Eqns.(5.27), (5.35) and (5.36) that, starting from the load-end, $l = 0$, as l is increased, a voltage maximum is encountered first, if $x_L > 0$ ($\phi_L > 0$); and a voltage minimum is encountered first, if $x_L < 0$ ($\phi_L < 0$). Thereafter, minima and maxima are encountered alternately. If $x_L = 0$, the first voltage maximum or minimum occurs at the load-end itself, according as $r_L >$ or < 1.

Substitution of Eqns.(5.35) and (5.36) into Eqn. (5.34) gives the amplitudes of the voltage at the antinodes and nodes to be

$$|\,V(l_{max})\,|_{max} = |\,A\,|\, e^{\alpha\, l_{max}}\,[1 + |\,\rho_L\,|\, e^{-2\,\alpha\, l_{max}}] \qquad (5.38)$$

and

$$|\,V(l_{min})\,|_{min} = |\,A\,|\, e^{\alpha\, l_{min}}\,[1 - |\,\rho_L\,|\, e^{-2\,\alpha\, l_{min}}] \qquad (5.39)$$

giving $$S = e^{\alpha\,(l_{max} - l_{min})}\left[\frac{1 + |\,\rho_L\,|\, e^{-2\,\alpha\, l_{max}}}{1 - |\,\rho_L\,|\, e^{-2\,\alpha\, l_{min}}}\right]$$

Or, $$S^{\pm} = e^{\pm\frac{\alpha\,\lambda_g}{4}}\left[\frac{1 + |\,\rho_L\,|\, e^{-2\,\alpha\, l_{min}}\, e^{\mp\frac{\alpha\,\lambda_g}{2}}}{1 - |\,\rho_L\,|\, e^{-2\,\alpha\, l_{min}}}\right]$$

taking
$$l_{max} = l_{min} \pm \frac{\lambda_g}{4}. \tag{5.40}$$

Note : If the VSWR at any point, P or l, is calculated strictly following its definition, namely, $S = \frac{|V|_{max}}{|V|_{min}}$, and keeping its experimental determination in mind, then it follows from simple logic that it must be the ratio of the value of the line voltage at the voltage maximum nearest to the point, P, and the value of the line voltage at the nearest voltage minimum. Obviously, the voltage maximum and the voltage minimum closest to P would be on opposite sides of the point, P. That is, if the nearest voltage minimum lies on the load-side of P, then the voltage maximum nearest to P must be on the generator-side of P [Ref. Fig. 5.6]. Mathematical expression of this statement is:

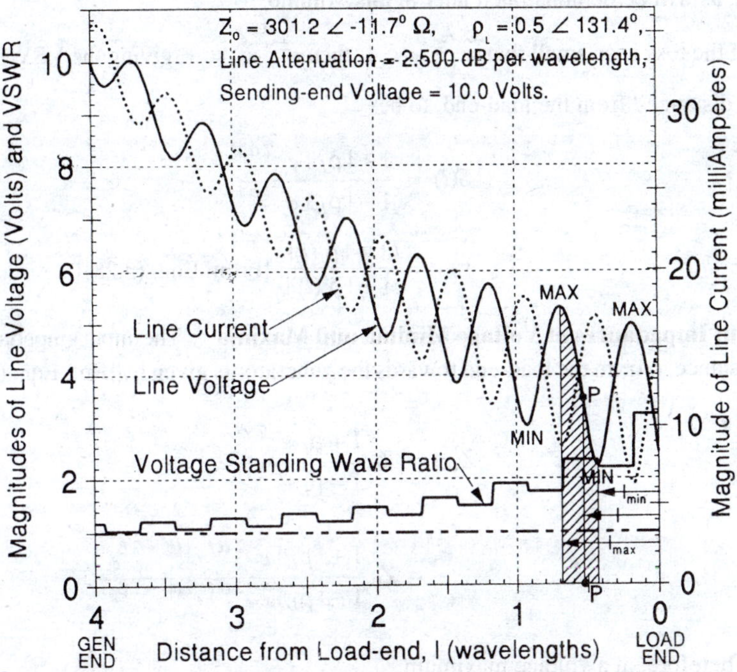

Fig. 5.6 Variation of VSWR along a Lossy Line.

$$l_{min} < l < l_{max} \tag{5.41}$$

If P is moved on the line towards the generator-end and at some point the condition (5.41) is satisfied, then it would remain satisfied over a $\lambda_g/4$ length of the line, shown shaded in Fig. 5.6. The value of the VSWR would be S^+, calculated from Eqn. (5.40) using the upper signs and it would remain constant for all positions of the point, P, on the $\lambda_g/4$ stretch of the line. As P continues its journey towards generator-end, for the next $\lambda_g/4$ length of the line, the following condition would hold.

$$l_{max} < l < l_{min} \tag{5.42}$$

Therefore, over this $\frac{\lambda_g}{4}$ stretch of the line, the VSWR would be S^-, calculated from Eqn. (5.40) using the lower signs. Hence, a plot of the VSWR, S, with l would be a staircase waveform with its largest value at the load-end and decreasing as l increases. At all nodes and antinodes, there would be discontinuities in the S vs. l graph: the magnitude of the discontinuity decreasing as l increases. When l is very large, S settles down to the value 1.

Notwithstanding the discussion under the Note given above, the VSWR, S, may be defined as an analytic function, as given below, for practical transmission lines which may be regarded as lossless or low-loss lines, particularly as high frequencies, as will be demonstrated later in this Volume.

If the loss is so small that $\frac{\alpha \lambda_g}{4} << 1$, then $e^{\pm\frac{\alpha \lambda_g}{4}} \approx 1$, giving the VSWR, $S(l)$, at a distance l from the load-end, to be

$$S(l) \approx \frac{1 + |\rho_L| e^{-2\alpha l}}{1 - |\rho_L| e^{-2\alpha l}} \tag{5.43}$$

$$= \frac{1 + |\rho(l)|}{1 + |\rho(l)|}. \text{ [From Eqn. (5.28)]} \tag{5.44}$$

Input Impedances at Voltage Minima and Maxima : The input impedance at a distance, l, from the load-end towards the generator is given by [Ref. Eqn. (2.53)]

$$Z(z) = Z_0 \frac{1 + \rho_L e^{-2\gamma l}}{1 - \rho_L e^{-2\gamma l}} \tag{5.45}$$

$$= Z_0 \frac{1 + |\rho_L| e^{-2\alpha l} e^{j(\phi_L - 2\beta_g l)}}{1 - |\rho_L| e^{-2\alpha l} e^{j(\phi_L - 2\beta_g l)}} \tag{5.46}$$

Therefore, at a voltage maximum,

$$Z(l_{max}) = Z_0 \frac{1 + |\rho_L| e^{-2\alpha l_{max}}}{1 - |\rho_L| e^{-2\alpha l_{max}}} \approx S Z_0 \text{ for low–loss lines} \tag{5.47}$$

and at a voltage minimum,

$$Z(l_{min}) = Z_0 \frac{1 + |\rho_L| e^{-2\alpha l_{min}}}{1 - |\rho_L| e^{-2\alpha l_{min}}} \approx \frac{1}{S} Z_0 \text{ for low–loss lines} \tag{5.48}$$

It may be observed from Eqn. (5.34), that the points where the line voltage is maximum, the line current is minimum and, therefore, the magnitude of the line impedance is maximum; and at the points where the line voltage is minimum, the line current is maximum and, therefore, the magnitude of the line impedance is minimum.

5.4 POWER FLOW

It would now be interesting to discuss the generator-end and load-end voltages and currents as well as the power flow along a transmission line on the basis of the voltage and current waves that are present on the line in the steady state. For reasons explained earlier, only time-harmonic voltages and currents (whose variation with respect to time can be represented by the factor, $e^{j\omega t}$) would be considered.

5.4.1 Wave Phenomena

When a generator is connected to the generator-end of a line of length l, a voltage wave travelling towards the load-end, is set up. Let this positively travelling wave be called the ***incident wave*** and let its complex amplitude or phasor at the generator-end be

$$V_{S1} = V_{inc}\, e^{j0} \tag{5.49}$$

The incident wave travels along the line with a phase constant, β_g. As it does so, its amplitude gets attenuated at the rate of α nepers per unit length so that the wave arrives at the load-end of the line with its amplitude attenuated by $\alpha\, l$ nepers and its phase retarded by $\beta_g\, l$ radians. Mathematically, these two effects may be taken into account by multiplying the complex amplitude of the wave at the generator-end by the factor, $e^{-\gamma l}$, where, $\gamma = \alpha + j\,\beta_g$. Note that every time the wave travels from one end of the line to the other, in either direction, its phasor or complex amplitude gets multiplied by the factor, $e^{-\gamma l}$. Therefore, the complex amplitude, V_{L1}, of the wave at the load-end is,

$$V_{L1} = e^{-\gamma l}\, V_{inc} \tag{5.50}$$

At the load-end, as the load impedance may not, in general, be equal to the characteristic impedance, that is, as $Z_L \neq Z_0$, a reflected wave is generated at the load-end. It travels towards the generator-end and as it does so, it gets attenuated. The complex amplitude or the phasor of this negatively travelling wave or the ***reflected wave***, as it may be called, at the load-end is

$$V_{L2} = \rho_L\, (e^{-\gamma l}\, V_{inc}) \tag{5.51}$$

where, ρ_L = the voltage reflection coefficient at the load-end $= \dfrac{Z_L - Z_0}{Z_L + Z_0}$.

When it arrives at the generator-end, its amplitude is

$$V_{S2} = e^{-\gamma l} [\rho_L (e^{-\gamma l} V_{inc})] = \rho_L e^{-2\gamma l} V_{inc} \qquad (5.52)$$

Case 1. To begin with, let it be assumed that the internal impedance of the generator is equal to the characteristic impedance of the transmission line so that the reflected wave coming from the load-end, does not get reflected at the generator-end.

Now, as has already been discussed, the net voltage at any point on the line is the sum of the complex amplitudes of

(1) the incident wave or the positively travelling wave, that is, the wave proceeding towards the load-end, and
(2) the reflected wave or the negatively travelling wave, that is, the wave proceeding towards the generator-end.

Therefore, the net voltages, V_S and V_L, appearing at the generator-end and at the load-end, respectively, are [Ref. Eqns.(5.49) - (5.52), and (5.50) - (5.51)]

$$V_S = V_{S1} + V_{S2} = V_{inc} + \rho_L e^{-2\gamma l} V_{inc} = (1 + \rho_S) V_{inc}, \qquad (5.53)$$

where, $\rho_S = \rho_L e^{-2\gamma l}$ = voltage reflection coefficient at the generator–end, looking towards the load-end

and, $$V_L = V_{L1} + V_{L2} = e^{-\gamma l} V_{inc} + \rho_L (e^{-\gamma l} V_{inc}) = (1 + \rho_L) e^{-\gamma l} V_{inc} \qquad (5.54)$$

The phenomena associated with the voltage waves on the line and described above, may be summarised in the tabular form given below:

Table 5.1 Voltage waves

Generator-end		*Load-end*
$V_{inc} \longrightarrow$	$\longrightarrow x\, e^{-\gamma l} \longrightarrow$	$\longrightarrow e^{-\gamma l} V_{inc}$ $\downarrow x\, \rho_L$
$\rho_L V_{inc} e^{-2\gamma l} \longleftarrow$	$\longleftarrow x\, e^{-\gamma l} \longleftarrow$	$\longleftarrow \rho_L e^{-\gamma l} V_{inc}$
net voltage : $V_S = (1 + \rho_S) V_{inc}$	$\rho_S = \rho_L e^{-2\gamma l}$	net voltage : $V_L = (1 + \rho_L) e^{-\gamma l} V_{inc}$

The corresponding current, waves behave exactly the same way as the voltage waves do. However, there is just one difference in that the current reflection coefficient is negative of the voltage reflection coefficient. Therefore, the description of the current waves may be obtained from that of the voltage waves by changing V's to I's and ρ's to $-\rho$'s, yielding.

Table 5.2 Current waves

Generator-end		*Load-end*
$I_{inc} \longrightarrow$	$\longrightarrow \times e^{-\gamma l} \longrightarrow$	$\longrightarrow e^{-\gamma l} I_{inc}$ $\downarrow \times - \rho_L$
$-\rho_L I_{inc} e^{-2\gamma l} \longleftarrow$	$\longleftarrow \times e^{-\gamma l} \longleftarrow$	$\longleftarrow -\rho_L e^{-\gamma l} I_{inc}$
net current :		net current :
$I_S = (1 - \rho_S) I_{inc}$	$\rho_S = \rho_L e^{-2\gamma l}$	$I_L = (1 - \rho_L) e^{-\gamma l} I_{inc}$

It may be noted that as V_{inc} and I_{inc} correspond to the travelling wave, therefore, I_{inc} is quantitatively related to V_{inc} through the relation,

$$I_{inc} = \frac{V_{inc}}{Z_0} \tag{5.55}$$

From Tables 5.1 and 5.2, given above, the line impedances, Z_S and Z_L, at the generator-end and the load-end, respectively, are,

$$Z_S = \frac{V_S}{I_S} = \frac{(1+\rho_S)\, V_{inc}}{(1-\rho_S)\, I_{inc}} = \frac{1+\rho_S}{1-\rho_S}\frac{V_{inc}}{I_{inc}} = \frac{1+\rho_S}{1-\rho_S} Z_0 \tag{5.56}$$

$$Z_L = \frac{V_L}{I_L} = \frac{(1+\rho_L)\, e^{-\gamma l}\, V_{inc}}{(1-\rho_L)\, e^{-\gamma l}\, I_{inc}} = \frac{1+\rho_L}{1-\rho_L}\frac{V_{inc}}{I_{inc}} = \frac{1+\rho_L}{1-\rho_L} Z_0 \tag{5.57}$$

Case 2. Consider now the general case in which the generator impedance is also not equal to the characteristic impedance of the line.

In this case, the reflected wave that gets generated at the load-end and travels down the line to the generator-end, suffers a reflection at the generator-end also. The complex amplitude of this reflected wave produced, is ρ_G times the amplitude of the reflected wave on its arrival at the generator-end from the load-end, where

$$\rho_G = \text{voltage reflection coefficient at the generator–end, looking towards the generator-end}$$

$$= \frac{Z_G - Z_0}{Z_G + Z_0}. \tag{5.58}$$

On arrival at the load-end from the generator-end, the reflected wave is *again* reflected. The complex amplitude of the reflected wave now generated at the load-end, is ρ_L times the amplitude of the incoming reflected wave at the load-end. This phenomenon of multiple reflections, theoretically, goes on indefinitely, and the net voltages and currents at the generator-end and at the load-end may be obtained by summing up the infinite series of voltages contributed by the incident and reflected waves.

Based on this, Table 5.1 may be extended for Case 2, as given below:

Table 5.3 Voltage waves

Generator-end		*Load-end*
$V_{inc} \longrightarrow$	$\longrightarrow x\, e^{-\gamma l} \longrightarrow$	$\longrightarrow e^{-\gamma l}\, V_{inc}$ $\downarrow x\, \rho_L$
$\rho_L\, e^{-2\gamma l}\, V_{inc} \longleftarrow$ $x\, \rho_G \downarrow$	$\longleftarrow x\, e^{-\gamma l} \longleftarrow$	$\longleftarrow \rho_L\, e^{-\gamma l}\, V_{inc}$
$\rho_G\, \rho_L\, e^{-2\gamma l}\, V_{inc} \longrightarrow$	$\longrightarrow x\, e^{-\gamma l} \longrightarrow$	$\longrightarrow \rho_G\, \rho_L\, e^{-3\gamma l}\, V_{inc}$ $\downarrow x\, \rho_L$
$\rho_G\, \rho_L^2\, e^{-4\gamma l}\, V_{inc} \longleftarrow$ $x\, \rho_G \downarrow$	$\longleftarrow x\, e^{-\gamma l} \longleftarrow$	$\longleftarrow \rho_G\, \rho_L^2\, e^{-3\gamma l}\, V_{inc}$
$\rho_G^2\, \rho_L^2\, e^{-4\gamma l}\, V_{inc} \longrightarrow$	$\longrightarrow x\, e^{-\gamma l} \longrightarrow$	$\longrightarrow \rho_G^2\, \rho_L^2\, e^{-5\gamma l}\, V_{inc}$ $\downarrow x\, \rho_L$
$\rho_G^2\, \rho_L^3\, e^{-6\gamma l}\, V_{inc} \longleftarrow$ $x\, \rho_G \downarrow$	$\longleftarrow x\, e^{-\gamma l} \longleftarrow$	$\longleftarrow \rho_G^2\, \rho_L^3\, e^{-5\gamma l}\, V_{inc}$
$\rho_G^3\, \rho_L^3\, e^{-6\gamma l}\, V_{inc} \longrightarrow$	$\longrightarrow x\, e^{-\gamma l} \longrightarrow$	$\longrightarrow \rho_G^3\, \rho_L^3\, e^{-7\gamma l}\, V_{inc}$ $\downarrow x\, \rho_L$
$\rho_G^3\, \rho_L^4\, e^{-8\gamma l}\, V_{inc} \longleftarrow$ $x\, \rho_G \downarrow$	$\longleftarrow x\, e^{-\gamma l} \longleftarrow$	$\longleftarrow \rho_G^3\, \rho_L^4\, e^{-7\gamma l}\, V_{inc}$
....	$\longrightarrow x\, e^{-\gamma l} \longrightarrow$	.... $\downarrow x\, \rho_L$
....	$\longleftarrow x\, e^{-\gamma l} \longleftarrow$	....

The net voltage appearing at the generator-end or the load-end at any other point on the line, may be obtained by summing up the complex amplitudes of the incident wave and the reflected waves arriving at that point after one, two, three, ... reflections.

Thus, the net voltage appearing at the generator-end is:

[It is convenient to sum up all odd-numbered contributions into one group and all even-numbered into another, and then to add these two group totals together.]

$$V_{S_{odd}} = V_{inc} + \rho_G \rho_L e^{-2\gamma l} V_{inc} + \rho_G^2 \rho_L^2 e^{-4\gamma l} V_{inc} + \rho_G^3 \rho_L^3 e^{-6\gamma l} V_{inc} + \ldots \text{ to } \infty$$

$$= V_{inc}\, [1 + \rho_G \rho_L e^{-2\gamma l} + \rho_G^2 \rho_L^2 e^{-4\gamma l} + \rho_G^3 \rho_L^3 e^{-6\gamma l} + \ldots \text{ to } \infty], \text{ and} \tag{5.59}$$

$$V_{S_{even}} = \rho_L e^{-2\gamma l} V_{inc} + \rho_G \rho_L^2 e^{-4\gamma l} V_{inc} + \rho_G^2 \rho_L^3 e^{-6\gamma l} V_{inc} + \rho_G^3 \rho_L^4 e^{-8\gamma l} V_{inc} + \ldots \text{ to } \infty$$

$$= \rho_L e^{-2\gamma l} V_{inc}\, [1 + \rho_G \rho_L e^{-2\gamma l} + \rho_G^2 \rho_L^2 e^{-4\gamma l} + \rho_G^3 \rho_L^3 e^{-6\gamma l} + \ldots \text{ to } \infty]. \tag{5.60}$$

The net voltage, V_S, appearing at the generator-end is now obtained by adding up Eqns. (5.59) and (5.60) to give,

$$V_S = V_{S_{odd}} + V_{S_{even}}$$

$$= (1 + \rho_L e^{-2\gamma l}) V_{inc} [1 + \rho_G \rho_L e^{-2\gamma l} + \rho_G^2 \rho_L^2 e^{-4\gamma l} + \rho_G^3 \rho_L^3 e^{-6\gamma l} + \ldots \text{ to } \infty] \quad (5.61)$$

The infinite series within the square brackets, [.], can be identified as a geometric series with $\rho_G \rho_L e^{-2\gamma l}$ as the common ratio. Its sum to ∞ is

$$1 + \rho_G \rho_L e^{-2\gamma l} + \rho_G^2 \rho_L^2 e^{-4\gamma l} + \rho_G^3 \rho_L^3 e^{-6\gamma l} + \ldots \text{ to } \infty = \frac{1}{1 - \rho_G \rho_L e^{-2\gamma l}} \quad (5.62)$$

Note from Eqn. (5.17) that

$$\rho_S = \rho_L e^{-2\gamma l} \quad (5.63)$$

Substitution of Eqns. (5.62) and (5.63) into Eqn. (5.61) yields the following expression for the generator-end voltage.

$$V_S = \frac{(1 + \rho_S)}{(1 - \rho_G \rho_L e^{-2\gamma l})} V_{inc} = \frac{(1 + \rho_S)}{(1 - \rho_G \rho_S)} V_{inc} \quad (5.64)$$

Similarly, the net voltage appearing at the load-end may be obtained by summing up all odd-numbered contributions into one group, and all even-numbered into another, and then by adding these two group totals together. From Table 5.3, the following are obtained

$$V_{L_{odd}} = e^{-\gamma l} V_{inc} + \rho_G \rho_L e^{-3\gamma l} V_{inc} + \rho_G^2 \rho_L^2 e^{-5\gamma l} V_{inc} + \rho_G^3 \rho_L^3 e^{-7\gamma l} V_{inc} + \ldots \text{ to } \infty$$

$$= e^{-\gamma l} V_{inc} [1 + \rho_G \rho_L e^{-2\gamma l} + \rho_G^2 \rho_L^2 e^{-4\gamma l} + \rho_G^3 \rho_L^3 e^{-6\gamma l} V_{inc} + \ldots \text{ to } \infty], \quad (5.65)$$

$$V_{L_{even}} = \rho_L e^{-\gamma l} V_{inc} + \rho_G \rho_L^2 e^{-3\gamma l} V_{inc} + \rho_G^2 \rho_L^3 e^{-5\gamma l} V_{inc} + \rho_G^3 \rho_L^4 e^{-7\gamma l} + \ldots \text{ to } \infty$$

$$= \rho_L e^{-\gamma l} V_{inc} [1 + \rho_G \rho_L e^{-2\gamma l} + \rho_G^2 \rho_L^2 e^{-4\gamma l} + \rho_G^3 \rho_L^3 e^{-6\gamma l} + \ldots \text{ to } \infty]. \quad (5.66)$$

The net voltage, V_L appearing at the load-end is now obtained by adding up Eqns. (5.65) and (5.66) to give,

$$V_L = V_{L_{odd}} + V_{L_{even}}$$

$$= (1 + \rho_L) e^{-\gamma l} V_{inc} [1 + \rho_G \rho_L e^{-2\gamma l} + \rho_G^2 \rho_L^2 e^{-4\gamma l} + \rho_G^3 \rho_L^3 e^{-6\gamma l} + \ldots \text{ to } \infty]$$

$$= \frac{(1 + \rho_L)}{(1 - \rho_G \rho_L e^{-2\gamma l})} e^{-\gamma l} V_{inc} = \frac{(1 + \rho_L)}{(1 - \rho_G \rho_S)} e^{-\gamma l} V_{inc} \text{ [From Eqn. (5.62)]} \quad (5.67)$$

Proceeding in the same way, it may easily be shown that the currents, I_S and I_L at the generator-end and the load-end, respectively, are

$$I_S = \frac{(1-\rho_S)}{(1-\rho_G\,\rho_S)}\, I_{inc} \tag{5.68}$$

$$I_L = \frac{(1-\rho_L)}{(1-\rho_G\,\rho_S)}\, e^{-\gamma l}\, I_{inc} \tag{5.69}$$

On comparing the equations giving the voltages and currents at the two ends of the line in case 2 with the corresponding equations for case 1, it may be observed that the effect of mismatch at the generator-end,

$$\left[Z_G \neq Z_0 \text{ or } \rho_G = \frac{Z_G - Z_0}{Z_G + Z_0} \neq 0 \right]$$

is to multiply the generator-end and load-end voltages and currents by the factor,

$$\frac{1}{1-\rho_G\,\rho_S}.$$

Notes :

1. This factor becomes unity even if ρ_L alone is zero. This is expected because if the load-end is matched ($\rho_L = 0$), there would be no reflected wave and therefore, even if there is mismatch ($\rho_G \neq 0$) at the generator-end, it would have no effect at all.
2. The expressions for the line impedances at the two ends of the line, remain the same as those for case 1.

5.4.2 Power Flow Calculations

Calculation of the power flow at any point on the line is quite straightforward if the voltage and current at that point are known. The (time) average power, P_L, ***delivered*** to, ***accepted*** by, ***dissipated*** in, or ***flowing*** into the terminals of a complex impedance, $Z_L = |Z_L|\, e^{j\theta_L} = R_L + j X_L$, having a voltage, V_L, across it and a current, $I_L = \frac{V_L}{Z_L}$ passing through it, is given by

$$P_L = \frac{1}{2} |I_L|^2 R_L \tag{5.70}$$

$$= \frac{1}{2} \left| \frac{V_L}{Z_L} \right|^2 R_L \qquad \left[\text{Since } I_L = \frac{V_L}{Z_L} \right] \tag{5.71}$$

$$= \frac{1}{2} |V_L|\,|I_L| \cos\theta_L \qquad \left[\text{Since} \left| \frac{V_L}{Z_L} \right|^2 R_L = |V_L| \left| \frac{V_L}{Z_L} \right| \frac{R_L}{|Z_L|} \text{ and } R_L = |Z_L| \cos\theta_L \right] \tag{5.72}$$

where $\cos\theta_L$ = ***power factor***, θ_L = difference in the (phase) angles of the phasors, V_L and I_L.

$$= \frac{1}{2}\,\text{Re}\,[V_L I_L^*] \tag{5.73}$$

Any of the above formulae may be used to calculate the power delivered to a load or to determine the power flow at any point on a transmission line. In the latter case, if the line voltage and current at the point are used, the above formulae give the power that is delivered to the line input impedance at that point, that is, the power flowing towards the load-end at that point.

For example, if generator-end voltage and current are used, the power input to the line at the generator-end may be calculated, and if the load-end voltage and current are used, the power delivered to the load may be determined.

Consider a (Z_0, γ) transmission line having a load of impedance, Z_L, connected at its load-end. Using Eqn. (5.70), the power, P_L, is

$$P_L = \frac{1}{2}|I_L|^2 R_L$$

$$= \frac{1}{2}|(1-\rho_L)\,e^{-\gamma l}\,I_{inc}|^2 R_L \qquad \text{[From Table 5.2]}$$

$$= \frac{1}{2}\left|\frac{2Z_0}{Z_L+Z_0}\frac{V_{inc}}{Z_0}\right|^2 R_L\,e^{-2\alpha l} \qquad \text{[Using Eqn. (5.8) for } \rho_L \text{ and Eqn. (5.55) for } I_{inc}\text{]}$$

$$= \frac{1}{2}\frac{4R_L}{|Z_L+Z_0|^2}\,e^{-2\alpha l}\,|V_{inc}|^2. \tag{5.74}$$

Now, using Eqn. (5.71), the power, P_L, is

$$P_L = \frac{1}{2}\left|\frac{V_L}{Z_L}\right|^2 R_L$$

$$= \frac{1}{2}\left|\frac{(1+\rho_L)\,e^{-\gamma l}\,V_{inc}}{Z_L}\right|^2 R_L \qquad [V_L \text{ from Eqn. (5.54)}]$$

$$= \frac{1}{2}\left|\frac{2Z_L}{Z_L+Z_0}\frac{V_{inc}}{Z_L}\right|^2 R_L\,e^{-2\alpha l} \qquad [\rho_L \text{ from Eqn. (5.8)}]$$

$$= \frac{1}{2}\frac{4R_L}{|Z_L+Z_0|^2}\,e^{-2\alpha l}\,|V_{inc}|^2. \tag{5.75}$$

Lastly, if Eqn. (5.73) is used, the power, P_L, is

$$P_L = \frac{1}{2}\,\text{Re}\left[V_L I_L^*\right]$$

$$= \frac{1}{2}\,\text{Re}\,[(1+\rho_L)e^{-\gamma l}V_{inc}(1-\rho_L^*)e^{-\gamma^* l}\,I_{inc}^*]\ \text{[Eqn. (5.54) for } V_L \text{ and Table 5.1 for } I_L]$$

$$= \frac{1}{2}\,\text{Re}\left[\frac{(1-|\rho_L|^2)+(\rho_L-\rho_L^*)}{Z_0^*}\right]e^{-2\alpha l}\,|V_{inc}|^2 \quad \text{[Eqn. (5.55) for } I_{inc}] \qquad (5.76)$$

Using the expression for ρ_L from Eqn. (5.8) and expressing $|\rho_L|^2$ as the product of ρ_L and its complex conjugate,

$$(1-|\rho_L|^2) = 1-\left[\frac{Z_L-Z_0}{Z_L+Z_0}\right]\left[\frac{Z_L^*-Z_0^*}{Z_L^*+Z_0^*}\right]$$

$$= \frac{(Z_L+Z_0)\,(Z_L^*+Z_0^*)-(Z_L-Z_0)\,(Z_L^*-Z_0^*)}{(Z_L+Z_0)\,(Z_L^*+Z_0^*)}$$

$$= \frac{2\,(Z_0\,Z_L^*+Z_0^*\,Z_L)}{|Z_L+Z_0|^2} \qquad (5.77)$$

and,
$$(\rho_L-\rho_L^*) = \frac{(Z_L-Z_0)\,(Z_L^*+Z_0^*)-(Z_L+Z_0)(Z_L^*-Z_0^*)}{(Z_L+Z_0)\,(Z_L^*+Z_0^*)}$$

$$= \frac{2\,(Z_0^*\,Z_L-Z_0\,Z_L^*)}{|Z_L+Z_0|^2} \qquad (5.78)$$

Substitution of Eqns. (5.77) and (5.78) into Eqn. (5.76) yields

$$P_L = \frac{1}{2}\,\text{Re}\left[\frac{4\,Z_0^*\,Z_L}{Z_0^*\,|Z_L+Z_0|^2}\right]e^{-2\alpha l}\,|V_{inc}|^2$$

$$= \frac{1}{2}\,\frac{4\,R_L}{|Z_L+Z_0|^2}\,e^{-2\alpha l}\,|V_{inc}|^2 \qquad (5.79)$$

5.4.3 Alternative Formulae for Power Flow

Power delivered to a load of complex impedance, Z_L, has been shown to be given by the formula,

$$P_L = \frac{1}{2}\,\frac{4\,R_L}{|Z_L+Z_0|^2}\,e^{-2\alpha l}\,|V_{inc}|^2. \qquad (5.80)$$

The right hand side of Eqn. (5.80) may be rewritten in alternative forms for compactness or for physical interpretation, as discussed below. The power associated with the incident wave at the input (generator-end) is

$$P_{inc} = \frac{1}{2}\left|\frac{V_{inc}}{Z_0}\right|^2 R_0 = \frac{1}{2}\frac{|V_{inc}|^2}{|Z_0|}\cos\theta_0$$

Therefore, $$|V_{inc}|^2 = 2\frac{|Z_0|}{\cos\theta_0}P_{inc} \tag{5.81}$$

Alternatively, $$|V_{inc}|^2 = 2\frac{|Z_0|^2}{R_0}P_{inc} \tag{5.82}$$

Substitution of Eqn. (5.81) into Eqn. (5.80), gives

$$P_L = \frac{1}{2}\frac{4R_L}{|Z_L+Z_0|^2}e^{-2\alpha l}\,2\frac{|Z_0|}{\cos\theta_0}P_{inc}.$$

$$= \frac{4|Z_L|\cos\theta_L}{|Z_L+Z_0|^2}\frac{|Z_0|}{\cos\theta_0}e^{-2\alpha l}P_{inc}.$$

Or, $$P_L = \left|\frac{2\sqrt{Z_L Z_0}}{Z_L+Z_0}\right|^2\frac{\cos\theta_L}{\cos\theta_0}[e^{-2\alpha l}P_{inc}] \tag{5.83}$$

Alternatively, if Eqn. (5.82) is substituted into Eqn. (5.80), the following formula for P_L is obtained.

$$P_L = \frac{1}{2}\frac{4R_L}{|Z_L+Z_0|^2}e^{-2\alpha l}\,2\frac{|Z_0|^2}{R_0}P_{inc}.$$

$$= \left|\frac{2Z_0}{Z_L+Z_0}\right|^2\frac{R_L}{R_0}e^{-2\alpha l}P_{inc}$$

Or, $$P_L = |1-\rho_L|^2\frac{R_L}{R_0}[e^{-2\alpha l}P_{inc}] \tag{5.84}$$

Yet another formula for P_L results, if Eqn. (5.73) is used and the expressions for V_L and I_L taken from Eqn. (5.54) and Table 5.2.

$$P_L = \frac{1}{2}\mathrm{Re}\,[V_L I_L^*]$$

$$= \frac{1}{2}\mathrm{Re}\,[(1+\rho_L)e^{-\gamma l}V_{inc}]\,[(1-\rho_L^*)e^{-\gamma^* l}I_{inc}^*]$$

$$= \frac{1}{2} \text{Re} \, [\{(1 - |\rho_L|^2) + 2j \, \text{Im} \, [\rho_L]\} \, |V_{inc} \, I_{inc}^*| \,] \, e^{-2\alpha l}$$

$$= \frac{1}{2} \text{Re} \left[\{(1 - |\rho_L|^2) + 2j \, \text{Im} \, [\rho_L]\} \left\{ 1 + j \frac{X_0}{R_0} \right\} \right] e^{-2\alpha l} \, (|I|^2 R_0)$$

$$= \left[(1 - |\rho_L|^2) - 2 \frac{X_0}{R_0} \text{Im} \, [\rho_L] \right] [e^{-2\alpha l} P_{inc}]$$

Or, $P_L = [(1 - |\rho_L|^2) - 2 \tan \theta_0 \, \text{Im} \, [\rho_L]] \, [e^{-2\alpha l} P_{inc}]$ (5.85)

Input Power : By expressing the generator-end voltage, V_S, and current, I_S, in terms of the incident wave voltage, V_{inc}, and current, I_{inc}, from Eqn. (5.53) and Table 5.1, the power entering the line or the power flow, P_s or P_{in}, at the input or the generator-end of the line may be calculated as given below.

$$P_{in} = P_S = \frac{1}{2} \text{Re} \, [V_S I_S^*]$$

$$= \frac{1}{2} \text{Re} \, [(1 + \rho_S) \, V_{inc} \, (1 - \rho_S^*) \, I_{inc}^*]$$

$$= \frac{1}{2} \text{Re} \left[\{(1 - |\rho_S|^2) + 2j \, \text{Im} \, [\rho_S]\} \left\{ 1 + j \frac{X_0}{R_0} \right\} \right] (|I_{inc}|^2 R_0)$$

$$= \left[(1 - |\rho_S|^2) - 2 \frac{X_0}{R_0} \text{Im} \, [\rho_S] \right] P_{inc}$$

Or, $P_{in} = P_S = [(1 - |\rho_S|^2) - 2 \tan \theta_0 \, \text{Im} \, [\rho_S]] \, P_{inc}$ (5.86)

By making use of Eqn. (5.86), the power flow, P_L, at the load-end may be expressed in terms of the input power, P_{in}.

Summarising, therefore, the three alternative formulae developed above are

$$\left.\begin{aligned} P_L &= \left| 2 \frac{\sqrt{Z_L Z_0}}{Z_L + Z_0} \right|^2 \frac{\cos \theta_L}{\cos \theta_0} [e^{-2\alpha l} P_{inc}] \\ \text{Or,} \quad & \\ P_L &= |1 - \rho_L|^2 \frac{R_L}{R_0} [e^{-2\alpha l} P_{inc}] \\ \text{Or,} \quad & \\ P_L &= [(1 - |\rho_L|^2) - 2 \tan \theta_0 \, \text{Im} \, [\rho_L]] \, [e^{-2\alpha l} P_{inc}] \end{aligned}\right\} \quad (5.87)$$

Note : From Eqn. (5.85), it may be observed that

the expression, $[(1 - |\rho_L|^2) - 2 \tan \theta_0 \operatorname{Im}[\rho_L]] = [(1 - |\rho_L|^2)]$, (5.88)

if and only if either

1. $\theta_0 = 0$, that is, if the characteristic impedance is a pure resistance, or,
2. $\operatorname{Im}[\rho_L] = 0$, that is, the normalised load impedance is a pure resistance (real).

5.5 SHORT-CIRCUITED LINE

An important case of termination is the one of a short-circuit. Consider, therefore, a line with a short-circuit as the load impedance. In this case,

$$z_L = 0, \rho_L = +1 \underline{/180^\circ}. \text{ and } V_L = 0. \tag{5.89}$$

Substitution of Eqn. (5.89), that is, $V_L = 0$, into Eqns. (2.50) and (2.51) gives the voltage and current on the short-circuited line to be

$$\begin{aligned} V(l) &= Z_0 I_L \sinh \gamma l \\ &= Z_0 I_L \sinh (\alpha l + j\beta_g l) \\ &= Z_0 I_L [\sinh (\alpha l) \cos (\beta_g l) + j \cosh (\alpha l) \sin (\beta_g l)] \\ &= |Z_0| I_L \sqrt{\sinh^2 (\alpha l) + \sin^2 (\beta_g l)} \\ &\qquad \underline{/ \tan^{-1} [\coth (\alpha l) \tan (\beta_g l)] + \text{Ang} [Z_0]}. \end{aligned} \tag{5.90}$$

and the line current,

$$\begin{aligned} I(l) &= I_L \cosh \gamma l \\ &= I_L \cosh (\alpha l + j \beta_g l) \\ &= I_L [\cosh (\alpha l) \cos (\beta_g l) + j \sinh (\alpha l) \sin (\beta_g l)] \\ &= I_L \sqrt{\sinh^2 (\alpha l) + \cos^2 (\beta_g l)} \; \underline{/ \tan^{-1} [\tanh (\alpha l) \tan (\beta_g l)]}. \end{aligned} \tag{5.91}$$

Note from Eqns. (5.90) and (5.91), the term $\sinh^2 (\alpha l)$ under the radical sign is zero at the load-end $l = 0$, and increases slowly and steadily. Therefore, initially both the voltage and current oscillate sinusoidally. But as l increases, $\sinh^2 (\alpha l)$ increases without any limit, whereas the variation in $\sin^2 (\beta_g l)$ remains confined to 1 with the result that, relatively, the amplitude of oscillation decreases with increase in l. At very large distances from the load-end, the $\sinh^2 (\alpha l)$ completely swamps the variation in $\sin^2(\beta_g l)$ or $\cos^2(\beta_g l)$ and the voltage increases practically exponentially with l. Near the load-end (short-circuit) the maxima and minima of $V(l)$ are practically given by the maxima and minima of $\cos(\beta_g l)$. Fig. 5.7 gives the variation of the line voltage with l for a short-circuited line.

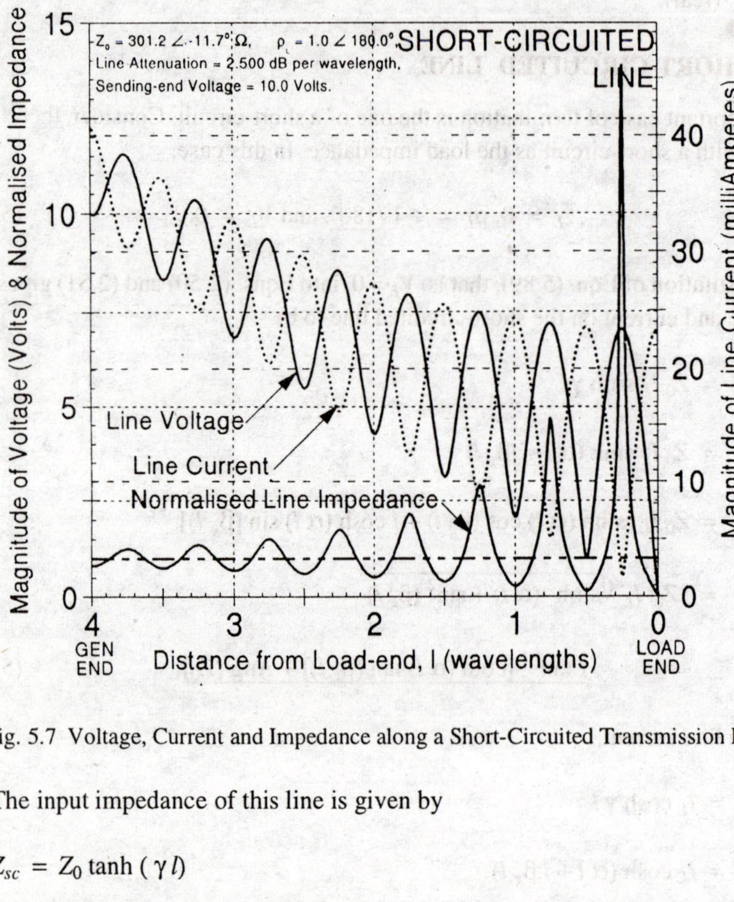

Fig. 5.7 Voltage, Current and Impedance along a Short-Circuited Transmission Line.

The input impedance of this line is given by

$$Z_{sc} = Z_0 \tanh(\gamma l)$$

$$= Z_0 \sqrt{\frac{\sinh^2(\alpha l) + \sin^2(\beta_g l)}{\sinh^2(\alpha l) + \cos^2(\beta_g l)}}$$

$$\underline{/ \tan^{-1}[\coth(\alpha l)\tan(\beta_g l)] - \tan^{-1}[\tanh(\alpha l)\tan(\beta_g l)]}. \qquad (5.92)$$

Now,

$$\tan^{-1}(\coth(\alpha l)\tan(\beta_g l)) - \tan^{-1}(\tanh(\alpha l)\tan(\beta_g l))$$

$$= \tan^{-1}\left[\frac{\coth(\alpha l)\tan(\beta_g l) - \tanh(\alpha l)\tan(\beta_g l)}{1+\tan^2(\beta_g l)}\right]$$

$$= \tan^{-1}\left[\frac{(1-\tanh^2(\alpha l))\tan(\beta_g l)}{\tanh(\alpha l)\sec^2(\beta_g l)}\right]$$

$$= \tan^{-1}\left[\frac{\sin(2\beta_g l)}{\sinh(2\alpha l)}\right] \tag{5.93}$$

Therefore, substitution of Eqn. (5.93) into Eqn. (5.92) gives

$$Z_{sc} = Z_0\sqrt{\frac{\sinh^2(\alpha l)+\sin^2(\beta_g l)}{\sinh^2(\alpha l)+\cos^2(\beta_g l)}}\ \underline{/\tan^{-1}\left[\frac{\sin(2\beta_g l)}{\sinh(2\alpha l)}\right]}. \tag{5.94}$$

The magnitude of the input impedance also shows similar variation with the length, l, of the line [Ref. Fig. 5.7]. As l becomes large, Z_{sc} becomes nearly equal to Z_0.

5.6 OPEN-CIRCUITED LINE

Another important case of termination is that of an open-circuit. Consider, therefore, a line with an open-circuit as the load impedance. In this case,

$$z_L \to \infty,\ \rho_L = +1\,\underline{/0^\circ}.\ \text{and}\ I_L = 0. \tag{5.95}$$

Substitution of Eqn. (5.95), that is, $I_L = 0$, into Eqns. (2.50) and (2.51) gives the voltage and current on the open-circuited line to be

$$\begin{aligned} V(l) &= V_L\cosh\gamma l \\ &= V_L\cosh(\alpha l + j\beta_g l) \\ &= V_L[\cosh(\alpha l)\cos(\beta_g l) + j\sinh(\alpha l)\sin(\beta_g l)] \\ &= V_L\sqrt{\sinh^2(\alpha l)+\cos^2(\beta_g l)}\ \underline{/\tan^{-1}[\tanh(\alpha l)\tan(\beta_g l)]}. \end{aligned} \tag{5.96}$$

and the line current,

$$\begin{aligned} I(l) &= \frac{V_L}{Z_0}\sinh\gamma l \\ &= \frac{V_L}{Z_0}\sinh(\alpha l + j\beta_g l) \\ &= \frac{V_L}{Z_0}[\sinh(\alpha l)\cos(\beta_g l) + j\cosh(\alpha l)\sin(\beta_g l)] \\ &= \frac{V_L}{|Z_0|}\sqrt{\sinh^2(\alpha l)+\sin^2(\beta_g l)}\ \underline{/\tan^{-1}\left[\frac{\tan[\beta_g l]}{\tanh(\alpha l)}\right] - \text{Ang}[Z_0]} \end{aligned} \tag{5.97}$$

Note from Eqns. (5.96) and (5.97) that the term, $\sinh^2 (\alpha l)$, under the radical sign is zero at the load-end, $l = 0$, and increases slowly and steadily. Therefore, initially both the voltage and current oscillate sinusoidally. But as l increases, $\sinh^2 (\alpha l)$ increases without any limit, whereas the variation in $\sin^2 (\beta_g l)$ remains confined to 1 with the result that, relatively, the amplitude of oscillation decreases with an increase in l. At very large distances from the load-end, the $\sinh^2 (\alpha l)$ completely swamps the variation in $\sin^2 (\beta_g l)$ or $\cos^2 (\beta_g l)$ and the voltage increases practically exponentially with l. Near the load-end (open-circuit) the maxima and minima of $V(l)$ are practically given by the maxima and minima of $\cos (\beta_g l)$. Fig. 5.8 gives the variation of the line voltage with l for an Open-Circuited line. The input impedance of this line is given by

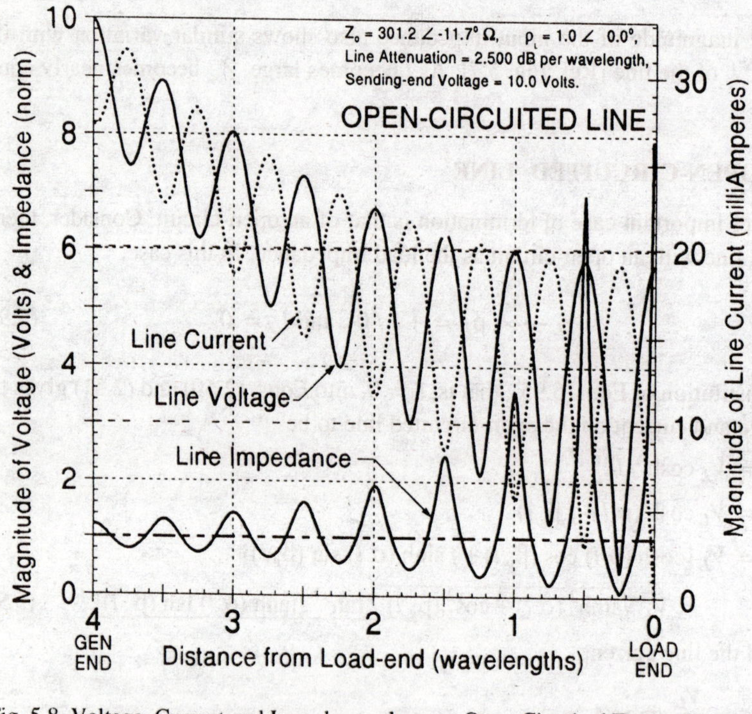

Fig. 5.8 Voltage, Current and Impedance along an Open-Circuited Transmission Line.

$$Z_{oc} = Z_0 \coth (\gamma l)$$

$$= Z_0 \sqrt{\frac{\sinh^2 (\alpha l) + \cos^2 (\beta_g l)}{\sinh^2 (\alpha l) + \sin^2 (\beta_g l)}} \angle \tan^{-1} [\tanh (\alpha l) \tan (\beta_g l)] - \tanh \left(\frac{\tan (\beta_g l)}{\tan (\alpha l)} \right) \quad (5.98)$$

Now,

$$\tan^{-1}[\tanh(\alpha l)\tan(\beta_g l)] - \tan^{-1}\left[\frac{\tan(\beta_g l)}{\tanh(\alpha l)}\right]$$

$$= \tan^{-1}\left[\frac{\tanh(\alpha l)\tan(\beta_g l) - \dfrac{\tan(\beta_g l)}{\tan(\alpha l)}}{1+\tan^2(\beta_g l)}\right]$$

$$= \tan^{-1}\left[\frac{\dfrac{\tanh^2(\alpha l)-1}{\tanh(\alpha l)}\tan(\beta_g l)}{\sec^2(\beta_g l)}\right]$$

$$= -\tan^{-1}\left[\frac{\sin(2\beta_g l)}{\sinh(2\alpha l)}\right] \tag{5.99}$$

Therefore, substitution of Eqn. (5.99) into Eqn. .(5.98) gives

$$Z_{oc} = Z_0\left[\frac{\sinh^2(\alpha l)+\cos^2(\beta_g l)}{\sinh^2(\alpha l)+\sin^2(\beta_g l)}\right]^{\frac{1}{2}} \angle \tan^{-1}\left[\frac{\sin(2\beta_g l)}{\sinh(2\alpha l)}\right] \tag{5.100}$$

The magnitude of the input impedance also shows similar variation with the length l of the line [Ref. Fig. 5.8]. As l becomes large, Z_{oc} differs very little from Z_0.

On comparison of Eqns. (5.96), (5.97) and (5.98) with Eqns.(5.90), (5.91) and (5.92), it is found that this case is a dual of the short-circuited line. Variation of the voltage and current with distance in the present case, is the same as the variation of the current and voltage, respectively, in the previous case. In fact, the discussion given above can be used here with the changes required by the duality.

5.7 FAULT LOCATION ON LINES

If a fault like a shorting or breaking of lines has occurred on a long telephone line, its position can be easily located by a simple experiment in which the magnitude of the input impedance of the line is studied as a function of frequency. The method is based on the fact that the fault acts as a discontinuity and, therefore, sends back a reflected wave [Ref. Fig. 5.9]. The fault thus gives rise to a voltage reflection coefficient in the line, whose value at the location of the fault may be given by

$$\rho_F = |\rho_F|\, e^{j\phi_F} \tag{5.101}$$

The input impedance, Z_{in} of the line at a distance, l, from the fault is given by [Ref. Eqn. (2.53)]

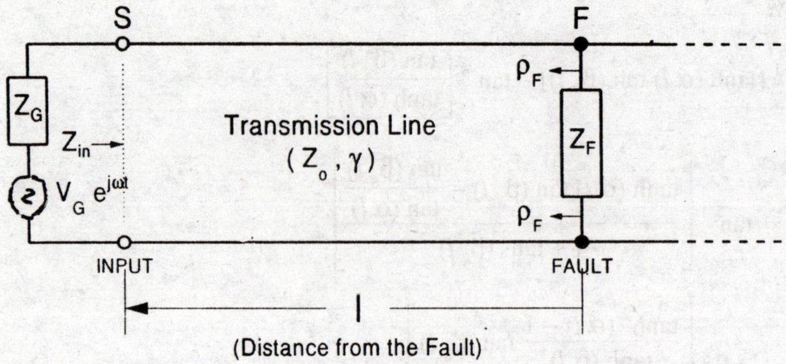

Fig. 5.9 Locating a Fault on a Line.

$$Z_{in} = Z_0 \frac{1 + \rho_F e^{-2\gamma l}}{1 - \rho_F e^{-2\gamma l}}$$

$$= Z_0 \frac{1 + |\rho_F| e^{-2\alpha l} e^{j(\phi_F - 2\beta_g l)}}{1 - |\rho_F| e^{-2\alpha l} e^{j(\phi_F - 2\beta_g l)}} \quad (5.102)$$

From this expression for the input impedance, Z_{in}, it is clear that if the frequency of operation is varied thereby varying $2\beta_g l$, $|Z_{in}|$ goes through maxima, $|Z_{in}|_{max}$, and minima, $|Z_{in}|_{min}$ [Ref. Fig. 5.10]. The value of $|Z_{in}|_{max}$ and $|Z_{in}|_{min}$ do not change with frequency provided α and ρ_F are constants with respect to frequency. Clearly, the frequency separation between any two adjacent maxima of $|Z_{in}|$ versus frequency curve corresponds to a change of 2π radians in the value of $2\beta_g l$.

Suppose, in an experiment, the magnitude of the input impedance is found to be maximum at a frequency, f_L, and the next maximum is observed to be at f_H. Then,

$$2\beta_{gH} l = 2\beta_{gL} l + 2\pi \quad (5.103)$$

or

$$l = \frac{\pi}{\beta_{gH} - \beta_{gL}} \quad (5.104)$$

Now,

$$\left.\begin{aligned} \beta_{gL} &= \frac{2\pi}{\lambda_{gL}} = \frac{2\pi f_L}{v_L} \\ \text{and} \quad \beta_{gH} &= \frac{2\pi}{\lambda_{gH}} = \frac{2\pi f_H}{v_H} \end{aligned}\right\} \quad (5.105)$$

where, v_L and v_H are the phase velocities of the waves along the line at the line at the lower frequency, f_L and the higher frequency f_H, respectively. Usually, f_L and f_H are close to each other, so that the velocity at these two frequencies, is practically the same.

That is,

$$v_L \sim v_H = v \quad (5.106)$$

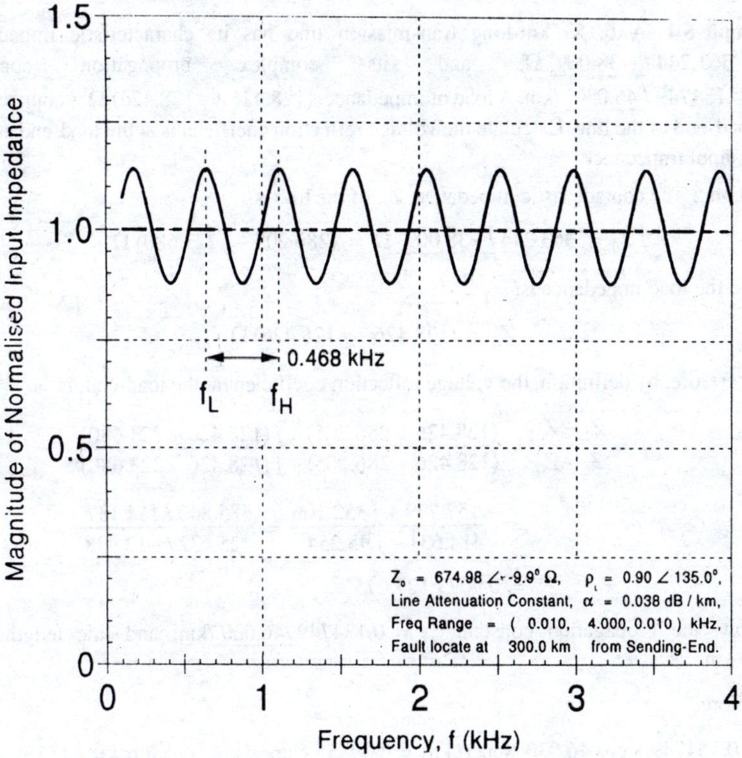

Fig. 5.10 Variation of Magnitude of the Input Line Impedance with Frequency.

Plugging Eqns.(5.105) and (5.106) into Eqn. (5.104) gives,

$$l = \frac{v}{2\,(f_H - f_L)} \tag{5.107}$$

It may be pointed out that the larger the reflection coefficient at the fault and the less the attenuation between the line fault and the point at which the impedance is measured, the greater will be the magnitude of the impedance variation.

These impedance measurements provide a method by which the time that the wave takes to travel to the reflecting point and return to the sending end can be accurately measured. This return time is

$$\frac{2\,l}{v} = \frac{1}{f_H - f_L} \tag{5.108}$$

and if the velocity, v, is accurately known, then the distance of the fault can be computed accurately.

SOLVED EXAMPLES

Example 5.1 A 6.153 km-long transmission line has its characteristic impedance, $Z_0 = 363.244 \underline{/-38.009°}\ \Omega$, and its complex propagation constant, $\gamma = 0.134749 \underline{/46.030°}$ /km. A load of impedance, $(128.426 + j\,128.426)\ \Omega$ is connected at the load-end of the line. Calculate the voltage reflection coefficients at the load-end as well as its input impedance.

Solution : The characteristic impedance, Z_0, of the line is,

$$Z_0 = 363.244 \underline{/-38.009°}\ \Omega = (286.205 - j\,223.680)\ \Omega,$$

and the load impedance is,

$$Z_L = (128.426 + j\,128.426)\ \Omega.$$

Therefore, by definition, the voltage reflection coefficient at the load-end, is

$$\rho_L = \frac{Z_L - Z_0}{Z_L + Z_0} = \frac{(128.426 - 286.205) + j\,(128.426 + 223.680)}{(128.426 + 286.205) + j\,(128.426 - 223.680)}$$

$$= \frac{-157.779 + j\,352.106}{414.631 - j\,95.254} = \frac{385.840 \underline{/114.137°}}{425.432 \underline{/-12.938°}}$$

$$= 0.907 \underline{/127.075°}$$

Now, the propagation constant, $\gamma = 0.134749 \underline{/46.030°}$/km, and line length, l = 6.153 km.

Therefore,

$$\alpha = 0.134749 \times \cos 46.030 \text{ Neper/ km.} = 0.09355 \text{ Neper/ km. and, } 2\,\alpha\, l = 1.1513 \text{Neper}$$

Also, $\beta_g = 0.134749 \times \frac{180}{\pi} \sin 46.030$ degrees/km $= 5.5565$ degrees/km, therefore, $2\beta_g\, l = 68.3784$ degrees.

Hence $e^{-2\gamma l} = e^{-1.1513} \underline{/-68.3784°}$ The voltage reflection coefficient, ρ_S, at the generator-end of the line may now be calculated using the formula,

$$\rho_S = \rho_L e^{-2\gamma l} = [0.907 \underline{/127.075°}]\,[e^{-1.1513} \underline{/-68.378°}]$$

$$= 0.2868 \underline{/58.697°} = 0.149 + j\,0.245$$

The input impedance, Z_S, of the line can now be calculated from the voltage reflection coefficient, ρ_S, as illustrated below:

$$Z_S = Z_0 \frac{1 + \rho_S}{1 - \rho_S} = Z_0 \frac{1 + (0.149 + j\,0.245)}{1 - (0.149 + j\,0.245)} = Z_0 \frac{1.149 + j\,0.245}{0.851 - j\,0.245}$$

$$= Z_0 \frac{1.1748 \underline{/12.0369°}}{0.8856 \underline{/-16.0609°}}$$

$$= 1.3266 Z_0 \underline{/28.0978°} = (1.3266 \times 363.244) \underline{/(28.0978 - 38.009)°}$$

$$= 481.8795 \underline{/-9.9112°}\ \Omega$$

Example 5.2 A 3.007 km long transmission line has its characteristic impedance, $Z_0 = 363.244 \underline{/-38.009°}\ \Omega$, and its complex propagation constant,

$(0.09355 + j\,0.09698)$/ km. The input impedance of the line is measured to be, $Z_S = 348.756 \underline{/15.997^\circ}\ \Omega$.

Calculate the voltage reflection coefficients at the generator-end and load-end as well as its load impedance.

Solution : This problem is an 'inverse' of those given above in Example 5.1, in that here the input impedance is given and the voltage reflection coefficient and impedance at the load-end are required to be calculated. But the procedure is very similar to the one used for solving Example 5.1. Therefore, only the steps of calculation, instead of the detailed work out, are being given.

As the first step, the voltage reflection coefficient, ρ_S, at the input or generator-end may be calculated using the data supplied. Then, by using the relation, $\rho_L = \rho_S \times e^{+2\gamma l}$ (Note the '+' sign in the exponent of ' '.), the voltage reflection coefficient, ρ_L, may be calculated. Finally, using the relationship between the line impedance and the voltage reflection coefficient, the load impedance may be obtained. The results are, however, given below.

[**Ans.** $\rho_S = 0.5100 \underline{/92.881^\circ}$, $\rho_L = 0.9069 \underline{/127.075^\circ}$ and $Z_L = 181.6213 \underline{/45.001^\circ}\ \Omega$.]

EXERCISES

1. If the length of the line in Example 5.1 is changed to 12.306 km, all other parameters remaining the same, what would the voltage reflection coefficients at the two ends and the input impedance be?
   [**Ans.** $\rho_L = 0.9069 \underline{/127.075^\circ}$, $\rho_S = 0.0907 \underline{/-9.681^\circ}$ and $Z_S = 434.554 \underline{/-39.771^\circ}\ \Omega$.]
2. If the length of the line in Example 5.1 is changed to 3.077 km, all other parameters remaining the same, what would the voltage reflection coefficients at the two ends and the input impedance be?
   [**Ans.** $\rho_L = 0.9069 \underline{/127.075^\circ}$, $\rho_S = 0.5100 \underline{/92.881^\circ}$ and $Z_S = 348.7564 \underline{/14.997^\circ}\ \Omega$.]
3. A 3.077 km long transmission line has its characteristic impedence, $Z_0 = 363.244 \underline{/-38.009^\circ}\ \Omega$, and its complex propagation constant, $\gamma = 0.134729 \underline{/46.030^\circ}$/km. A load of impedance, $(363.244 + j\,0)\ \Omega$ is connected at the load-end of the line. Calculate the voltage reflection coefficients at the load-end and generator-end as well as its input impedance.
   [**Ans.** $\rho_L = 0.3444 \underline{/90.000^\circ}$, $\rho_S = 0.1937 \underline{/55.805^\circ}$ and $Z_S = 449.4602 \underline{/-19.599^\circ}\ \Omega$.]
4. If the length of the line in Exercise 3 is changed to 6.153 km, all other parameters remaining the same, what would the voltage reflection coefficients at the two ends and the input impedance be?
   [**Ans.** $\rho_L = 0.3444 \underline{/90.000^\circ}$, $\rho_S = 0.1089 \underline{/21.622^\circ}$ and $Z_S = 444.9409 \underline{/-33.365^\circ}\ \Omega$.]
5. If the length of the line in Exercise 3 is changed to 12.306 km, all other parameters remaining the same, what would the voltage reflection coefficients at the two ends and the input impedance be?
   [**Ans.** $\rho_L = 0.3444 \underline{/90.000^\circ}$, $\rho_S = 0.0344 \underline{/-46.757^\circ}$ and $Z_S = 380.7894 \underline{/-40.885^\circ}\ \Omega$.]
6. If the length of the line in Exercise 2 is changed to 6.153 km and the input line impedance to $Z_S = 481.904 \underline{/-9.905^\circ}\ \Omega$, all other parameters remaining the same, what would the voltage reflection coefficients at the two ends and the load impedance be?
   [**Ans.** $\rho_S = 0.2868 \underline{/58.697^\circ}$, $\rho_L = 0.9069 \underline{/127.075^\circ}$ and $Z_L = 181.6217 \underline{/45.000^\circ}\ \Omega$.]
7. If the length of the line in Exercise 2 is changed to 12.306 km and the input line impedance to $Z_S = 434.554 \underline{/-39.771^\circ}\ \Omega$, all other parameters remaining the same,

what would the voltage reflection coefficients at the two ends and the load impedance be?

[**Ans.** $\rho_S = 0.0907 \underline{/-9.683^\circ}$, $\rho_L = 0.9069 \underline{/127.075^\circ}$ and $Z_L = 181.6292 \underline{/45.000^\circ}$ Ω.]

8. If the input impedance of the line in Exercise 2 is changed to $Z_S = 380.789 \underline{/-40.885^\circ}$ Ω, all other parameters remaining the same, what would the voltage reflection coefficients at the two ends and the load impedance be?

[**Ans.** $\rho_S = 0.0344 \underline{/-46.757^\circ}$, $\rho_L = 0.3444 \underline{/89.999^\circ}$ and $Z_L = 363.2464 \underline{/0^\circ}$ Ω]

9. In order to locate a line fault, the variation in the magnitude of the input impedance of the line is studied as a function of the frequency of operation. In one case, the magnitude of the input impedance is found to have its maximum at a frequency of 456 Hz and it is found to reach its next maximum at 1654 Hz. If the velocity of propagation along the line is assumed to be constant and equal to 28.604×10^4 km/sec over this frequency range, estimate the distance of the fault.

[**Ans.** Fault location = 119.4 km approximately from the sending end.]

10. For a short-circuited line of fixed length l, and assuming α and Z_0 to be frequency independent, derive a formula giving the difference between two consecutive frequencies, f_H and f_L, at which the magnitude of the input impedance of the line is maximum.

11. Check whether the formula derived in Exercise 10 is applicable to the case of open-circuited lines also.

12. Finally, check whether the formula derived in Exercise 10 is applicable to the case of lines terminated in any arbitrarily prescribed impedance also.

Chapter 6

LINE AS A TWO-PORT NETWORK

6.1 INTRODUCTION

A transmission line of a finite length terminated in an arbitrary, load, Z_L, has, after all, an input port—its sending-end or the generator-end, and an output port—its receiving-end or the load-end. Therefore, it may as well be regarded as a two-port communication network. Like any other two-port network, its terminal behaviour may, for some applications, best be described in terms of its two-port network parameters like the *z*-parameters, *y*-parameters, *h*-parameters, or *ABCD* parameters. Before taking up for discussion the terminal behaviour of a transmission line of a finite length, regarding it as a two-port network, the definitions, properties and inter-relationships of the various network parameters of a general two-port network are reviewed in this chapter.

6.1.1 Two-port Network Parameters

Consider a general two-port network. Let the voltage and current at its input port be V_S and I_S, respectively, and those at the output port be V_L and I_L^-,* as shown in Fig. 6.1. Out of these four quantities, only two may be treated as independent and

---

* In the parlance of network theory, the current at a port is regarded positive, if it is directed ***into*** the network. Therefore, to conform to this convention, I_L^- [which is equal to $-I_L$, where I_L is conventional line current (at the load-end)] has been used as the current at the output port throughout this chapter.

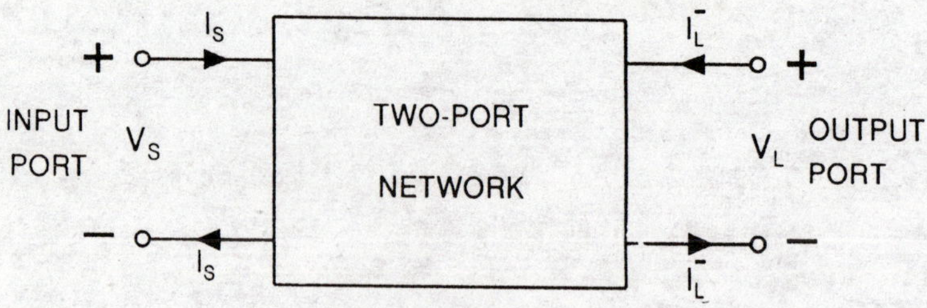

Fig. 6.1 Definitions of Voltages & Currents for a Two-Port Network.

the other two may be expressed as linear combinations of the independent ones. The two independent quantities may be chosen out of the four available, in six different ways, giving rise to six sets of parameters [Ref. Fig. 6.2]. These are described below:

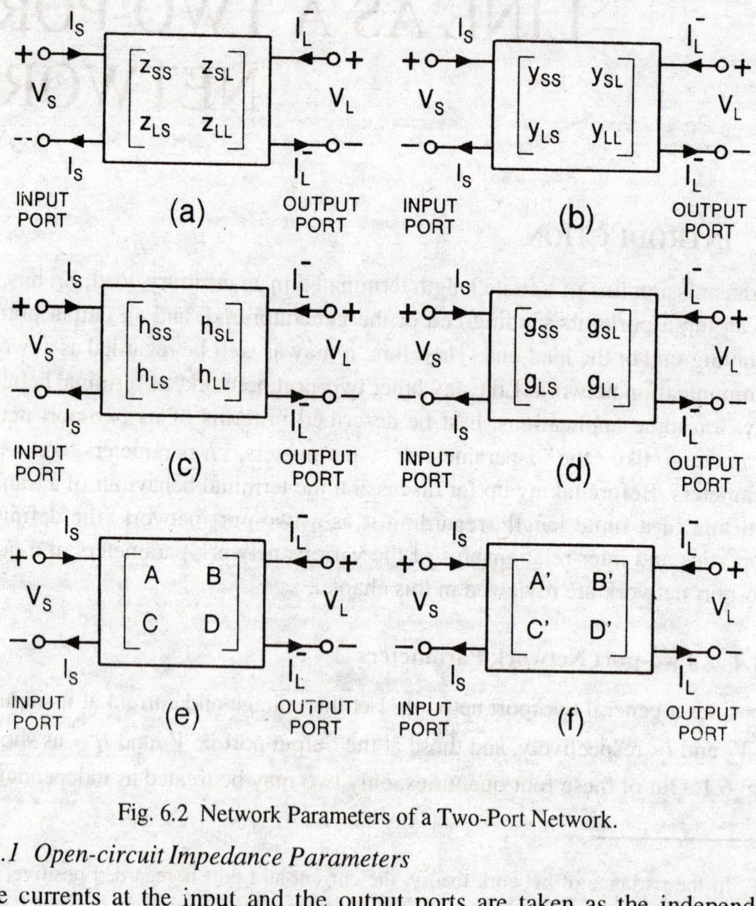

Fig. 6.2 Network Parameters of a Two-Port Network.

6.1.1.1 Open-circuit Impedance Parameters

If the currents at the input and the output ports are taken as the independent variables, the voltages at the two ports may be expressed in terms of I_S and I_L by the relations,

$$V_S = z_{SS} I_S + z_{SL} I_L, \tag{6.1}$$

and

$$V_L = z_{LS} I_S + z_{LL} I_L. \tag{6.2}$$

All the four z-parameters, z_{SS}, z_{SL}, z_{LS} and z_{SS}, have the dimensions of an impedance. Therefore, they are called the ***Open-circuit Impedance Parameters*** or ***z-parameters*** of the two-port network. Define the matrix, Z, of the network having these four z-parameters as its elements, that is,

$$Z = \begin{bmatrix} z_{SS} & z_{SL} \\ z_{LS} & z_{LL} \end{bmatrix}. \tag{6.3}$$

Then, in the matrix notation, Eqns. (6.1) and (6.2) may be written as

$$\begin{bmatrix} V_S \\ V_L \end{bmatrix} = \begin{bmatrix} z_{SS} & z_{SL} \\ z_{LS} & z_{LL} \end{bmatrix} \begin{bmatrix} I_S \\ I_L \end{bmatrix}, \tag{6.4}$$

Or,

$$[V] = [z]\,[I] \tag{6.5}$$

An inspection of the defining equations, Eqns. (6.1) and (6.2) and (6.4), shows that the z-parameters may be determined experimentally by measuring the input and transfer impedances at the port, S, when the port L, is open-circuted and then making these measurements at the port L, with the port, S, open-circuited. Physical interpretation of the z-parameters is given by the following equations:

$$\left.\begin{aligned}
z_{SS} &= \left.\frac{V_S}{I_S}\right|_{I_L=0} = \text{open-circuit input impedance.} \\
z_{LS} &= \left.\frac{V_L}{I_S}\right|_{I_L=0} = \text{open-circuit transfer impedance.} \\
z_{LL} &= \left.\frac{V_L}{I_L}\right|_{I_S=0} = \text{open-circuit output impedance.} \\
z_{SL} &= \left.\frac{V_S}{I_L}\right|_{I_S=0} = \text{open-circuit transfer impedance.}
\end{aligned}\right\} \tag{6.6}$$

From the definitions given above, it follows that if the network is ***reciprocal***,

$$z_{LS} = z_{SL}. \tag{6.7}$$

Further, if the network is electrically ***symmetrical,***

$$z_{LL} = z_{SS}. \tag{6.8}$$

6.1.1.2 Short-circuited Admittance Parameters

Similarly, if the voltages at the input and the output ports are taken as the independent variables, the currents at the two ports may be expressed in terms of V_S and V_L by the relations,

$$I_S = y_{SS} V_S + y_{SL} V_L, \tag{6.9}$$

and $$I_L = y_{LS}\, V_S + y_{LL}\, V_L\,. \tag{6.10}$$

All the four *y*-parameters, y_{SS}, y_{SL}, y_{LS} and y_{SS}, have the dimensions of an admittance. Therefore, they are called the ***Short*-circuit Admittance Parameters** or ***y-parameters*** of the two-port network. Define the matrix, *Y*, of the network having these four *y*-parameters as its elements, that is,

$$Y = \begin{bmatrix} y_{SS} & y_{SL} \\ y_{LS} & y_{LL} \end{bmatrix} \tag{6.11}$$

Then, in the matrix notation, Eqns.(6.9) and (6.10) may be written as

$$\begin{bmatrix} I_S \\ I_L \end{bmatrix} = \begin{bmatrix} y_{SS} & y_{SL} \\ y_{LS} & y_{LL} \end{bmatrix} \begin{bmatrix} V_S \\ V_L \end{bmatrix} \tag{6.12}$$

Or, $$[I] = [y]\,[V]. \tag{6.13}$$

An inspection of the defining equations, Eqns. (6.9) and (6.10) or (6.12) shows that the parameters may be determined experimentally by measuring the input and transfer admittances at the port, *S*, when the port, *L*, is short-circuited and then making these measurements at the port, *L*, with the port, *S*, short-circuited. Physical interpretation of the *y*-parameters is given by the following equations:

$$\left.\begin{aligned} y_{SS} &= \left.\frac{I_S}{V_S}\right|_{V_L=0} = \text{short-circuit input admittance.} \\ y_{LS} &= \left.\frac{I_L}{V_S}\right|_{V_L=0} = \text{short-circuit transfer admittance.} \\ y_{LL} &= \left.\frac{I_L}{V_L}\right|_{V_S=0} = \text{short-circuit output admittance.} \\ y_{SL} &= \left.\frac{I_S}{V_L}\right|_{V_S=0} = \text{short-circuit transfer admittance.} \end{aligned}\right\} \tag{6.14}$$

From the definitions given above, it follows that if the network is ***reciprocal,***

$$y_{LS} = y_{SL} \tag{6.15}$$

Further, if the network is electrically ***symmetrical***,

$$y_{LL} = y_{SS} \tag{6.16}$$

6.1.1.3 Hybrid Parameters

The performance of a two-port network may also be expressed in terms of its ***h-parameters*** which give the input port voltage and the output port current as linear combinations of the input port current and the output port voltage, as given below:

$$V_S = h_{SS}\, I_S + h_{SL}\, V_L\,, \tag{6.17}$$

and $$I_L = h_{LS}\, I_S + h_{LL}\, V_L. \tag{6.18}$$

Define the matrix, H, of the network having these four h-parameters as its elements, that is

$$H = \begin{bmatrix} h_{SS} & h_{SL} \\ h_{LS} & h_{LL} \end{bmatrix}. \tag{6.19}$$

Using the matrix notation, Eqns. (6.17) and (6.18) may be written as

$$\begin{bmatrix} V_S \\ I_L^- \end{bmatrix} = \begin{bmatrix} h_{SS} & h_{SL} \\ h_{LS} & h_{LL} \end{bmatrix} \begin{bmatrix} I_S \\ V_L \end{bmatrix} \tag{6.20}$$

It may be noted from the defining equations, Eqns. (6.17) and (6.18) or (6.20), that whereas h_{SS} has the dimension of an impedance and h_{LL} that of an admittance, h_{SL} and h_{LS} are dimensionless, being the ratios of two voltages and of two currents, respectively. Thus, the h-parameters form a mixed or hybrid set: Therefore, they are called the ***hybrid parameters.*** Physical interpretation of the h-parameters is given by the following equations.

$$\left.\begin{aligned} h_{SS} &= \left.\frac{V_S}{I_S}\right|_{V_L=0} = \text{short-circuit input impedance.} \\ h_{LS} &= \left.\frac{I_L^-}{I_S}\right|_{V_L=0} = \text{short-circuit current gain.} \\ h_{SL} &= \left.\frac{V_S}{V_L}\right|_{I_S=0} = \text{open-circuit reverse voltage gain.} \\ h_{LL} &= \left.\frac{I_L^-}{V_L}\right|_{I_S=0} = \text{open-circuit output admittance.} \end{aligned}\right\} \tag{6.21}$$

It may be easily shown that if the network is ***reciprocal***,

$$h_{LS} = -\,h_{SL}. \tag{6.22}$$

and, if the network is electrically ***symmetrical***,

$$\det[h] = h_{LL}\,h_{SS} - h_{SL}\,h_{LS} = \Delta h = 1. \tag{6.23}$$

6.1.1.4 Inverse Hybrid Parameters

Another set of mixed or hybrid parameters of a two-port network is formed by its ***Inverse hybrid parameters*** or the **g-*parameters*** which give the input port current and the output port voltage as linear combinations of the input port voltage and the output port current, as given below:

$$I_S = g_{SS}\,V_S + g_{SL}\,I_L^-, \tag{6.24}$$

and

$$V_L = g_{LS}\,V_S = g_{LL}\,I_L^-. \tag{6.25}$$

Define the matrix, ***G***, of the network having these four *g*-parameters as its elements, that is,

$$G = \begin{bmatrix} g_{SS} & g_{SL} \\ g_{LS} & g_{LL} \end{bmatrix} \tag{6.26}$$

Using the matrix notation, Eqns. (6.24) and (6.25), may be written as

$$\begin{bmatrix} I_S \\ V_L \end{bmatrix} = \begin{bmatrix} g_{SS} & g_{SL} \\ g_{LS} & g_{LL} \end{bmatrix} \begin{bmatrix} V_S \\ I_L^- \end{bmatrix}. \tag{6.27}$$

Physical interpretation of the *g*-parameters is given by the following equations:

$$\left.\begin{aligned} g_{SS} &= \left.\frac{I_S}{V_S}\right|_{I_L^-=0} = \text{open-circuit input admittance.} \\ g_{LS} &= \left.\frac{V_L}{V_S}\right|_{I_L^-=0} = \text{open-circuit voltage ratio.} \\ g_{LL} &= \left.\frac{V_L}{I_L^-}\right|_{V_S=0} = \text{short-circuit output impedance.} \\ g_{SL} &= \left.\frac{I_S}{I_L^-}\right|_{V_S=0} = \text{short-circuit current ratio.} \end{aligned}\right\} \tag{6.28}$$

It may be easily shown that if the network is ***reciprocal,***

$$g_{LS} = -\,g_{SL} \tag{6.29}$$

and, if the network is electrically symmetrical,

$$\det[g] = g_{LL}\,g_{SS} - g_{SL}\,g_{LS} = \Delta_g = 1. \tag{6.30}$$

6.1.1.5 Transmission Parameters

Properties of two-port networks connected in cascade or in chain or in tandem, may be described best in terms of ***Transmission Parameters*** or **ABCD-*parameters***, also called the ***Chain Parameters*** or ***General Circuit Parameters*** which express the voltage and current at the input port in terms of those at the output port, as given below:

$$V_S = AV_L - B\,I_L^-, \tag{6.31}$$

and

$$I_S = CV_L - DI_L^-. \tag{6.32}$$

Define the matrix, *T*, of the network having *A, B, C* and *D* as its elements, that is,

$$T = \begin{bmatrix} A & B \\ C & D \end{bmatrix} \tag{6.33}$$

Using the matrix notation, Eqns. (6.31) and (6.32), may be written as

$$\begin{bmatrix} V_S \\ I_S \end{bmatrix} = \begin{bmatrix} A & B \\ C & D \end{bmatrix} \begin{bmatrix} V_L \\ -I_L \end{bmatrix}. \tag{6.34}$$

Physical interpretation of the *ABCD*-parameters is given by the following equations:

$$\left.\begin{aligned} \frac{1}{A} &= \left.\frac{V_L}{V_S}\right|_{I_L=0} = \text{open-circuit voltage gain.} \\ \frac{1}{C} &= \left.\frac{V_L}{I_S}\right|_{I_L=0} = \text{open-circuit transfer impedance.} \\ -\frac{1}{B} &= \left.\frac{I_L}{V_S}\right|_{V_L=0} = \text{short-circuit transfer admittance.} \\ -\frac{1}{D} &= \left.\frac{I_L}{I_S}\right|_{V_L=0} = \text{short-circuit current gain.} \end{aligned}\right\} \tag{6.35}$$

It may be easily shown that if the network is ***reciprocal,***

$$AD - BC = 1 \tag{6.36}$$

and, if the network is electrically ***symmetrical,***

$$A = D \tag{6.37}$$

6.1.1.6 Inverse Transmission Parameters

For studying propagation through cascaded two-port networks in the reverse direction, it may be convenient to use the ***Inverse Transmission Parameters*** or ***A′B′C′D′-parameters***, which express the voltage and current at the output port in terms of those at the input port, as given below:

$$V_L = A' V_S - B' I_S, \tag{6.38}$$

and

$$I_L = C' V_S - D' I_S. \tag{6.39}$$

Define the matrix, T', of the network having A', B', C' and D' as its elements, that is,

$$T' = \begin{bmatrix} A' & B' \\ C' & D' \end{bmatrix} \tag{6.40}$$

Using the matrix notation, Eqns. (6.38) and (6.39), may be written as

$$\begin{bmatrix} V_L \\ I_L \end{bmatrix} = \begin{bmatrix} A' & B' \\ C' & D' \end{bmatrix} \begin{bmatrix} V_S \\ -I_S \end{bmatrix} \tag{6.41}$$

Physical interpretation of the $A'B'C'D'$-parameters is given by the following equations:

$$\left.\begin{aligned}
\frac{1}{A'} &= \left.\frac{V_S}{V_L}\right|_{I_S=0} = \text{open-circuit voltage gain.}\\
\frac{1}{C'} &= \left.\frac{V_S}{V_L}\right|_{I_S=0} = \text{open-circuit transfer impedance.}\\
-\frac{1}{B'} &= \left.\frac{I_S}{V_L}\right|_{V_S=0} = \text{short-circuit transfer admittance.}\\
-\frac{1}{D'} &= \left.\frac{I_S}{I_L}\right|_{V_S=0} = \text{short-circuit current gain.}
\end{aligned}\right\} \tag{6.42}$$

It may be easily shown that if the network is ***reciprocal***,

$$A'\,D' - B'\,C' = 1. \tag{6.43}$$

and, if the network is electrically ***symmetrical***,

$$A' = D' \tag{6.44}$$

6.1.2 Relationships among Network Parameters

In the last Section, various network parameters have been described. From the equations defining these parameters, relationships among these parameters may easily be deduced. For instance, from the matrix equations (6.4) and (6.12), it immediately follows that

$$Y = Z^{-1} \tag{6.45}$$

Relationships among all the network parameters are presented in a tabular form[2] in Table 6.1. In this Table, all the matrices along a row are equivalent to each other.

6.1.3 Applications of Network Parameters

Although, in principle, any one set of network parameters is sufficient for describing a two-port network yet six sets of parameters have been defined above. The reason for doing so is mainly the convenience in carrying out the analysis of a network. Depending on the nature of the network to be analysed, the analysis may become very simple and easy if an appropriate set of parameters is employed. In practice, the two-port network required to be analysed may be either given to be

---

2 M.E. van Valkenburg, *Network Analysis*, Prentice Hall of India Pvt. Ltd, 3rd Edition, 1985, Chapter 11.

Table 6.1 Relationships among Network Parameters

	[z]	[y]	[T]	[T′]	[h]	[g]
[z]	$\begin{matrix} z_{SS} & z_{SL} \\ z_{LS} & z_{LL} \end{matrix}$	$\begin{matrix} \frac{y_{LL}}{\Delta_y} & \frac{-y_{SL}}{\Delta_y} \\ \frac{-y_{LS}}{\Delta_y} & \frac{y_{SS}}{\Delta_y} \end{matrix}$	$\begin{matrix} \frac{A}{C} & \frac{\Delta_T}{C} \\ \frac{1}{C} & \frac{D}{C} \end{matrix}$	$\begin{matrix} \frac{D'}{C'} & \frac{1}{C'} \\ \frac{\Delta_T'}{C'} & \frac{A'}{C'} \end{matrix}$	$\begin{matrix} \frac{\Delta_h}{h_{LL}} & \frac{h_{SL}}{h_{LL}} \\ \frac{-h_{LS}}{h_{LL}} & \frac{1}{h_{LL}} \end{matrix}$	$\begin{matrix} \frac{1}{g_{SS}} & \frac{-g_{SL}}{g_{SS}} \\ \frac{g_{LS}}{g_{SS}} & \frac{\Delta_g}{g_{SS}} \end{matrix}$
[y]	$\begin{matrix} \frac{z_{LL}}{\Delta_z} & \frac{-z_{SL}}{\Delta_z} \\ \frac{-z_{LS}}{\Delta_z} & \frac{z_{SS}}{\Delta_z} \end{matrix}$	$\begin{matrix} y_{SS} & y_{SL} \\ y_{LS} & y_{LL} \end{matrix}$	$\begin{matrix} \frac{D}{B} & \frac{-\Delta_T}{B} \\ \frac{-1}{B} & \frac{A}{B} \end{matrix}$	$\begin{matrix} \frac{A'}{B'} & \frac{-1}{B'} \\ \frac{-\Delta_{T'}}{B'} & \frac{D'}{B'} \end{matrix}$	$\begin{matrix} \frac{1}{h_{SS}} & \frac{-h_{SL}}{h_{SS}} \\ \frac{-h_{LS}}{h_{SS}} & \frac{\Delta_h}{h_{SS}} \end{matrix}$	$\begin{matrix} \frac{\Delta_g}{g_{LL}} & \frac{-g_{SL}}{g_{LL}} \\ \frac{-g_{LS}}{g_{LL}} & \frac{1}{g_{LL}} \end{matrix}$
[T]	$\begin{matrix} \frac{z_{SS}}{z_{LS}} & \frac{\Delta_z}{z_{LS}} \\ \frac{1}{z_{LS}} & \frac{z_{LL}}{z_{LS}} \end{matrix}$	$\begin{matrix} \frac{-y_{LL}}{y_{LS}} & \frac{-1}{y_{LS}} \\ \frac{-\Delta_y}{y_{LS}} & \frac{-y_{SS}}{y_{LS}} \end{matrix}$	$\begin{matrix} A & B \\ C & D \end{matrix}$	$\begin{matrix} \frac{D'}{\Delta_{T'}} & \frac{D'}{\Delta_T} \\ \frac{C'}{\Delta_{T'}} & \frac{A'}{\Delta_{T'}} \end{matrix}$	$\begin{matrix} \frac{-\Delta_h}{h_{LS}} & \frac{-h_{SS}}{h_{LS}} \\ \frac{-h_{LL}}{h_{LS}} & \frac{-1}{h_{LS}} \end{matrix}$	$\begin{matrix} \frac{1}{g_{LS}} & \frac{g_{LL}}{g_{LS}} \\ \frac{g_{SS}}{g_{LS}} & \frac{\Delta_g}{g_{LS}} \end{matrix}$
[T′]	$\begin{matrix} \frac{z_{LL}}{z_{SL}} & \frac{\Delta_z}{z_{SL}} \\ \frac{1}{z_{SL}} & \frac{z_{SS}}{z_{SL}} \end{matrix}$	$\begin{matrix} \frac{-y_{SS}}{y_{SL}} & \frac{-1}{y_{SL}} \\ \frac{-\Delta_y}{y_{SL}} & \frac{-y_{LL}}{y_{SL}} \end{matrix}$	$\begin{matrix} \frac{D}{\Delta_T} & \frac{B}{\Delta_T} \\ \frac{C}{\Delta_T} & \frac{A}{\Delta_T} \end{matrix}$	$\begin{matrix} A' & B' \\ C' & D' \end{matrix}$	$\begin{matrix} \frac{1}{h_{SL}} & \frac{h_{SS}}{h_{SL}} \\ \frac{h_{LL}}{h_{SL}} & \frac{\Delta_h}{h_{SL}} \end{matrix}$	$\begin{matrix} \frac{-\Delta_g}{g_{SL}} & \frac{-g_{LL}}{g_{SL}} \\ \frac{-g_{SS}}{g_{SL}} & \frac{-1}{g_{SL}} \end{matrix}$
[h]	$\begin{matrix} \frac{\Delta_z}{z_{LL}} & \frac{z_{SL}}{z_{LL}} \\ \frac{-z_{LS}}{z_{LL}} & \frac{1}{z_{LL}} \end{matrix}$	$\begin{matrix} \frac{1}{y_{SS}} & \frac{-y_{SL}}{y_{SS}} \\ \frac{y_{LS}}{y_{SS}} & \frac{\Delta_y}{y_{SS}} \end{matrix}$	$\begin{matrix} \frac{B}{D} & \frac{\Delta_T}{D} \\ \frac{-1}{D} & \frac{C}{D} \end{matrix}$	$\begin{matrix} \frac{B'}{A'} & \frac{1}{A'} \\ \frac{-\Delta_{T'}}{A'} & \frac{C'}{A'} \end{matrix}$	$\begin{matrix} h_{SS} & h_{SL} \\ h_{LS} & h_{LL} \end{matrix}$	$\begin{matrix} \frac{g_{LL}}{\Delta_g} & \frac{g_{SL}}{\Delta_g} \\ \frac{-g_{LS}}{\Delta_g} & \frac{g_{SS}}{\Delta_g} \end{matrix}$
[g]	$\begin{matrix} \frac{1}{z_{SS}} & \frac{-z_{SL}}{z_{SS}} \\ \frac{z_{LS}}{z_{SS}} & \frac{\Delta_z}{z_{SS}} \end{matrix}$	$\begin{matrix} \frac{\Delta_y}{y_{LL}} & \frac{y_{SL}}{y_{LL}} \\ \frac{-y_{LS}}{y_{LL}} & \frac{1}{y_{LL}} \end{matrix}$	$\begin{matrix} \frac{C}{A} & \frac{-\Delta_T}{A} \\ \frac{1}{A} & \frac{B}{A} \end{matrix}$	$\begin{matrix} \frac{C'}{D'} & \frac{-1}{D'} \\ \frac{\Delta_{T'}}{D'} & \frac{B'}{D'} \end{matrix}$	$\begin{matrix} \frac{h_{LL}}{\Delta_h} & \frac{-h_{SL}}{\Delta_h} \\ \frac{-h_{LS}}{\Delta_h} & \frac{h_{SS}}{\Delta_h} \end{matrix}$	$\begin{matrix} g_{SS} & g_{SL} \\ g_{LS} & g_{LL} \end{matrix}$

an interconnection of two two-port networks or it may be possible to regard it as an interconnection of two simpler two-port networks. Now, two two-port networks, N_1 and N_2, may be interconnected in a variety of ways like the ones given below:

1. Series-Series Connection : The input ports of the two networks are connected in series and their output ports are also connected in series with each other. [Ref. Fig. 6.3]

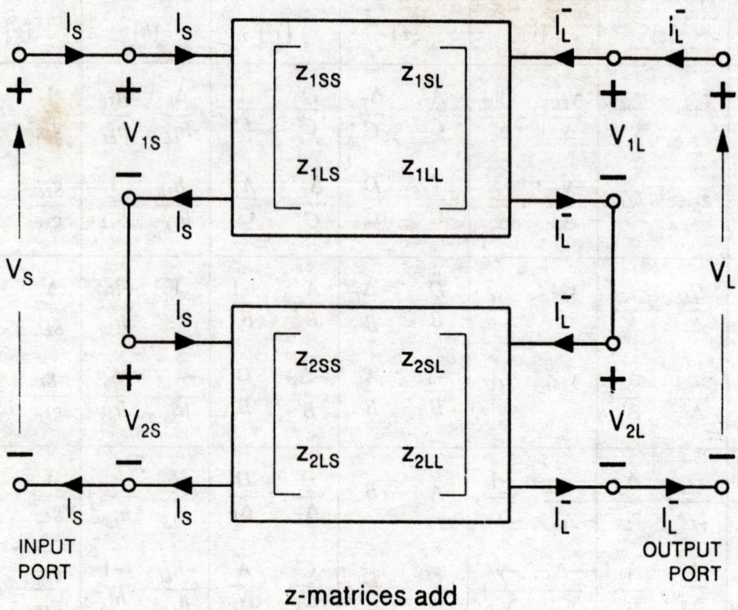

Fig. 6.3 Series-Series Connection of two Two-Port Networks.

In this case, the *z*-matrix of the combination is the sum of the *z*-matrices of the individual networks. Therefore, in this case, it would be convenient to carry out the analysis in terms of the *z*-parameters.

2. Parallel-Parallel Connection : The input ports of the two networks are connected in parallel (or shunt) and their output ports are also connected in parallel (or shunt) with each other. [Ref. Fig. 6.4]

In this case, the *y*-matrix of the combination is the sum of the *y*-matrices of the individual networks. Therefore, in this case, it would be convenient to carry out the analysis in terms of the *y*-parameters.

3. Series-Parallel Connection : The input ports of the two networks are connected in series, but their output ports are connected in parallel (or shunt) with each other. [Ref. Fig. 6.5]

In this case, the *h*-matrix of the combination is the sum of the *h*-matrices of the individual networks. Therefore, in this case, it would be convenient to carry out the analysis in terms of the *h*-parameters.

4. Parallel-Series Connection : The input ports of the two networks are connected in parallel (or shunt) but their output ports are connected in series with each other. [Ref. Fig. 6.6]

In this case, the *g*-matrix of the combination is the sum of the *g*-matrices of the individual networks. Therefore, in this case, it would be convenient to carry out the analysis in terms of the *g*-parameters.

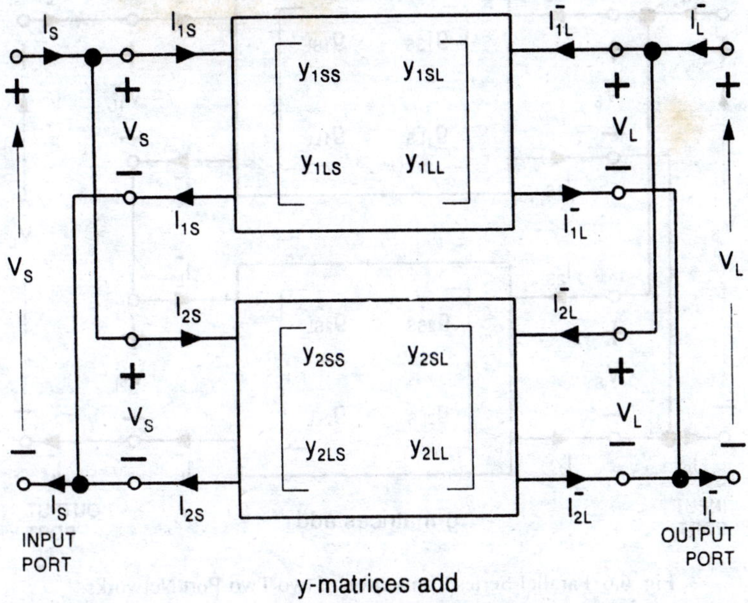

Fig. 6.4 Parallel-Parallel Connection of two Two-Port Networks.

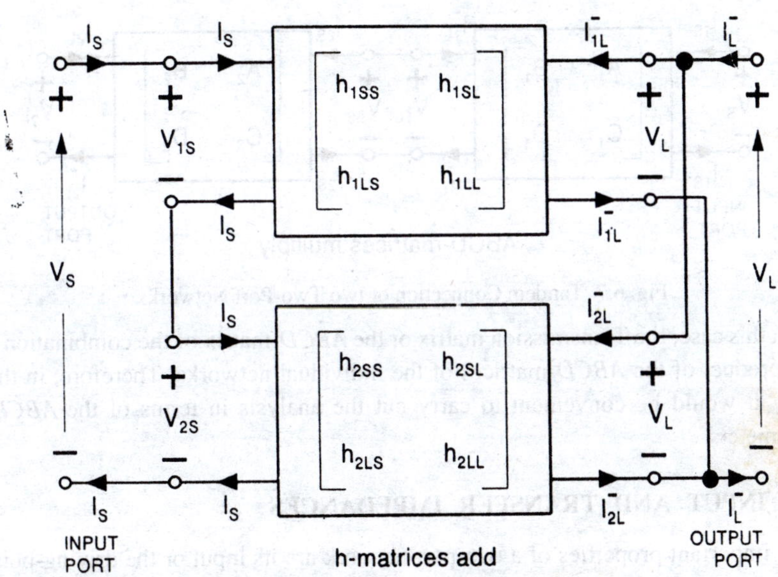

Fig. 6.5 Series-Parallel Connection of two Two-Port Networks.

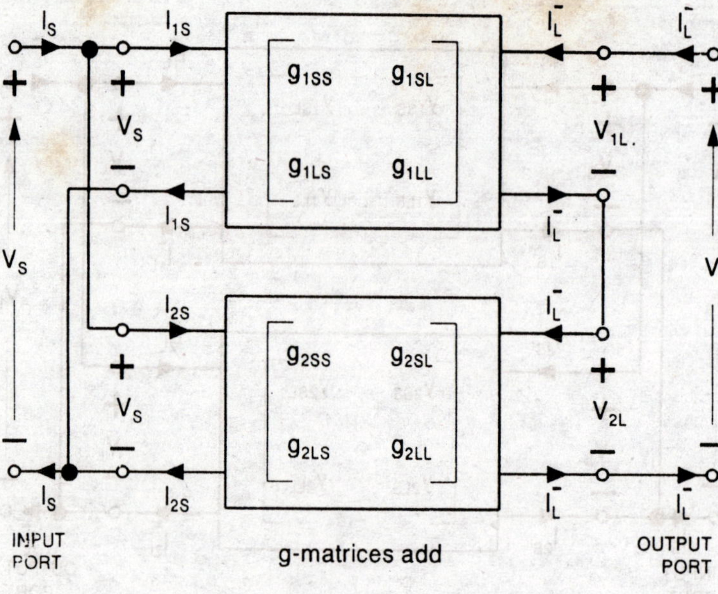

Fig. 6.6 Parallel-Series Connection of two Two-Port Networks.

5. *Cascade or Tandem Connection* : The output port of one network is connected to the input port of the other network, [Ref. Fig. 6.7]

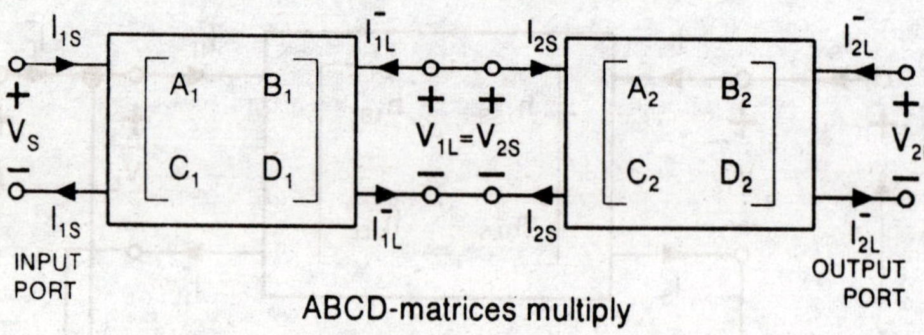

Fig. 6.7 Tandem Connection of two Two-Port Networks.

In this case, the Transmission matrix or the *ABCD*-matrix of the combination is the product of the *ABCD*-matrices of the individual networks. Therefore, in this case, it would be convenient to carry out the analysis in terms of the *ABCD*-parameters.

6.2 INPUT AND TRANSFER IMPEDANCES

Two important properties of a two-port network are its input or the driving-point impedance and the transfer impedance. For a transmission line, the input impedance, Z_{in}, is obtained from Eqn. (2.53) by setting $l = L$ in it, so that

$$Z_{in} = \frac{V_S}{I_S} = Z_o \frac{Z_L + Z_o \tanh(\gamma L)}{Z_o + Z_L \tanh(\gamma L)}. \qquad (6.46)$$

Or, alternatively,

$$Z_{in} = Z_o \frac{1 + \rho_L e^{-2\gamma L}}{1 - \rho_L e^{-2\gamma L}} \tag{6.47}$$

The transfer impedance, Z_{Tjk}, is defined as the ratio of voltage applied in the jth port to the current flowing in any other port k with all other emf's removed. For a transmission line, therefore, the transfer impedance may be defined as

$$Z_T = \frac{V_S}{I_L} \tag{6.48}$$

The expression for this may be obtained from Eqn. (2.50) with $l = L$ that is,

$$Z_T = Z_L \cosh(\gamma L) + Z_o \sinh(\gamma L). \tag{6.49}$$

The $ABCD$ matrix for a transmission line of length L has already been obtained in Eqns. (2.50) and (2.51). It is explicitly given by

$$\begin{bmatrix} V_S \\ I_S \end{bmatrix} = \begin{bmatrix} \cosh(\gamma L) & Z_o \sinh(\gamma L) \\ \frac{1}{Z_o} \sinh(\gamma L) & \cosh(\gamma L) \end{bmatrix} \begin{bmatrix} V_L \\ I_L \end{bmatrix} \tag{6.50}$$

6.3 REFLECTION LOSS

When a line is terminated in Z_o, a wave travelling down the line is completely absorbed at the load-end. If it is not so terminated, all the power that arrives at the load-end, will not be absorbed but a return or a reflected wave is set up. This idea of reflection at a junction is carried over to lumped networks also. If the load impedance Z_L is different from the internal impedance Z_G of the generator, the power delivered to the load may be less than that delivered if the impedance were matched; a reflection loss is said to have occurred. The ratio of the magnitude of the current actually flowing in the load to the magnitude of the current which would flow if the load impedance were matched, is the measure of the reflection loss.

Consider the load Z_L to be fed by the generator (V_G, Z_G) as in Fig. 6.8. If the load is directly connected to the generator, the current through the load is

$$I_L = \frac{V_G}{Z_G + Z_L} \tag{6.51}$$

Let the load be now matched, on the image basis, to the generator by an ideal transformer and a phase shifter. Then the current through the primary of the transformer under image matching is

$$I_G = \frac{V_G}{2Z_G} \tag{6.52}$$

and, therefore, the current through the load would be

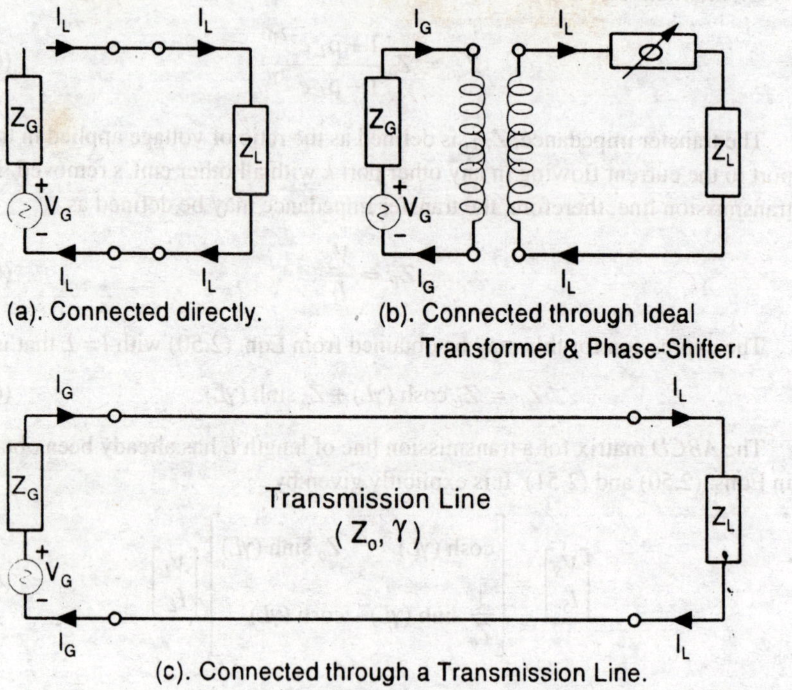

Fig. 6.8 A Load connected to a Generator.

$$I_{Lm} = I_G \sqrt{\frac{Z_G}{Z_L}} = \frac{V_G}{2\sqrt{Z_G Z_L}} \tag{6.53}$$

Hence, the ratio of the magnitude of the current actually flowing in the load to the magnitude of the current which would flow under image-matched conditions is

$$\left|\frac{I_L}{I_{Lm}}\right| = \left|\frac{2\sqrt{Z_G Z_L}}{Z_G + Z_L}\right|$$

$$= \left|\frac{\sqrt{Z_G Z_L}}{\frac{1}{2}(Z_G + Z_L)}\right|$$

$$= \left|\frac{\text{Geometric Mean}}{\text{Arithmetic Mean}} \text{ of } Z_L \text{ and } Z_G\right| \tag{6.54}$$

This ratio gives the change in the magnitude of the load current due to reflection at the mismatched junction and is called the ***reflection factor***. It is denoted by k and is given by

$$k = \left|\frac{2\sqrt{Z_G Z_L}}{Z_G + Z_L}\right| \tag{6.55}$$

The reflection loss, RL, in nepers and decibels, is given respectively by

$$RL = \ln\left(\frac{1}{k}\right)$$

$$= \ln\left|\frac{Z_G + Z_L}{2\sqrt{Z_G Z_L}}\right| \text{ nepers} \tag{6.56}$$

and

$$RL = 20\log\left(\frac{1}{k}\right)$$

$$= 20\log\left|\frac{Z_G + Z_L}{2\sqrt{Z_G Z_L}}\right| \text{ dB} \tag{6.57}$$

6.4 INSERTION LOSS OF A LINE

When a two-port network or a line is inserted between the generator and the load, the load current and the power delivered to the load would change from what they would have been, had the load been connected directly to the generator. This change in power or the power loss which may be positive or negative, due to the insertion of the network or the line is called the ***insertion loss**** of the network or the line. In what follows, an expression for the insertion loss of a line fed by a generator of internal impedance, Z_G, and having Z_L as its load impedance, is derived directly following the definition of the insertion loss.

The load current when the load is directly connected to the generator, is given by

$$I_{LO} = \frac{V_G}{Z_G + Z_L} \tag{6.58}$$

The load current when the line is inserted between the generator and the load, is obtained from Eqn. (2.60) by setting $l = 0$. That is,

$$I_L = \left(\frac{V_G}{Z_o + Z_G}\right)\left[\frac{1 - \rho_L}{e^{\gamma L} - \rho_G \rho_L e^{-\gamma L}}\right] \tag{6.59}$$

$$= \frac{V_G}{Z_o + Z_G} e^{-\gamma L} \frac{\dfrac{2Z_o}{Z_L + Z_o}}{1 - \rho_G \rho_L e^{-2\gamma L}} \tag{6.60}$$

Therefore,

$$\frac{I_{LO}}{I_L} = \frac{Z_o + Z_G}{Z_L + Z_G} e^{\gamma L} \frac{Z_L + Z_o}{2Z_o} [1 - \rho_G \rho_L e^{-2\gamma L}]$$

$$= \frac{Z_o + Z_G}{2\sqrt{Z_o Z_G}} \frac{Z_L + Z_o}{2\sqrt{Z_L Z_o}} \frac{2\sqrt{Z_L Z_G}}{Z_L + Z_G} e^{\gamma L} [1 - \rho_G \rho_L e^{-2\gamma L}] \tag{6.61}$$

---

* W.L. Everitt and G.E. Anner, *Communication Engineering*, McGraw-Hill Book Co Inc, 1956.

If $\alpha\, l >> 2.3$, $\left|\rho_G\, \rho_L\, e^{-2\gamma L}\right| << 1$ and may be neglected in comparison with 1, yielding

$$\left|\frac{I_{LO}}{I_L}\right| = \frac{1}{k_G}\frac{1}{k_L}\, k_{LG}\, e^{\alpha L} \tag{6.62}$$

Therefore, the insertion loss, *IL*, of the line is

$$IL = \ln\left(\frac{1}{k_G}\right) + \ln\left(\frac{1}{k_L}\right) - \ln\left(\frac{1}{k_{GL}}\right) + \alpha L \text{ nepers} \tag{6.63}$$

or $$IL = 20\left[\log\left(\frac{1}{k_G}\right) + \log\left(\frac{1}{k_L}\right) - \log\left(\frac{1}{k_{GL}}\right) + 0.4343\alpha L\right] \text{dB} \tag{6.64}$$

Reconciliation:

As a reconciliation or physical interpretation of the above expression for the insertion loss of the line, the following may be stated:

As $e^{-2\gamma L}$ has been neglected in comparison with 1, the impedance of the transmission line at its terminals will be Z_o irrespective of the terminating impedance at the other end of the line. Therefore, when the line of characteristic impedance, Z_o, is inserted between the generator of internal impedance, Z_G, and the load of impedance, Z_L, a mismatch is created at the generator-end of the line, namely, the one between Z_G and Z_o (between the generator and the line). Correspondingly, a reflection loss of 20 log $\left(\frac{1}{k_G}\right)$ dB results. Similarly, a mismatch occurs at the load-end of the line, namely, that between Z_o and Z_L (between the line and the load) giving rise to a reflection loss of 20 log $\left(\frac{1}{k_L}\right)$ dB. To these two components of the loss, the ohmic or ***dissipation loss*** = $20 \times 0.4343\, \alpha\, L$ dB, occurring on the line, must be added to give the total loss due to the insertion of the line. The total loss, IL_{tot}, is, therefore, given by

$$IL_{tot} = 20\left[\log\left(\frac{1}{k_G}\right) + \log\left(\frac{1}{k_L}\right) + 0.4343\, l\, L\right] \text{dB} \tag{6.65}$$

However, it may be pointed out that even if the line were not inserted and the load directly connected to the generator, a reflection loss of 20 log $\left(\frac{1}{k_{GL}}\right)$ dB would have resulted. This means that this loss is not caused by the insertion of the line. Hence, to compute the loss due truly to the insertion of the line, from the total loss already computed above, the inherent reflection loss between the generator and the load, given by 20 log $\left(\frac{1}{k_{GL}}\right)$ dB, must be subtracted to give the Insertion Loss, *IL*, of the line. Hence, the insertion loss or the loss resulting only from the insertion of the line between the generator and the load, is precisely what Eqns. (6.63) and (6.64) give.

Lastly, it may be mentioned that if αL is not large, simplification of Eqn. (6.61) is not possible. The insertion loss may then be determined by direct computation of the current in the load, with and without the line being present.

6.5 TRANSMISSION LOSS OF A LINE

In general, the power P_L, delivered to a load, Z_L, connected at the load-end of a transmission line of characteristic impedance, Z_o, and length, L, is less than the power, P_{in}, input at the sending-end of the line. Therefore, $10 \log \left(\dfrac{P_{in}}{P_L}\right)$ dB may be defined as the ***transmission loss***, *TL, of the line in decibels. An* expression for the transmission loss may be derived as follows:

From Eqns. (2.50) and (2.51), the line voltage and current at a distance, l, from the load are given by

$$V(l) = I_L [Z_L \cosh \gamma l + \sinh \gamma l] \tag{6.66}$$

and

$$I(l) = I_L \left[\cosh \gamma l + \frac{Z_L}{Z_o} \sinh \gamma l\right] \tag{6.67}$$

These equations may be re-written as

$$V(l) = I_L e^{\gamma l} \frac{Z_L + Z_o}{2} [1 + \rho_L e^{-2\gamma l}] \tag{6.68}$$

and

$$I(l) = I_L e^{\gamma l} \frac{Z_L + Z_o}{2 Z_o} [1 - \rho_L e^{-2\gamma l}] \tag{6.69}$$

The power, P_{in}, input to the line at the sending-end is

$$P_{in} = \frac{1}{2} \text{Re}\,[V_S I_S^*] = \frac{1}{2} \text{Re}\,[V(L) I^*(L)] \tag{6.70}$$

From Eqns. (6.68) and (6.69),

$$V_S I_S^* = |I_L|^2 e^{2\alpha L} \left|\frac{Z_L + Z_o}{2 Z_o}\right|^2 Z_o [1 + \rho_L e^{-2\gamma L}][1 - \rho_L^* e^{-2\gamma^* L}]$$

$$= |I_L|^2 e^{2\alpha L} \left| \frac{Z_L + Z_o}{2 Z_o} \right|^2 Z_o [1 - |\rho_L|^2 e^{-4\alpha L} + 2j \operatorname{Im} (\rho_L e^{-2\gamma L}) \quad (6.71)$$

Substitution of Eqn. (6.71) into Eqn. (6.70) gives, on using the relations:

$$\left. \begin{array}{ll} Z_o = R_o + jX_o = |Z_o| e^{\theta_o}, & Z_L = R_L + jX_L = |Z_L| e^{\theta_L}, \\ \rho_L = |\rho_L| e^{\phi_L} \quad \text{and} & \gamma = \alpha + j\beta_g \end{array} \right\} \quad (6.72)$$

$$P_{in} = \frac{1}{2} |I_L|^2 e^{2\alpha L} \left| \frac{Z_L + Z_O}{2Z_0} \right|^2 R_o$$
$$\left[\left(1 - |\rho_L|^2 e^{-4\alpha L} \right) - 2 |\rho_L| e^{-2\alpha L} \tan \theta_O \sin (\phi_L - 2\beta_g L) \right] \quad (6.73)$$

The power delivered to the load is,

$$P_L = \frac{1}{2} |I_L|^2 R_L \quad (6.74)$$

Therefore, from Eqns. (6.74) and (6.73), the **Transmission loss (ratio)** is given by

$$TL = \frac{P_{in}}{P_L} = \left[\left| \frac{Z_L + Z_o}{2 Z_o} \right|^2 \frac{R_o}{R_L} \right] [e^{2\alpha L}]$$
$$\left[\left(1 - |\rho_L|^2 e^{-4\alpha L} \right) - 2 |\rho_L| e^{-2\alpha L} \tan \theta_o \sin (\phi_L - 2\beta_g L) \right] \quad (6.75)$$

$$\left[\left| \frac{1}{K_L} \right|^2 \frac{\cos \theta_o}{\cos \theta_L} \right] [e^{2\alpha L}] [(1 - |\rho_L|^2 e^{-4\alpha L}) - 2 |\rho_L| e^{-2\alpha L} \tan \theta_o \sin(\phi_L - 2\beta_g L))]$$
$$(6.75a)$$

By taking $10 \log_{10}$ of the above equation, the Transmission Loss in dB's may be obtained as a sum of three terms arising from the quantities appearing inside the three square brackets. Each of these has its own physical significance and may respectively be defined as

Transmission Loss = Mismatch Loss (Power) + Dissipation Loss + Interaction Loss

where

Mismatch Loss = Reflection Loss + Power Factor Loss (6.76)

Consider a special case of a line having pure resistances as its characteristic and load impedances, that is, $\theta_o = \theta_L = 0$, and very small attenuation constant. On this line, the voltage standing wave ratio, S, equals $\frac{R_L}{R_o}$ or $\frac{R_o}{R_L}$, according as $R_L > R_o$ or $R_o > R_L$ and the magnitude of the voltage reflection coefficient, $|\rho_L| = \frac{S-1}{S+1}$.

Under these conditions, Eqn. (6.75) simplifies to yield

$$TL = \cosh (2\alpha L) + \frac{1}{2} \left(S + \frac{1}{S} \right) \sinh (2\alpha L) \quad (6.77)$$

It is clear from the above equation that for this transmission line, for a given value of $2\alpha L$, the transmission loss is minimum when $\left(S + \frac{1}{S} \right)$ is minimum.

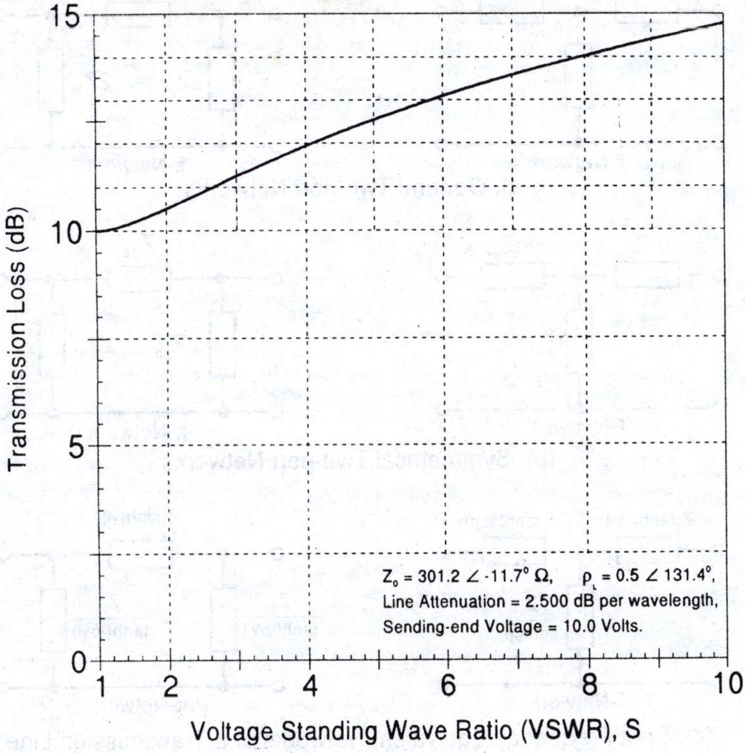

Fig. 6.9 Transmission Loss vs VSWR.

Therefore, the transmission loss is minimum and equal to 8.686 αL *dB* for $S = 1$, and it increases as S increases [Ref. Fig. 6.9.]

6.6 ARTIFICIAL LINES

In communication engineering, it is often desired that various types of transmission equipment and apparatus be tested under conditions that simulate those in the field. Again, in radar engineering, a finite time delay such as produced by a very long length of a transmission line is required. It is very inconvenient to use long lengths of cable required for these purposes. Therefore, lumped-element networks called ***artificial lines*** which simulate the behaviour of transmission lines, are designed and because of their compactness, they are used in laboratories.

If simulation of a given length, l, of the transmission lines having the parameters R, L, G and C, is required at a single spot frequency, f, the design of the artificial line is a straightforward one as only a ***single symmetrical T-section*** or a ***single symmetrical*** π***-section*** [Ref. fig. 6.10 (a)] is sufficient.

In general, the components of a T-network or a π-network that is equivalent, at a single frequency, to a given linear, bilateral, passive, two-port network, may be

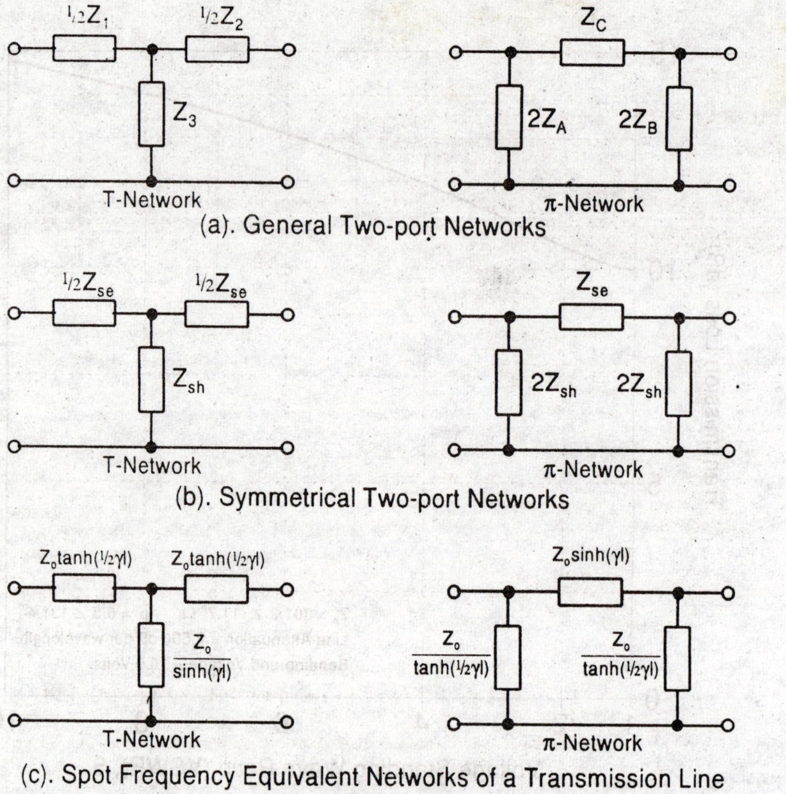

Fig. 6.10 Spot Frequency Equivalents of a Line.

determined[4] in terms of the open- and short-circuit impedances measured at the two ports of the given network calculated using the formulae: [Ref. Fig. 6.10 (a)]

Equivalent *T*-network: **Equivalent π-network:**

$$\left.\begin{aligned}
\frac{Z_1}{2} &= Z_{1\,oc} - \sqrt{Z_{2\,oc}\,(Z_{2\,oc} - Z_{1\,sc})}, & 2\,Z_A &= \frac{Z_{2\,oc}\,Z_{1\,sc}}{Z_{2\,oc} - \sqrt{Z_{2\,oc}\,(Z_{1\,oc} - Z_{1\,sc})}} \\
\frac{Z_2}{2} &= Z_{2\,oc} - \sqrt{Z_{2\,oc}\,(Z_{2\,oc} - Z_{1\,sc})}, & 2\,Z_B &= \frac{Z_{2\,oc}\,Z_{1\,sc}}{Z_{1\,oc} - \sqrt{Z_{2\,oc}\,(Z_{1\,oc} - Z_{1\,sc})}} \\
Z_3 &= \sqrt{Z_{2\,oc}\,(Z_{1\,oc} - Z_{1\,sc})}, & Z_C &= \frac{Z_{2\,oc}\,Z_{1\,sc}}{\sqrt{Z_{2\,oc}\,(Z_{1\,oc} - Z_{1\,sc})}}
\end{aligned}\right\} \quad (6.78)$$

---

4 John D. Ryder, *Networks, Lines and Fields,* Prentice-Hall of India Private Limited, New Delhi, 2nd Edition, 1988, Chap. 6 and 1.

Now, if the given network is known to be symmetrical like a transmission line of length, l, then [Ref. Fig. 6.10 (b)]

$$\left.\begin{aligned} Z_{2\,sc} &= Z_{1\,sc} = Z_{sc} = Z_o \tanh(\gamma l) \\ Z_{2\,oc} &= Z_{1\,oc} = Z_{oc} = \frac{Z_o}{\tanh(\gamma l)} \\ &\text{then, for a } T\text{-network,} \\ \frac{Z_1}{2} &= \frac{Z_2}{2} = \frac{Z_{Tse}}{2} \\ \text{and, let} \quad Z_3 &= Z_{Tsh}. \\ &\text{Also, for a } \pi\text{-network,} \\ 2\,Z_A &= 2\,Z_B = 2\,Z_{\pi\,sh} \\ \text{and, let} \quad Z_C &= Z_{\pi\,se} \end{aligned}\right\} \qquad (6.79)$$

Substitution of Eqns. (6.79) into Eqns. (6.78) gives [Ref. Fig. 6.10 (c)]

Equivalent T-Network: **Equivalent π-Network:**

$$\left.\begin{aligned} \frac{Z_{Tse}}{2} &= Z_o \tanh\left(\frac{\gamma l}{2}\right), & Z_{\pi\,se} &= Z_o \sinh(\gamma l). \\ \text{and} \\ Z_{Tsh} &= \frac{Z_o}{\sinh(\gamma l)}, & 2\,Z_{\pi\,sh} &= \frac{Z_o}{\tanh\left(\frac{\gamma l}{2}\right)}. \end{aligned}\right\} \qquad (6.80)$$

A knowledge of R, L, G and C permits the calculation of the characteristic impedance, Z_o, and the propagation constant, γ, at the frequency of interest. However, as Z_o and γ are frequency-dependent, all the impedances, (Z_{Tse}, Z_{Tsh}) and ($Z_{\pi\,se}$, $Z_{\pi\,sh}$) are dependent on the frequency and hence, theoretically, no single set of (Z_{Tse}, Z_{Tsh}) or ($Z_{\pi\,se}$, $Z_{\pi\,sh}$) would be able to simulate a transmission line over even a limited band of frequencies. On the other hand, in communication engineering, simulation of the line is always required at least over a band of frequencies and not at a spot frequency.

An approximate simulation over a band of frequencies may be obtained by breaking up the given length, l, of the transmission line into a large number, N, of sections so that the length of each section, $\Delta l = \frac{l}{N}$, is such that both $\alpha \Delta l$ and $\beta_g \Delta l$ are less than 0.1 radian or so. The components of the T-network and the T-network representing each of these sections, called the ***nominal T-section*** and ***nominal π-section***, are given by [Ref. Fig. 6.11]

Nominal *T*-section:

$$\frac{Z_{Tse}}{2} = Z_o \tanh\left(\frac{\gamma \Delta l}{2}\right) \approx Z_o \frac{\gamma \Delta l}{2} = \frac{1}{2}[(R\,\Delta l) + j\,\omega\,(L\,\Delta l)].$$

and,

$$Z_{Tsh} = \frac{Z_o}{\sinh(\gamma \Delta l)} \approx \frac{1}{\dfrac{\gamma}{Z_o}\Delta l} = \frac{1}{(G + j\,\omega\, C)\,\Delta l} = \frac{1}{(G\,\Delta l) + j\,\omega\,(C\,\Delta l)}.$$

Nominal π-section:

$$\left.\begin{aligned} Z_{\pi\, se} &= Z_o \sinh(\gamma\, l) \approx Z_o\, \gamma\, \Delta l = (R\,\Delta l) + j\,\omega\,(L\,\Delta l).\\ 2\,Z_{\pi\, sh} &= \frac{Z_o}{\tanh\left(\dfrac{\gamma \Delta l}{2}\right)} \approx \frac{1}{\dfrac{\gamma}{Z_o}\dfrac{\Delta l}{2}} = 2\,\frac{1}{(G + j\,\omega\, C)\,\Delta l} = 2\,\frac{1}{(G\,\Delta l) + j\,\omega\,(C\,\Delta l)}. \end{aligned}\right\} \quad (6.78)$$

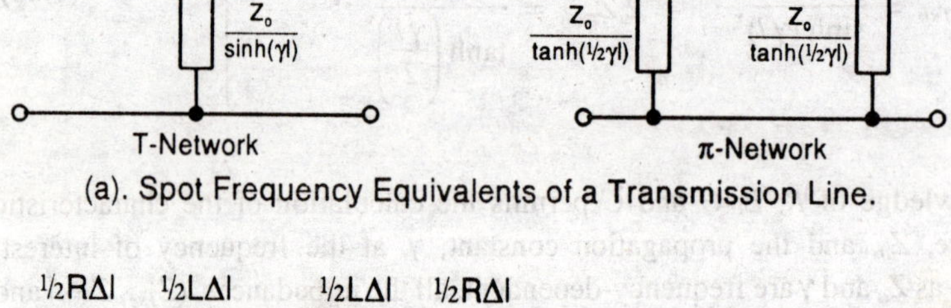

(a). Spot Frequency Equivalents of a Transmission Line.

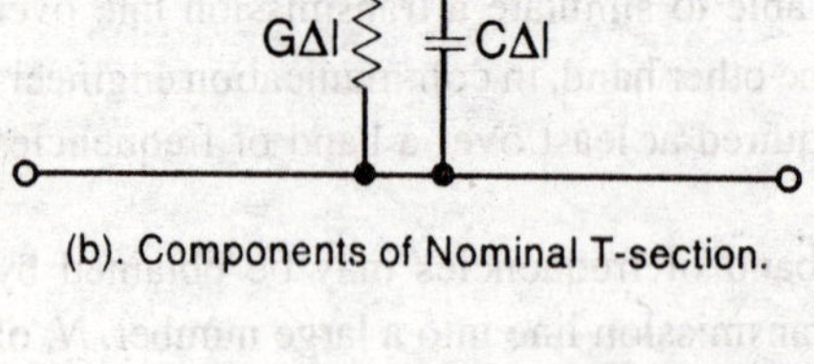

(b). Components of Nominal T-section.

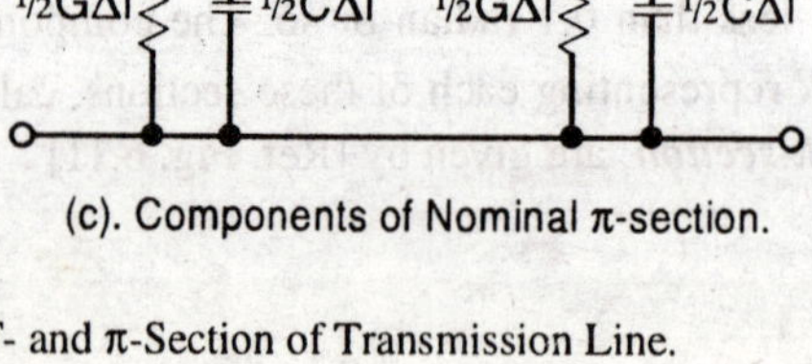

(c). Components of Nominal π-section.

Fig. 6.11 Nominal T- and π-Section of Transmission Line.

Note that under the assumptions made ($\gamma\,\Delta\,l << 1$),

$$\left.\begin{aligned} Z_{Tse} &= Z_{\pi\,se} = Z_{se} \\ Z_{Tsh} &= Z_{\pi sh} = Z_{sh} \end{aligned}\right\} \tag{6.82}$$

The expressions for Z_{se} and Z_{sh} suggest that the series impedance, Z_{se}, is a series combination of resistance, $R\,\Delta\,l$, and an inductance, $L\,\Delta\,l$, which are nothing but resistance and inductance lumped up over the length, $\Delta\,l$, of the line! Similarly, the shunt impedance, Z_{sh}, is a parallel combination of the conductance, $G\,\Delta\,l$, and the capacitance, $C\,\Delta\,l$, which are, once again the conductance and the capacitance lumped up over the length, $\Delta\,l$, of the line!

The length, l, of the transmission line is, then, simulated by N numbers of such T-sections or π-sections connected in tandem.

It may be pointed out that both α and β_g increase with frequency for a transmission line. Hence, the design of the nominal T-section must be carried out for the highest frequency of interest. It may be noted that the design restriction that $\alpha\,\Delta\,l$, $\beta_g\,\Delta\,l$ be less than 0.1, is more conservative than is usually necessary. In practice, a nominal T-section is designed to simulate much longer length, $\Delta\,l$, of the line than $\dfrac{0.1}{\beta_g}$.

SOLVED EXAMPLES

Example 6.1 (a) Derive the ($ABCD$)-matrix of a transmission line of length, l.

(b) What would this matrix be, if the line is lossless and a quarter wavelength long?

(c) Hence, determine the ($ABCD$)-matrix of the halfwave, lossless ($\alpha = 0$) line with a shunt admittance, Y_M, connected at its centre.

Solution : : The lossless halfwave line with a shunt admittance, Y_M, connected at its centre, may be regarded as three two-port networks connected in tandem. The first of these two-port networks is a lossless quarterwave line, the second is the shunt admittance, Y_M, and the third two-port network is again a lossless quarterwave line. Hence, the required $ABCD$-matrix (of the halfwave line) may be obtained by multiplying the $ABCD$-matrices of each of the three component two-port networks described above.

With reference to Fig. 6.12, the $ABCD$-matrix of the shunt admittance, Y_M, is given by the equations,

$$V_{S2} = V_{L2}$$

$$I_{S2} = Y_M\,V_{L2} + I_{L2}$$

or, in the matrix form,

$$\begin{bmatrix} V_{S2} \\ I_{S2} \end{bmatrix} = \begin{bmatrix} 1 & 0 \\ Y_M & 1 \end{bmatrix} \begin{bmatrix} V_{L2} \\ I_{L2} \end{bmatrix}$$

Now, the $ABCD$ matrix for a transmission line of length, l, has already been obtained in Eqns. (2.50) and (2.51). It is explicitly given by

$$\begin{bmatrix} V_S \\ I_S \end{bmatrix} = \begin{bmatrix} \cosh(\gamma\,l) & Z_o \sinh(\gamma\,l) \\ \dfrac{1}{Z_o}\sinh(\gamma\,l) & \cosh(\gamma\,l) \end{bmatrix} \begin{bmatrix} V_L \\ I_L \end{bmatrix}$$

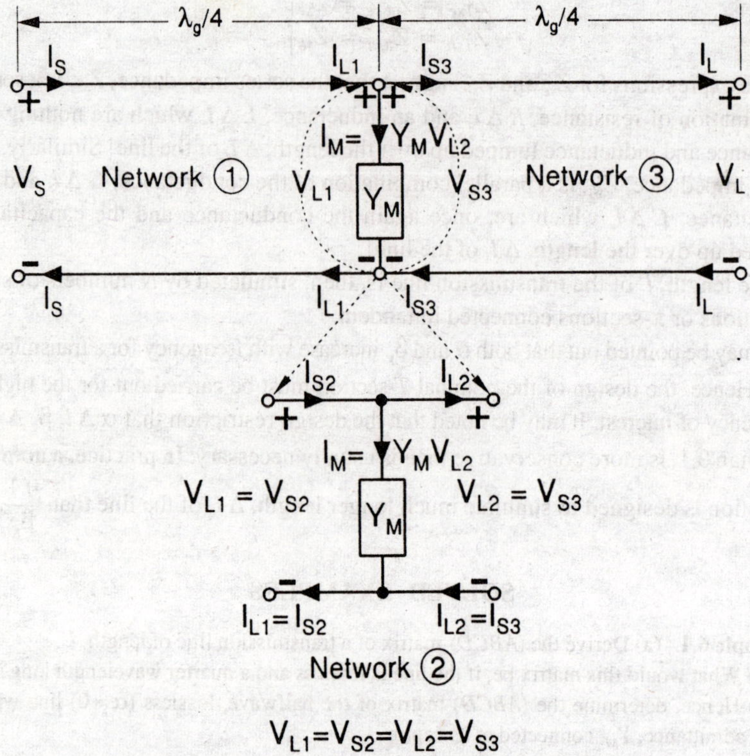

Fig. 6.12 Halfwave Line with Shunt Admittance at Centre.

Therefore, for the first quarterwave lossless line, it becomes, (by putting $\gamma l = j\dfrac{\pi}{2}$ and $Z_o = R_o$)

$$\begin{bmatrix} V_S \\ I_S \end{bmatrix} = \begin{bmatrix} 0 & jR_o \\ \dfrac{j}{R_o} & 0 \end{bmatrix} \begin{bmatrix} V_{L1} \\ I_{L1} \end{bmatrix}$$

And, for the other quarterwave lossless line, it is,

$$\begin{bmatrix} V_{S3} \\ I_{S3} \end{bmatrix} = \begin{bmatrix} 0 & jR_o \\ \dfrac{j}{R_o} & 0 \end{bmatrix} \begin{bmatrix} V_L \\ I_L \end{bmatrix}$$

Hence, the required *ABCD*-matrix of the given halfwave lossless line with the given shunt admittance at the centre is obtained by multiplying the *ABCD*-matrices of the *three two-port networks, as stated above, giving*

$$\begin{bmatrix} V_S \\ I_S \end{bmatrix} = \begin{bmatrix} 0 & jR_o \\ \dfrac{j}{R_o} & 0 \end{bmatrix} \begin{bmatrix} 1 & 0 \\ Y_M & 1 \end{bmatrix} \begin{bmatrix} 0 & jR_o \\ \dfrac{j}{R_o} & 0 \end{bmatrix} \begin{bmatrix} V_L \\ I_{L1} \end{bmatrix}$$

Simplification of the above matrix equation gives,

$$\begin{bmatrix} V_S \\ I_S \end{bmatrix} = \begin{bmatrix} -1 & -R_o^2 Y_M \\ 0 & -1 \end{bmatrix} \begin{bmatrix} V_L \\ I_L \end{bmatrix}$$

Example 6.2 A sinusoidal generator operating at a frequency of 1 kHz, applies 10 Volts (peak) to one end of a transmission line whose data is given below:

$R = 6.66\ \Omega$/km, $L = 2.22$ mH/km, $G = 0.55\ \mu\ \Omega$/km, $C = 5.55$ m μ F/km, and length, $l = 108.5$ km.

A load of impedance, $Z_L = (222.0 + j\,0)\ \Omega$ is connected at the other end of the line. Calculate the voltages, currents, powers, and reflection and mismatch losses at the two ends of the line. Also determine the transmission loss and the power transmission efficiency of the transmission line.

Solution : For $f = 1$ kHz, angular frequency $\omega = 6.2832 \times 10^3$ radians/sec.

$$R + j\omega L = (6.66 + j13.9487)\ \Omega/\text{km} = 15.4571 \underline{/ 64.477^\circ}\ \Omega/\text{km},$$

$$G + j\omega C = (0.5500 + j\,34.8717)\ \mu\ \text{mhos/km} = 34.8760 \underline{/ 89.096^\circ}\ \text{mhos/km},$$

Substitution of these into Eqns. (2.10) and (2.6), gives, at $f = 1$ kHz,

$$Z_o = (650.4278 - j\,141.9306)\ \Omega = 665.7331 \underline{/ - 12.310^\circ}\ \Omega,$$

and,

$$\gamma = (0.005307 + j\,0.02260)/\text{km} = 0.02322 \underline{/ 76.787^\circ}/\text{km}.$$

Before proceeding further with the solution of the problem, it would be better to recapitulate some of the formulas which may be required here:

Formulae :

For ready reference, some of the formulae which are required for an analysis of transmission lines, are reproduced below. [Ref. Eqns. (5.53), (5.54), Tables 5.1 and 5.2, (5.56) to (5.60)]

$$\rho_L = \frac{Z_L - Z_o}{Z_L + Z_o} \quad (1), \qquad \rho_S = \frac{Z_S - Z_o}{Z_S + Z_o} \quad (2), \qquad \rho_S = \rho_L e^{-2\gamma l} \quad (3).$$

$$V_L = (1 + \rho_L) e^{-\gamma l} V_{inc} \quad (4), \qquad I_L = (1 - \rho_L) e^{-\gamma l} I_{inc} \quad (5), \qquad Z_L = \frac{1 + \rho_L}{1 - \rho_L} Z_o \quad (6).$$

$$V_S = (1 + \rho_S) V_{inc} \quad (7), \qquad I_S = (1 - \rho_S) I_{inc} \quad (8), \qquad Z_S = \frac{1 + \rho_S}{1 - \rho_S} Z_o \quad (9).$$

$$V_{inc} = Z_o I_{inc} \quad (a), \qquad V_{inc} = \frac{e^{\gamma l} V_L}{1 + \rho_L} \quad (b), \qquad V_{inc} = \frac{V_S}{1 + \rho_S} \quad (c)$$

$$P = \frac{1}{2} \text{Re}\,[V I^*] \quad (d), \qquad V_L = \frac{1 + \rho_L}{1 + \rho_S} e^{-\gamma l} V_S \quad (e), \qquad V_S = \frac{1 + \rho_S}{1 + \rho_L} e^{\gamma l} V_L \quad (f).$$

Some definitions:

Mismatch Loss = Reflection Loss + Power Factor Loss.

Transmission Loss = [Mismatch Loss at the Load-end + Dissipation Loss – Mismatch Loss at the Gen-end.]

Now, the load impedance is given to be $Z_L = (222.0 + j0)\ \Omega$. Therefore, the voltage reflection coefficient, ρ_L, at the load-end is obtained as

$$\rho_L = \frac{(222.0 + j0) - (650.4278 - j141.9306)}{(222.0 + j0) + (650.4278 - j141.9306)} = \frac{-428.4278 + j141.9306}{872.4278 - j141.9306}$$

$$= \frac{451.3254}{883.8974} \underline{/\ 161.671^\circ + 9.240^\circ} = 0.5106 \underline{/\ 170.911^\circ} = -0.5042 + j\,0.08066$$

The length of the line is given to be $l = 108.5$ km. Therefore,

$$e^{-2\gamma l} = e^{-(2 \times 0.005307 \times 108.5)} \underline{/\ (-2 \times 0.02260 \times (180/\pi) \times 108.5)^\circ}$$
$$= 0.3161 \underline{/\ -281.033^\circ}$$

and, $$e^{-\gamma l} = 0.5622 \underline{/\ -140.516^\circ}$$

From these results, the voltage reflection coefficient, ρ_S, at the generator-end, may be calculated as

$$\rho_S = \rho_L e^{-2\gamma l} = (0.5106 \times 0.3161) \underline{/\ 170.911^\circ - 281.033^\circ}$$
$$= 0.1614 \underline{/\ -110.122^\circ} = -0.05553 - j\,0.1516$$
$$\rho_L = -0.5042 + j\,0.08066 = 0.5106 \underline{/\ 170.911^\circ}$$
$$1 + \rho_L = 0.4958 + j\,0.08066 = 0.5023 \underline{/\ 9.240^\circ}$$
$$1 - \rho_L = 1.5042 + j\,0.08066 = 1.5064 \underline{/\ -3.069^\circ}$$
$$\rho_S = -0.05553 - j\,0.1516 = 0.1614 \underline{/\ -110.12^\circ}$$
$$1 + \rho_S = 0.9445 - j\,0.1516 = 0.9566 \underline{/\ -0.9117^\circ}$$
$$1 - \rho_S = 1.0555 + j0.1516 = 1.0664 \underline{/\ 8.171^\circ}$$

LOAD-END Quantities:

Impedance, $Z_L = 222.0 \underline{/\ 0^\circ}\ \Omega$.

Voltage, $$V_L = \frac{1 + \rho_L}{1 + \rho_S} e^{-\gamma l} V_S$$

$$= \frac{0.5023 \times 0.5622 \times 10.0}{0.9566} \underline{/\ (9.240 - 140.516 + 0 + 9.117)^\circ} \text{ Volts}.$$

$$= 2.9525 \underline{/\ -122.160^\circ} \text{ Volts}.$$

Current, $$\frac{V_L}{Z_L} = I_L = \frac{2.9525}{222.0} \underline{/\ -122.160^\circ - 0^\circ} \text{ Amp}$$

$$= 13.2997 \underline{/\ -122.160^\circ} \text{ milliamperes.}$$

Output power, $$\rho_L = \frac{1}{2} \text{Re}\,[V_L I_L^*]$$

$$= \frac{1}{2} \text{Re}\,[(2.9525 \times 13.2997) \underline{/\ -122.160^\circ + 122.160^\circ}] \text{ mW}$$

$$= 19.6340 \text{ mW}.$$

Reflection Loss:

$$\frac{Z_L + Z_o}{2} = 436.2139 - j\,70.9653\ \Omega = 441.9487 \underline{/-9.240^\circ}\ \Omega$$

$$\sqrt{Z_L Z_o} = \sqrt{222.0 \times 665.7331}\ \underline{/0.5 \times (0 - 12.310)^\circ}\ \Omega$$

$$= 384.4382 \underline{/-6.155^\circ}\ \Omega$$

Therefore, $$k_{LO} = \frac{2\sqrt{Z_L Z_o}}{Z_L + Z_o} = \frac{384.4382}{441.9487} \underline{/(-6.155 + 9.240)^\circ}\ \Omega$$

$$= 0.8699 \underline{/3.085^\circ}$$

Reflection Loss, $$RL = 20 \times \log \left| \frac{1}{k_{LO}} \right| = 20 \times \log \frac{1}{0.8699}\ \text{dB} = 1.2209\ \text{dB} \quad (6.83)$$

Power Factor Loss:

Power Factor of $Z_o = \cos\theta_o = \cos(-12.310^\circ) = 0.9770$,

Power Factor of $Z_L = \cos\theta_L = \cosh 0^\circ = 1.0000$,

Therefore, Power Factor Loss $$PL = 10 \times \log \frac{\cos\theta_o}{\cos\theta_L} = 10 \times \log (0.9770)\ \text{dB}$$

$$= -0.1010\ \text{dB}$$

Dissipation Loss:

Dissipation Loss $$DL = 10 \times \log e^{2\alpha L} = 20 \times (\log e)\,\alpha L$$

$$= 8.6859 \times 0.005307 \times 108.5\ \text{dB}$$

$$= 5.0015\ \text{dB}$$

Mismatch Loss, $$ML = RL + PL = 1.2109 + (-0.1010)\ \text{dB}$$

$$= 1.1099\ \text{dB}$$

GEN-END Quantities:

Voltage, $V_S = 10.0 \underline{/0^\circ}$ volts .

Impedance, $$Z_S = Z_o \frac{1 + \rho_S}{1 - \rho_S}$$

$$= 665.7331 \times \frac{0.9566}{1.0664} \underline{/(-12.310 - 9.117 - 8.171^\circ)^0}\ \Omega,$$

$$= 597.1845 \underline{/-29.597^\circ}\ \Omega = 519.2627 - j\,294.9502\ \Omega$$

Current, $$I_S = \frac{V_S}{Z_S} = \frac{10.0 \angle 0^\circ}{597.1845 \underline{/-29.597^\circ}}\ \text{Amp} = 16.7452 \underline{/29.597^\circ}\ \text{mA}.$$

Power, $$P_S = \frac{1}{2} \text{Re}\,[V_S I_S^*] = \frac{1}{2} \text{Re}\,[10.0 \times 16.7452] \underline{/0.0^\circ - 29.597^\circ}\ \text{mW}$$

$$= 72.8015\ \text{mW}.$$

Reflection Loss:

$$\frac{Z_S + Z_o}{2} = 584.8452 - j218.4404\ \Omega = 624.3077 \underline{/-20.481^\circ}\ \Omega$$

$$\sqrt{Z_S Z_o} = \sqrt{222.0 \times 597.1845} \;/\; 0.5 \times (-12.310 - 29.597)^\circ \; \Omega$$

$$= 630.5280 \;/\; -20.953^\circ \; \Omega$$

Therefore, $$k_{SO} = \frac{2\sqrt{Z_S Z_o}}{Z_S + Z_o} = \frac{630.5280}{624.3077} \;/\; (-20.953 + 10.481)^\circ$$

$$= 1.0100 \;/\; -0.473^\circ$$

Reflection Loss $$RL = 20 \times \log \left| \frac{1}{k_{SO}} \right| = 20 \times \log \frac{1}{1.0100} \text{ dB}$$

$$= -0.0861 \text{ dB}$$

Power Factor Loss:

Power Factor of $Z_o = \cos\theta_o = \cos(-12.310^\circ) = 0.9770$,

Power Factor of $Z_S = \cos\theta_L = \cos(-29.597^\circ) = 0.8695$,

Therefore, Power Factor Loss $$PL = 10 \times \log \frac{\cos\theta_o}{\cos\theta_L} = 10 \times \log (1.1236) \text{ dB}$$

$$= 0.5062 \text{ dB}$$

Mismatch loss, $$ML = RL + PL = -0{,}0861 + 0.5062 \text{ dB}$$

$$= 0.4201 \text{ dB}$$

INCIDENT WAVE Quantities:

Voltage, $$V_{inc} = \frac{V_S}{1 + \rho_S} = \frac{10.0 \;/\; 0^\circ}{0.9566 \;/\; -9.117^\circ} \text{ Volts}$$

$$= 10.4542 \;/\; 9.117^\circ \text{ Volts}$$

Current, $$I_{inc} = \frac{V_{inc}}{Z_o} = \frac{10.4542 \;/\; 9.117^\circ}{665.7331 \;/\; -12.310^\circ} \text{ Amps}$$

$$= 15.7033 \;/\; 21.426^\circ \text{ milliamperes}$$

Power, $$P_{inc} = \frac{1}{2} \text{Re}\,[V_{inc} I^*_{inc}] = 10.4542 \times 15.7033 \cos(-12.310^\circ)$$

$$= 80.1953 \text{ milliwatts.}$$

SUMMARY: Power Flow expressed	Power associated with the Incident wave, P_{inc}	Power entering the line at the Gen-end, P_S	Power delivered to the load at the load-end, P_L
in milliwatts	80.1953 mW	72.8015 mW	19.6340 mW
in dBs wrt P_{inc}	0 dB (Ref)	– 0.4201 dB	– 6.1114 dB
dBs below P_s	Transmission Loss:	→ 0 dB (Ref)	5.6913 dB

Power Transmission Efficiency of the line $= \frac{19.6340}{72.8015} \times 100 + = 26.969\ \%$

EXERCISES

1. (a) Derive the Impedance matrix of a transmission line of length, l.

(b) What form does the Z-matrix take, if the line is a lossless one (that is, $\gamma = 0 + j\,\beta_g$)?

(c) From the solution of Example 6.1(b), deduce the Z-matrices for Quarterwave $\left(l = \frac{\lambda_g}{4}\right)$ and Halfwave $\left(l = \frac{\lambda_g}{2}\right)$ lossless transmission lines.

2. (a) Derive the Admittance matrix of a transmission line of length, l.

(b) What form does the Y-matrix take, if the line is a lossless one (that is, $\gamma = 0 + j\,\beta_g$) ?

(c) From the solution of Example 6.2(b), deduce the Y-matrices for Quarterwave $\left(l = \frac{\lambda_g}{4}\right)$ and Halfwave $\left(l = \frac{\lambda_g}{2}\right)$ lossless transmission lines.

3. A 400 Hz sinusoidal voltage of 10.0 volts is applied to a 152.189 km long transmission line having $Z_o = 614.133 \underline{/-10.0^\circ}\ \Omega$, and $\gamma = 0.021624 \underline{/77.883^\circ}$/km. Calculate the power received by a $307.066 \underline{/-60.5^\circ}\ \Omega$ impedance connected at the other end of the line. [**Ans.** $P_L = 13.4836$ millwatts.]

4. Calculate the line voltages, currents, impedances and voltage reflection coefficients at the two ends of the line given in Exercise 1 above.

[**Ans.** $V_S = 10.000 \underline{/0.000^\circ}$ Volts, $I_S = 20.450 \underline{/20.014^\circ}$ mAmps.,
$Z_S = 489.008 \underline{/-20.014^\circ}\ \Omega$, $V_L = 4.101 \underline{/147.040^\circ}$ Volts,
$I_L = 13.355 \underline{/-152.460^\circ}$ mAmps., $Z_L = 307.066 \underline{/-60.500^\circ}\ \Omega$
$\rho_S = 0.143 \underline{/-142.890^\circ}$, $\rho_L = 0.571 \underline{/-134.186^\circ}$.]

5. Calculate the reflection and mismatch losses at the two ends of the above line (Exercise 1) ends and the transmission loss. If the line is matched at the load-end, what power would be delivered to the load?

[**Ans.** Reflection Loss: at the Gen-end = 0.0234 dB, and at the Load-end = – 0.2549 dB.
Mismatch Loss: at the Gen-end = 0.2275 dB, and at the Load-end = 2.7552 dB.
Transmission Loss = 8.5280 dB. Matched Power, $P_L = 25.4300$ milliwatts.]

6. Design an artificial line to simulate, over the frequency range of 150 Hz to 6 kHz, a 3 km length of the transmission line whose primary constants are given below:

$R = 6.543\ \Omega$/km, $L = 2.345$ mH/km, $G = 0.456\ \mu\ \Omega$/km, $C = 5.432$ mμF/km.

[**Ans.** $R_{se} = 4.907\ \Omega$, $L_{se} = 1.759$ mH, $R_{sh} = 2.924\ M\ \Omega$, $C_{sh} = 4.074$ mμF.]

7. Calculate and compare the characteristic impedance and the propagation constants of the actual line and the artificial line designed in Exercise 6 above at operating frequencies of 150 Hz, 1 kHz and 6 kHz.

[**Ans.** At 6 kHz: $Z_o = 657.1451 \underline{/-2.047^\circ}\ \Omega$,
$\gamma = (0.003849$ neper $+ j\,5.7884$ deg)/Nominal T–section
At 150 Hz: $Z_o = 1159.1664 \underline{/-33.123^\circ}\ \Omega$,
$\gamma = (0.002764$ neper $+ j\,0.2012$ deg)/T–section]

One nominal T-section represents 0.75 km of the given line. Correspondingly, for comparison, calculated values of γ per 0.75 km length of the actual line are :

At 6 kHz: $Z_0 = 657.9367 \underline{/-2.053^\circ}\ \Omega$,
$\gamma = (0.003844$ neper $+ j\,5.7856$ deg)/ 0.75 km.
At 150 Hz: $Z_o = 1159.1664 \underline{/-33.123^\circ}\ \Omega$,
$\gamma = (0.002764$ neper $+ j\,0.2012$ deg)/ 0.75 km.]

8. A 1 kHz sinusoidal voltage of 10.0 volts is applied to 6.153 km long transmission line having $Z_o = 363.2439 \underline{/-38.009^\circ}\ \Omega$, and $\gamma = 0.134749 \underline{/46.030^\circ}$ /km. Calculate the power received by a $363.244 \underline{/0^\circ}\ \Omega$ impedance connected at the other end of the line. Also calculate the mismatch loss at the load-end and the transmission loss of the line.

[Ans. $P_L = 40.097$ mW, Mismatch Loss = – 1.5220 dB, Transmission Loss = 3.6934 dB.]

Chapter 7

LOSSLESS LINES

7.1 LOW-LOSS LINES

As the frequency of operation is raised, the series reactance, ωL, of a transmission line increases directly as ω whereas the series resistance, R, increases as $\sqrt{\omega}$ due to the skin-effect. Therefore, at high frequencies, $\omega L >> R$. Also, for lines of practical interest, the shunt conductance, G, is negligible compared to ωC. Thus, at high frequencies, the characteristic impedance of a line is given by

$$Z_0 = \sqrt{\frac{j\,\omega L\left(1 - j\dfrac{R}{\omega L}\right)}{j\,\omega C\left(1 - j\dfrac{G}{\omega C}\right)}} \tag{7.1}$$

$$\approx Z_{0\,\infty}\left[1 + \frac{1}{8}\left(\frac{R}{\omega L}\right)^2\right] \angle \tan^{-1}\left\{\left(\frac{R}{2\,\omega L}\right)\left[1 - \frac{1}{4}\left(\frac{R}{\omega L}\right)^2\right]\right\}. \tag{7.2}$$

where,

$$Z_{0\,\infty} = \sqrt{\frac{L}{C}}, \tag{7.3}$$

and the propagation constant is given by

$$\gamma = \alpha + j\,\beta_g = \sqrt{j\,\omega L\left(1 - j\frac{R}{\omega L}\right) j\,\omega C\left(1 - j\frac{G}{\omega C}\right)} \tag{7.4}$$

$$\approx \frac{1}{2}\left(\frac{R}{Z_{0\,\infty}} + GZ_{0\,\infty}\right)\left(1 + \frac{RG}{8\omega^2 LC}\right) + j\,\omega\sqrt{LC}\left[\frac{1 + (RC - GL)^2}{8\,\omega^2 L^2 C^2}\right]. \tag{7.5}$$

Therefore,

$$\alpha = \frac{1}{2}\left(\frac{R}{Z_{o\infty}} + GZ_{o\infty}\right)\left[1 + \frac{RG}{8\,\omega^2 LC}\right] \tag{7.6}$$

and,

$$\beta_g = \omega\sqrt{LC}\left[1 + \frac{(RC - GL)^2}{8\,\omega^2 L^2 C^2}\right]. \tag{7.7}$$

If the frequency of operation is raised further, $\frac{R}{\omega L}$ and $\frac{G}{\omega C}$ will become negligibly small compared to 1 and may, therefore, be neglected altogether in Eqns. (7.1) and (7.4), giving

$$Z_o = \sqrt{\frac{j\,\omega L}{j\,\omega C}} = \sqrt{\frac{L}{C}} = Z_{o\infty} = R_o = \text{Real}, \tag{7.8}$$

and,

$$\gamma = \alpha + j\beta_g = \sqrt{j\,\omega L\, j\,\omega C} = j\,\omega\sqrt{LC} = \text{Imaginary}, \tag{7.9}$$

which gives,

$$\left.\begin{aligned} \alpha &= 0 \\ \beta_g &= \omega\sqrt{LC} \end{aligned}\right\} \tag{7.10}$$

Eqns. (7.8) and (7.10) show that at high frequencies, all lines may be regarded as lossless, having

$$\left.\begin{aligned} \text{Characteristic impedance, } Z_o &= \sqrt{\frac{L}{L}} = R_o \\ \text{Attenuation constant, } \alpha &= 0 \\ \text{Phase Constant, } \beta_g &= \omega\sqrt{LC} \\ \text{Phase Velocity, } v_p &= \frac{1}{\sqrt{LC}} \\ \text{Group Velocity, } v_g &= \frac{1}{\sqrt{LC}} = v_p. \end{aligned}\right\} \tag{7.11}$$

7.2 VOLTAGE AND CURRENT ON LOSSLESS LINES

Substitution of Eqns. (7.11) into the solutions of the line equations, given in Sec.(2.2.3), gives the distribution of the voltage and current along a lossless line. For instance, Eqns. (2.20), (2.24) and (2.26) yield,

$$V(z) = V_S \cos(\beta_g z) - j\,R_o I_S \sin(\beta_g z), \tag{7.12}$$

$$I(z) = I_S \cos(\beta_g z) - j\frac{V_S}{R_o}\sin(\beta_g z), \tag{7.13}$$

$$Z(z) = R_o\left[\frac{Z_S - j\,R_o \tan(\beta_g z)}{R_o - j\,Z_S \tan(\beta_g z)}\right] \tag{7.14}$$

and

$$V(l) = V_L \cos(\beta_g l) + j R_o I_L \sin(\beta_g l) \tag{7.15}$$

$$= \left(\frac{Z_L + R_o}{2Z_L}\right) V_L e^{j\beta_g l} [1 + \rho_L e^{-j2\beta_g l}], \tag{7.16}$$

$$I(l) = I_L \cos(\beta_g l) + j \frac{V_L}{R_o} \sin(\beta_g l) \tag{7.17}$$

$$= \left(\frac{Z_L + R_o}{2R_o}\right) I_L e^{j\beta_g l} [1 - \rho_L e^{-j2\beta_g l}], \tag{7.18}$$

and

$$Z(l) = R_o \left[\frac{Z_L + j R_o \tan(\beta_g l)}{R_o + j Z_L \tan(\beta_g l)}\right] \tag{7.19}$$

$$= R_o \left[\frac{1 + \rho_L e^{-j2\beta_g l}}{1 - \rho_L e^{-j2\beta_g l}}\right]. \tag{7.20}$$

Variations in the magnitude of the line voltage, $V(l)$, line current, $I(l)$ and line impedance, $\frac{Z(l)}{R_o}$ along the length of a typical lossless line, are plotted in Fig. 7.1. It may be observed from the figure that the maximum and minimum values of $|V(l)|$, $|I(l)|$ and $|Z(l)|$ remain constant throughout the length of the line. Further, it may be observed that $|V(l)|$, $|I(l)|$ and $\frac{|Z(l)|}{R_o}$ are periodic function of l or z with a period of $\frac{\lambda_g}{2}$. At points where the line voltage, $|V(l)|$, attains its maximum value, the line current, $|I(l)|$, reaches its minimum value; and, therefore, the line impedance, $|Z(l)|$, attains its maximum value at that point.

7.3 REFLECTION COEFFICIENT

The voltage reflection coefficient at the load terminals, ρ_L, is given by [Ref. Eqn. (5.8)]

$$\rho_L = |\rho_L| e^{j\phi_L} = \frac{Z_L - R_o}{Z_L + R_o}. \tag{7.21}$$

The reflection coefficient, $\rho(l)$, at a distance l from the load towards the generator, is given by [Ref. Eqn. (5.28)]

$$\rho(l) = \rho_L e^{-j2\beta_g l} \tag{7.22}$$

$$= |\rho_L| e^{j(\phi_L - 2\beta_g l)} \tag{7.23}$$

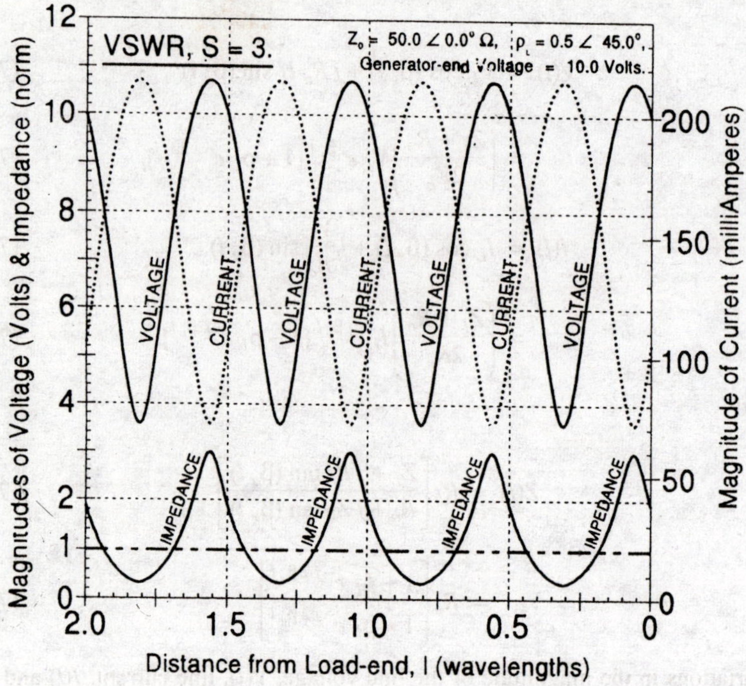

Fig. 7.1 Magnitudes of Voltage, Current and Impedance along a Lossless Line.

Note that in the case of a lossless line, the magnitude of the reflection coefficient, $|\rho(l)|$, is constant (equal to $|\rho_L|$) everywhere on the line. The phase angle of the voltage reflection coefficient, ϕ, however, goes on decreasing from its value ϕ_L at the load terminals, linearly as $\phi = \phi_L - 2\beta_g l$, as l increases from zero. In the complex ρ-plane, therefore, the point representing ρ starts from its position $|\rho_L| e^{j\phi L}$ corresponding to the voltage reflection coefficient at the load terminals and moves clockwise on the ($|\rho|$ = constant)-circle, as l is increased from zero, or, as the observation point moves on the physical line from the load-end towards the generator. This is illustrated in Fig. 7.2. Note that the phase angle of the reflection coefficient varies along the length of a line at double the rate at which the phase angle of a wave travelling along the line does.

7.4 VOLTAGE STANDING WAVE RATIO

For a lossless line, $\alpha = 0$, therefore, amplitudes of both the forward wave and the reflected wave remain constant all along the line but their phases change linearly with l. At points on the line where these two waves are in phase with each other (the phase difference being 0 or an even multiple of π), the net line voltage is the sum of the two amplitudes giving the maximum voltage (antinode); and at points where the two waves, are in antiphase with each other (the phase difference being

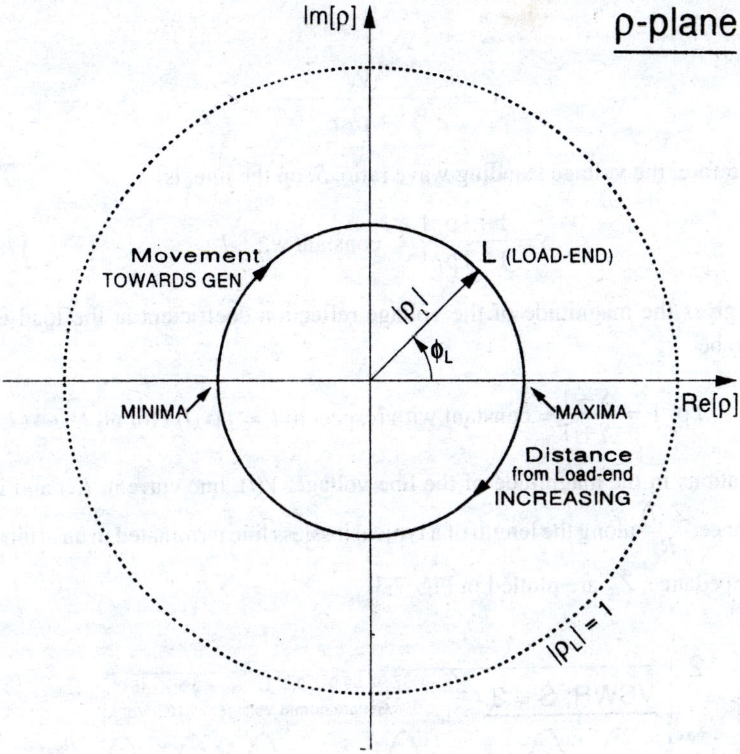

Fig. 7.2 Variation of Voltage Reflection Coefficient along a Lossless Transmission Line.

π or an odd multiple of π), the net line voltage is the difference of the two amplitudes giving the minimum voltage (nodes). As the two amplitudes are constant all along the line, the values of the maximum and the minimum voltages are the same everywhere resulting in the *VSWR*, *S*, being constant all along the line. The location of the antinodes and nodes of the voltage are given by Eqns. (5.35) and 5.36).

That is,

$$\phi_L - 2\beta_g \, l_{max} = -2\,k\,\pi, \quad k = 0, 1, 2, \ldots$$

or

$$l_{max} = \left(\frac{\phi_L}{\pi}\frac{\lambda_g}{4}\right) + k\frac{\lambda_g}{2} \tag{7.24}$$

and

$$\phi_L - 2\,\beta_g \, l_{min} = -(2\,k+1)\,\pi, \quad k = 0, 1, 2, \ldots$$

or

$$l_{min} = \left(\frac{\phi_L}{\pi}\frac{\lambda_g}{4}\right) + k\frac{\lambda_g}{2} + \frac{\lambda_g}{4} \tag{7.25}$$

As $\alpha = 0$, Eqns. (5.40) give

$$|V(l_{max})|_{max} = |A|(1 + |\rho_L|)$$

and $$|V(l_{\min})|_{\min} = |A|(1 - |\rho_L|) \tag{7.26}$$

where,

$$A = \frac{V_S}{e^{j\beta_g L} + \rho_L e^{-j\beta_g L}}.$$

Therefore, the voltage standing wave ratio, S, on the line, is

$$S = \frac{1 + |\rho_L|}{1 - |\rho_L|} = \text{constant w.r.t. } l \tag{7.27}$$

which gives the magnitude of the voltage reflection coefficient at the load-end, $|\rho_L|$, to be

$$|\rho_L| = \frac{S-1}{S+1} = \text{constant with respect to } l = |\rho(l)| \text{ for all } l. \tag{7.28}$$

Variations in the magnitude of the line voltage, $V(l)$, line current, $I(l)$ and line impedance, $\frac{Z(l)}{R_o}$ along the length of a typical lossless line terminated in an arbitrary load impedance, Z_L, are plotted in Fig. 7.3.

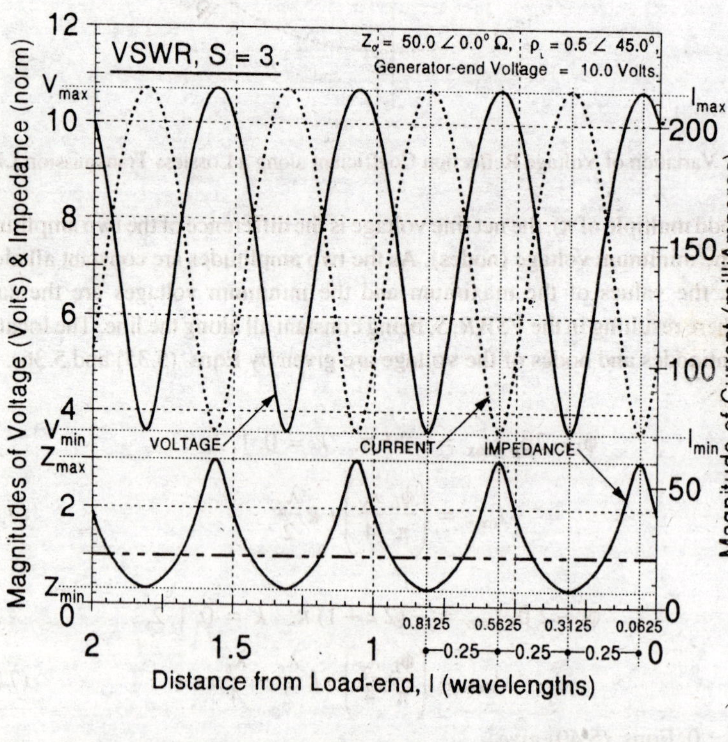

Fig. 7.3 Magnitudes of Voltage, Current and Impedance along Lossless Line (Arbitrary Load Impedance).

Now, the input impedance of the line at a distance l from the load-end towards the generator is [Ref. Eqns. (7.20) and (7.23)]

$$Z(l) = R_o \left[\frac{1 + |\rho_L| e^{-j(\phi_L - 2\beta_g l)}}{1 - |\rho_L| e^{-j(\phi_L - 2\beta_g l)}} \right] \tag{7.29}$$

Using Eqns. (7.29), (7.24) and (7.25), the impedances at the voltage maxima and minima are obtained to be

$$Z(l_{max}) = R_o \frac{1 + |\rho_L|}{1 - |\rho_L|} = SR_o = \text{a resistance} > R_o \tag{7.30}$$

and

$$Z(l_{min}) = R_o \frac{1 - |\rho_L|}{1 + |\rho_L|} = \frac{1}{S} R_o = \text{a resistance} < R_o \tag{7.31}$$

That is, the input impedance of a lossless line terminated in an arbitrary load impedance Z_L, at a voltage maximum turns out to be a pure resistance of value, $SR_o > R_o$, and that at a voltage minimum, it is again a pure resistance, $\frac{1}{S} R_o < R_o$. If $Z_L = Z_o$, there is no reflected wave and there are no maxima and minima, the voltage is constant everywhere, and *VSWR*, $S = 1$. The input impedance of the line at all points is then the resistance R_o. As discussed in Sec. 5.3., if $X_L > 0$, then as l is increased from zero, a voltage maximum is encountered first. Thereafter, voltage minima and maxima occur alternately, separated from each other at distances of $\frac{\lambda_g}{4}$. Similarly, if $X_L < 0$, then as l is increased from zero, a voltage minimum is encountered first. Thereafter, voltage maxima and minima occur alternately, separated from each other by $\frac{\lambda_g}{4}$. If $X_L = 0$ ($Z_L = R_L$ = a pure resistance), then the first voltage maximum or minimum occurs at the load-end itself according as $R_L > R_o$ or $R_L < R_o$.

7.4.1 *VSWR* produced by Arbitrary Load

If an arbitrary load impedance, $Z_L = R_L + j X_L$, is connected at the load-end of a lossless line of characteristic impedance, R_o, the voltage reflection coefficient, ρ_L, introduced at the load-end is given by

$$\rho_L = \frac{Z_L - R_o}{Z_L + R_o} \tag{7.32}$$

and the resulting *VSWR*, S, produced in the line is given by

$$S = \frac{1 + |\rho_L|}{1 - |\rho_L|} \tag{7.33}$$

Thus, if the load impedance, Z_L, is given, the voltage reflection coefficient, ρ_L, can be calculated [Eqn. (7.32)], and once ρ_L is known, the voltage standing wave ratio (*VSWR*), *S*, may be calculated using Eqn. (7.33). *S* may be expressed in terms of the load impedance, Z_L, as derived below:

$$\rho_L = \frac{Z_L - R_o}{Z_L + R_o} = \frac{z_L - 1}{z_L + 1} \tag{7.34}$$

where,

$$z_L = \frac{Z_L}{R_o} = r_L + j\,x_L = \frac{R_L}{R_o} + j\frac{X_L}{R_o} = \text{normalised load-impedance.} \tag{7.35}$$

From Eqn. (7.34),

$$|\rho_L|^2 = \rho_L \rho_L^* = \frac{(z_L - 1)(z_L^* - 1)}{(z_L + 1)(z_L^* + 1)}$$

$$= \frac{|z_L|^2 + 1 - 2r_L}{|z_L|^2 + 1 + 2r_L}$$

or

$$|\rho_L| = \sqrt{\frac{|z_L|^2 + 1 - 2r_L}{|z_L|^2 + 1 + 2r_L}} \tag{7.36}$$

Substitution of Eqn. (7.36) into Eqn. (7.33) gives

$$S = \frac{\sqrt{|z_L|^2 + 1 + 2\,r_L} + \sqrt{|z_L|^2 + 1 - 2\,r_L}}{\sqrt{|z_L|^2 + 1 + 2r_L} - \sqrt{|z_L|^2 + 1 - 2r_L}}$$

rationalising the denominator gives

$$S = \frac{2\,|z_L|^2 + 2 + 2\,\sqrt{(|z_L|^2 + 1)^2 - 4\,r_L^2}}{4r_L}$$

$$= \frac{\left[\dfrac{|z_L|^2 + 1}{r_L}\right] + \sqrt{\left[\dfrac{|z_L|^2 + 1}{r_L}\right]^2 - 4}}{2} \tag{7.37}$$

This expression for *S* suggests that *S* is one of the roots of the quadratic equation,

$$S^2 - \left[\frac{|z_L|^2 + 1}{r_L}\right] S + 1 = 0 \tag{7.38}$$

Note that the other root of this quadratic equation is $\frac{1}{S}$ (as the product of the roots is unity). For all real values of r_L and x_L, the discriminant of the quadratic equation may be shown to be positive. Therefore, Eqn. (7.38) has both of its roots real, namely, *S* and $\frac{1}{S}$; one of them being greater than unity (to be identified with *S*), and the other being less than unity. From Eqn. (7.38), it may be noted that

$$S = 1, \text{ if } \frac{|z_L|^2 + 1}{r_L} = 2$$

or if

$$(r_L - 1)^2 + x_L^2 = 0 \tag{7.39}$$

As both r_L and x_L are real, Eqn. (7.39) can be true, if $r_L = 1$ **and** $x_L = 0$

Hence, $$S = 1, \text{ if } r_L = 1\ (R_L = R_o) \textbf{ and } x_L = 0\ (X_L = 0) \tag{7.40}$$

7.5 SPECIAL CASES OF LINE

There are some cases of a transmission line of practical importance depending on either the terminating or the load impedances connected to the line or depending on the length of the line. These are discussed in this Section. First, consider the special cases arising out of the load impedance connected to the line.

7.5.1 Short-Circuited Line

A case of practical importance is the one in which the terminating or the load impedance, $Z_L = 0$. In this case, the voltage at the load-end, V_L, must be zero and, therefore, for this case, Eqns. (7.15), (7.17), and (7.19) give the voltage, current and impedance, respectively, at a distance, l, from the load-end to be

$$V(l) = j R_o I_L \sin(\beta_g l)$$

$$I(l) = I_L \cos(\beta_g l)$$

and $$Z_{sc}(l) = j R_o \tan(\beta_g l)$$

Voltage Reflection Coefficient,

$$\rho(l) = |\rho_L| e^{j(\phi_L - 2\beta_g L)} \tag{7.41}$$

where, $$|\rho_L| = 1.0$$

and, $$\phi_L = 180°$$

Voltage Standing Wave Ratio,
VSWR, $$S = \infty.$$

The voltage reflection coefficient at the load-end, ρ_L is equal to -1, producing a *VSWR*, $S = \infty$ on the line. Eqns. (7.41) show that the maximum amplitudes of the line voltages and line current are constant and the minimum values of both the voltage and current are, zero. Note further that the voltage and current on the line are in time and space quadrature, that is, when the line voltage is maximum, the line current is zero and vice versa; and the antinodes of the voltage wave are the nodes of the current wave and vice versa. The net power flow, therefore, at all points and at all times is zero giving purely stationary waves. From these, it follows that the input impedance of the line at all points, $Z_{sc}(l)$, is a pure reactance. This is corroborated by Eqn. (7.41). This input impedance is a positive reactance (inductive) for $l < \frac{\lambda_g}{4}$, and it is a negative reactance (capacitive) for $\frac{\lambda_g}{4} < l < \frac{\lambda_g}{2}$. The situation repeats itself with a period of $\frac{\lambda_g}{2}$. From (7.41), it is observed that the line is resonant ($Z_{sc} = 0$) at even multiples of a quarter wavelength and antiresonant ($Z_{sc} = \infty$) at odd multiples of a quarter wavelength. It may be pointed out that the input impedance of a lossless line is a function of the

$$\textit{electrical length}; \quad \theta = \beta_g l = 2\pi \frac{l}{\lambda_g} = \frac{2\pi}{v_p} f l,$$

and not of the physical length, l alone. Therefore, if, instead of varying the physical length of the line, it is kept constant and the frequency is varied, the electrical length of the line will vary as effectively as when the frequency is kept constant and the physical length of the line is varied. Hence, a short-circuited lossless line of a given length, l, exhibits resonance at frequencies which are even multiples of $\frac{v_p}{4l}$, and it is antiresonant at frequencies which are odd multiples of $\frac{v_p}{4l}$.

Eqn. (7.41) shows that at a given frequency of operation, it is possible to obtain any, positive or negative, pure reactance desired, by varying the length of the line.

Variations in the magnitude of the line voltage, $V(l)$, line current, $I(l)$ and the normalised line impedance, $\frac{Z(l)}{R_o}$ along the length of a typical lossless short-circuited line, are plotted in Fig. 7.4. Note that the amplitude of the line voltage and

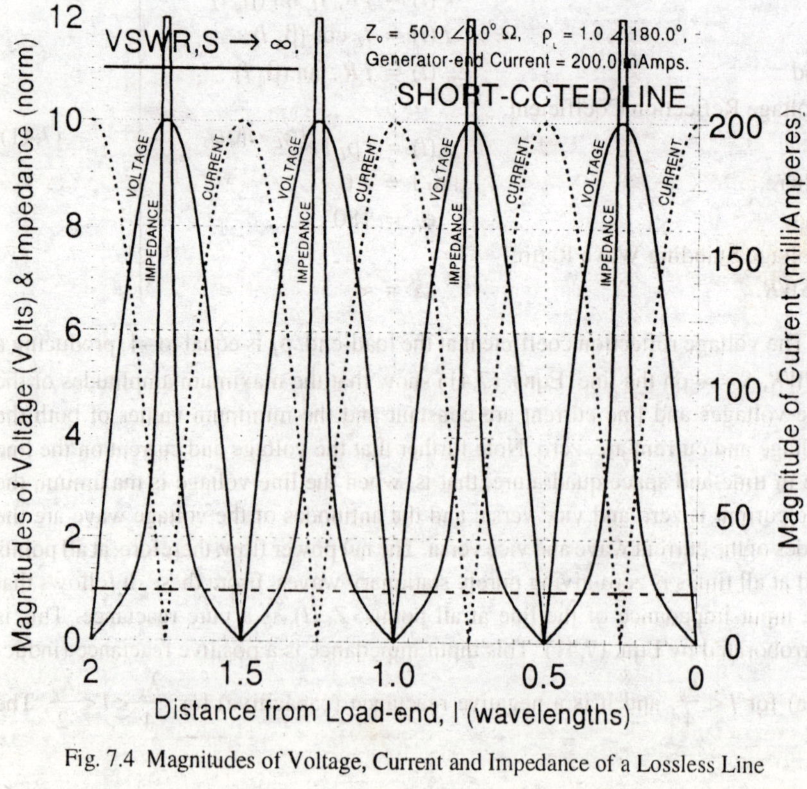

Fig. 7.4 Magnitudes of Voltage, Current and Impedance of a Lossless Line

hence the impedance at the load-end are zero (minimum), but the current amplitude is maximum. Thereafter, the maxima and minima of these quantities occur along the line alternately every quarter wavelength.

7.5.2 Open-Circuited Line

Another case of practical importance is that of a lossless terminated by an open-circuit so that the current at the load-end, I_L, is zero. The voltage, current and impedance of the line at a distance, l, from the load-end towards the generator, are obtained from Eqns. (7.15), (7.17), and (7.19), respectively, by setting $I_L = 0$.

$$\left.\begin{aligned} V(l) &= V_L \cos(\beta_g l) \\ I(l) &= j\frac{V_L}{R_o}\sin(\beta_g l) \\ \text{and} \quad z_{sc}(l) &= -j R_o \cot(\beta_g l) \\ \text{Voltage Reflection Coefficient,} \quad \rho(l) &= |\rho_L| e^{j(\phi_L - 2\beta_g l)} \\ \text{where,} \quad |\rho_L| &= 1.0 \\ \text{and,} \quad \phi_L &= 0^\circ \\ \text{Voltage Standing Wave Ratio, } VSWR, \quad S &= \infty. \end{aligned}\right\} \quad (7.42)$$

From Eqn. (7.21), the voltage reflection coefficient at the load-end, $\rho_L = 1$, producing a *VSWR* of $s = \infty$ on the line.

A comparison of these equations with Eqns. (7.41) shows that the voltage on the open-circuited line varies along the line like the current along the short-circuited line does, and *vice versa*. Further, the admittance of the open-circuited line behaves as the impedance of the short-circuited line. This case is, therefore, a dual of the short-circuited line and hence, the discussion given above may be used here with the changes required by the duality.

Like the short-circuited lossless line, the open-circuited line may be employed, in practice, to simulate at a given operating frequency, any positive or negative, pure susceptance desired, by varying the length of the line. At high frequencies, it becomes difficult to achieve $Z_L = \infty$ because there would always be some radiation from the open-end of a transmission line and hence Z_L is finite and not infinite. However, a short-circuit can be quite effectively produced, and therefore, in practice, open-circuit may be easily simulated by employing a short-circuited line whose length is an odd multiple of $\frac{\lambda_g}{4}$ at the operating frequency.

Variations in the magnitude of the line voltage, $V(l)$, line current, $i(l)$ and the normalised line impedance, $\frac{Z(l)}{R_o}$ along the length of a typical lossless open-circuited line, are plotted in Fig. 7.5. Note that variations in the magnitudes of the line

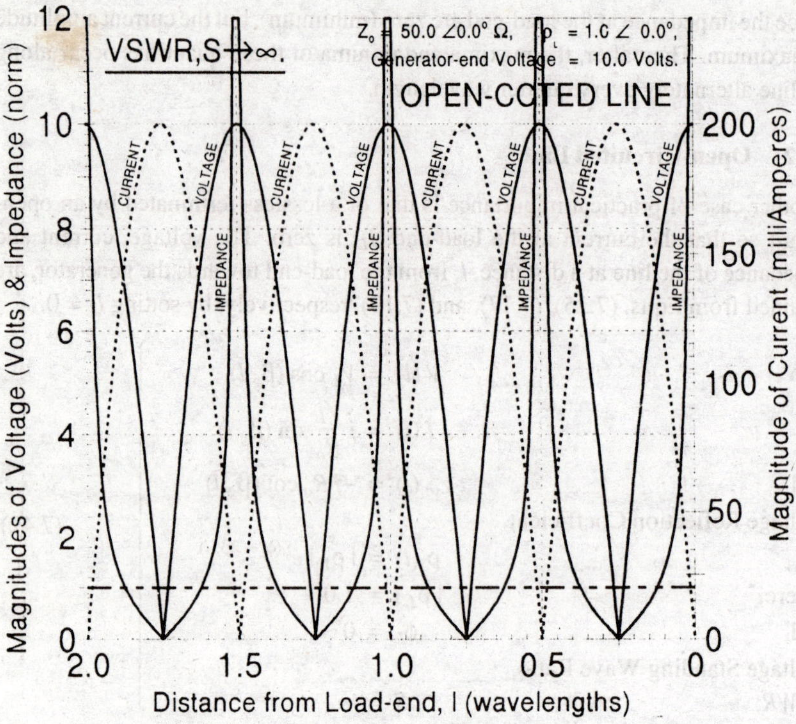

Fig. 7.5 Magnitudes of Voltage, Current and Impedance of Lossless Line Open-Circuited at Load-end.

voltage, current and impedance are very much the same as in the case of a short-circuited line except that the amplitude of the line current at the load-end is zero (minimum) but the voltage amplitude and hence the line impedance are maximum. Thereafter, the maxima and minima of these quantities occur along the line alternately every quarter wavelength. It may be observed that the patterns or distributions of the line voltage, current and impedance along the length of the line is essentially the same as that in the case of short-circuited line except that the patterns get shifted to the left (towards the generator) through a quarter of a wavelength.

7.5.3 Matched Line

Yet another important and interesting special case arises when the load impedance equals the characteristic impedance of the line. In this case, there is no reflection at the load-end and consequently, there are no reflected voltage and current waves on the line; only the travelling waves are present. As there are no losses on the line, the magnitudes of the line voltage and current are constant; only their phases vary along the line. Line impedance, in this case, is constant and equal to the characteristic impedance of the line everywhere along the line. This case is referred to as the ***matched line.***

$$
\left.
\begin{aligned}
& & V(l) &= V_L e^{(\beta_g l)} \\
& & I(l) &= \frac{V_L}{R_o} e^{(\beta_g l)} \\
&\text{and} & Z_{sc}(l) &= R_o \\
&\text{Voltage Reflection Coefficient,} & & \\
& & \rho(l) &= |\rho_L| e^{j(\phi_L - 2\beta_g l)} \\
&\text{where,} & |\rho_L| &= 0.0 \\
&\text{and,} & \phi_L &= 0^\circ \\
&\text{Voltage Standing Wave Ratio,} & & \\
&VSWR, & S &= 1.0
\end{aligned}
\right\} \quad (7.43)
$$

7.5.4 Quarterwave Line

Practically important cases also arise when a line terminated in an arbitrary load, has some specific lengths, for instance, like a quarter or half of the guided wavelength along the line.

Consider a line terminated in an arbitrary load impedance, Z_L, but having a length equal to a quarter of the wavelength. The most important property of this line is its impedance inversion capability. For a transmission line of length,

$$
\left.
\begin{aligned}
& l = \frac{\lambda_g}{4} \\
&\text{therefore,} \quad \beta_g l = \frac{\pi}{2} \\
&\text{Substituting this in Eqns. (7.15) to (7.20) gives} \\
& V_{in} = j\frac{R_o}{Z_L} V_L = \nu V_L, \text{ where } \nu = j\frac{R_o}{Z_L} \\
& I_{in} = j\frac{Z_L}{R_o} I_L = \frac{I_L}{\nu}, \\
& Z_{in} = \frac{R_o^2}{Z_L^2} Z_L = -\nu^2 Z_L, \text{ that is, } z_{in} = \frac{1}{z_L} \\
&\text{where, } z_{in} \text{ and } z_L \text{ are normalised impedances,} \\
&\text{defined by} \quad z_{in} = \frac{Z_{in}}{R_o} \text{ and } z_L = \frac{Z_L}{R_o}
\end{aligned}
\right\} \quad (7.44)
$$

Note that

1. the voltage at the input leads the load current by 90° in phase and its magnitude is $|z_L|$ times smaller than the voltage at the load end.
2. the current at the input leads the load voltage by 90° in phase and its magnitude is $|z_L|$ times larger than the current at the load end.
3. the normalised input impedance of the Quarterwave line is the reciprocal of the normalised load impedance.

or, the product of the input impedance and the load impedance equals the square of the characteristic impedance.
or, the characteristic impedance is the geometric means of the load impedance and the input impedance.

7.5.5 Halfwave Line

Next, consider a line terminated in an arbitrary load impedance, Z_L but having a length equal to a half of the wavelength. The most important property of this line is its phase reversal capability. For a transmission line of length,

$$\left.\begin{aligned} l &= \frac{\lambda_g}{2}, \\ \text{the electrical length, } \beta_g\, l &= \pi \end{aligned}\right\}. \tag{7.45}$$

Substituting this in Eqns. (7.15) to (7.20) gives

$$\left.\begin{aligned} V_{in} &= -Z_L, \\ I_{in} &= -I_L, \\ Z_{in} &= Z_L \end{aligned}\right\} \tag{7.46}$$

Note that

1. The voltage and current at the input are 180° out of phase with the voltage and current, respectively, at the load end.
2. The input impedance of the halfwave line is equal to the load impedance itself.

7.6 POWER RELATIONS

The Power, $P(z)$, flowing along the line at any point is given by

$$P(z) = \frac{1}{2}\,\mathrm{Re}\,[V(z)\, I^*(z)] \tag{7.47}$$

Substitution of Eqns. (7.12) and (7.13) into this equation yields

$$\begin{aligned} P(z) &= \frac{1}{2}\,\mathrm{Re}\left[\{V_S \cos(\beta_g z) - j R_o \sin(\beta_g z)\}\left\{I_S^* \cos(\beta_g z) + j\frac{V_S^*}{R_o}\sin(\beta_g z)\right\}\right] \\ &= \frac{1}{2}\,\mathrm{Re}\Bigg[V_S I_S^* \cos^2(\beta_g z) + V_S^* I_S \sin^2(\beta_g z) \\ &\qquad - j\left\{R_o\,|I_S|^2 - \frac{|V_S|^2}{R_o}\right\}\cos(\beta_g z)\sin(\beta_g z)\Bigg] \\ &= \frac{1}{2}\,\mathrm{Re}\Bigg[V_S I_S^* + 2j\,\mathrm{Im}\,(V_S^* I_S)\sin^2(\beta_g z) \\ &\qquad + j\left\{\frac{|V_S|^2}{R_o} - R_o\,|I_S|^2\right\}\cos(\beta_g z)\sin(\beta_g z)\Bigg] \\ &= \frac{1}{2}\,\mathrm{Re}\,[V_S I_S^*] = P(0) \end{aligned} \tag{7.48}$$

Note that $P(z)$ is independent of z. This is expected because the line is lossless and, therefore, there cannot be a change in the amount of power flow with distance along the line.

Using the expressions for $V(z)$ and $I(z)$ given by Eqns. (7.16) and (7.18) in Eqn. (7.47) yields

$$\begin{aligned}
P(l) &= \text{Re}\,[V(l)\,I^*(l)] \\
&= \frac{1}{2}\,\text{Re}\left[\frac{1}{2}\,|I_L|^2\,\frac{|Z_L+R_o|^2}{R_o}\,\{1-|\rho_L|^2+2j\,\text{Im}\,(\rho_L\,e^{-j2\beta_g l})\}\right] \\
&= \frac{|I_L|^2}{8}\,\frac{|Z_L+R_o|^2}{R_o}\,(1-|\rho_L|^2) \\
&= \frac{|I_L|^2}{2}\,\frac{|Z_L+R_o|^2}{4R_o}\left[1-\frac{Z_L-R_o}{Z_L+R_o}\,\frac{Z_L^*-R_o}{Z_L^*+R_o}\right]
\end{aligned}$$

$$\begin{aligned}
P(l) &= \frac{|I_L|^2}{2}\,\frac{|Z_L+R_o|^2}{4R_o}\,\frac{4R_oR_L}{|Z_L+R_o|^2} \\
&= \frac{1}{2}\,|I_L|^2\,R_L
\end{aligned} \tag{7.49}$$

From Eqns. (7.48) and (7.49), it is clear that at all points on a lossless transmission line the power flow is constant and equal to the power delivered to or dissipated in the load.

In the light of discussions under Sec 4.1 and 5.2, the propagation of waves on a transmission line may be described as follows:

When a generator of internal impedance, R_o, is connected to a lossless transmission line of characteristic impedance, R_o, length l, and terminated in a load impedance Z_L, a voltage wave of amplitude V_{inc} is set up on the line, which travels towards the load. The power associated with this wave, which flows towards the load-end, is

$$P_{inc} = \frac{1}{2}\,\frac{|V_{inc}|^2}{R_o} \tag{7.50}$$

On reaching the load-end, the amplitude of the voltage wave is $V_{inc}\,e^{-j\beta_g l}$ and a fraction ρ_L of this gets reflected and starts moving towards the generator. The total voltage at the load-end is

$$V_L = (1+\rho_L)\,V_{inc}\,e^{-j\beta_g l} \tag{7.51}$$

The power, P_{refl}, associated with the wave which travels towards the generator-end is

$$P_{refl} = |\rho_L|^2\,P_{inc}\,. \tag{7.52}$$

In practice, particularly at high frequencies, a quick and convenient measure of the degree of Reflectionless Match existing at the load-end, is provided by the ***Return Loss*** (in dBs) which is defined as

$$\text{Return Loss} = 10 \log_{10} \frac{P_{inc}}{P_{refl}} = -20 \log_{10} \rho_L = -20 \log_{10} \frac{(S-1)}{(S+1)} \tag{7.53}$$

From this definition, it is clear that the return loss is nothing but the number of decibels by which the reflected power is below the level of the incident power at the load-end. Therefore, for perfect (reflectionless) match,

$$P_{refl} = 0, \rho_L = 0 \text{ or } S = 1, \text{ the Return Loss} \to \infty$$

and, for the worst case (complete mismatch)

$$P_{refl} = P_{inc}, \ \rho_L = 1 \text{ or } S \to \infty, \text{ the Return Loss} = 0$$

The balance of power, $P_L = P_{inc} - P_{refl}$, is absorbed by the load. On reaching the generator-end, the total voltage, V_s, at the generator-end is given by

$$V_S = (1 + \rho_L e^{-j2\beta_g l}) V_{inc}, \tag{7.54}$$

assuming that no reflection takes place at the generator-end. The power delivered to the load is

$$P_L = P_{inc} - P_{refl} \tag{7.55}$$

$$= (1 - |\rho_L|^2) P_{inc}$$

$$= (1 - |\rho_L|^2) \frac{|V_{inc}|^2}{2R_o} \tag{7.56}$$

$$= \frac{|V_S|^2}{|1 + \rho_L e^{-j2\beta_g l}|^2} \frac{1}{2R_o} (1 - |\rho_L|^2) \tag{7.57}$$

It is clear that the maximum power that could be delivered to the load is P_{inc}, that is, when no power is reflected at the load-end, or

for $$|\rho_L| = 0 \text{ or } S = 1, P_{L_{max}} = P_{inc} \tag{7.58}$$

in general, there may be some mismatch, that is, $|\rho_L| \neq 0$ or $S \neq 1$, and some power gets reflected back giving rise to ***reflection loss*** which may be defined as

$$L_{ref} = 10 \log \left(\frac{P_{L_{max}}}{P_L} \right) = \left(\frac{1}{1 - |\rho_L|^2} \right) \text{dB}$$

$$= 20 \log \left[\frac{1}{2} \left(\sqrt{S} + \frac{1}{\sqrt{S}} \right) \right] \tag{7.59}$$

7.6.1 Power Carrying Capacity of a Line

It is clear from Eqn. (7.60) that the power carried by a line goes on increasing indefinitely as the line voltage is increased. But in practice, this cannot be done because when the line voltage exceeds a value V_m, the electric field strength in the dielectric is such as to produce dielectric breakdown. Thus, the maximum power that can be carried by the line is limited by the dielectric breakdown.

Consider lossless line having, in general, *VSWR*, *S*, on it. The voltage on the line is varying and its maximum value is given by [Ref. Sec. 5.3]

$$|V_{max}| = |A|(1+|\rho_L|) \tag{7.60}$$

and the current at that point is minimum, given by

$$|I_{min}| = \frac{|A|}{R_o}(1-|\rho_L|) \tag{7.61}$$

for a lossless line, as the line impedance at a voltage maximum or minimum is a pure resistance, the voltage and current at a voltage maximum or minimum will be in phase with each other, so that the power carried by the line is

$$P_L = \frac{1}{2}|V_{max}||I_{min}|$$

$$= \frac{|A|^2}{2R_o}(1-|\rho_L|^2) \tag{7.62}$$

The power carrying capacity of the transmission line is obtained from Eqn. (7.61) by using in it the maximum permitted value of $|A|$ which, in turn, is determined by the condition that the maximum voltage on the line does not exceed v_m, that is,

$$|V_{max}| = |A|(1+|\rho_L|) < V_m$$

or

$$|A| < \frac{V_m}{1+|\rho_L|} \tag{7.63}$$

Hence, from Eqns. (7.61) and (7.62),

$$P_{L_{max}} = \frac{V_m^2}{2R_o}\frac{1-|\rho_L{}^2}{(1+|\rho_L|)^2}$$

$$= \frac{V_m^2}{2R_o}\frac{1}{\dfrac{1+|\rho_L|}{1-|\rho_L|}}$$

$$= \frac{V_m^2}{2R_o}\frac{1}{S} \tag{7.64}$$

The power carrying capacity of the transmission line, given by Eqn. (7.64), is inversely proportional to the *VSWR* on the line. Therefore, the power carrying capacity of a line is largest for $S = 1$, and it decreases monotonically as S increases.

SOLVED EXAMPLES

Example 7.1 A 60 Ω lossless line feeds an antenna of an input impedance of $(76.5 + j76.5)\ \Omega$ at a frequency of 1200 MHz. Calculate the voltage reflection coefficient at the load-end and determine the *VSWR* produced. How far from the load-end is the first voltage minimum located?

Solution : Characteristic Impedance, $Z_o = 60\ \Omega$ (given), and Load Impedance, $Z_L = (76.5 + j76.5)\ \Omega$ (given).

Therefore, normalised load impedance,

$$z_L = \frac{(76.5 + j76.5)}{60} = 1.275 + j1.275$$

By def. the voltage transmission coefficient at the load-end,

$$\rho_L = \frac{(1.275 - 1) + j1.275}{(1.275 + 1) + j1.275}$$

or, $$\rho_L = \frac{(1.275 - 1) + j1.275}{(1.275 + 1) + j1.275} = \frac{1.3043 \angle 77.829^\circ}{2.6079 \angle 29.2680^\circ} = 0.5001 \angle 48.5605^\circ$$

VSWR, $$S = \frac{1 + 0.5001}{1 - 0.5001} = 3.001.$$

Angle of the voltage reflection coefficient,

$$\phi = \phi_L - 720^\circ \frac{l}{\lambda_g}, \text{ where, } \phi_L = 48.5605^\circ$$

As l is increased from zero, the angle of the voltage transmission coefficient, ϕ decreases from its value, 48.5605 °, at the load-end, becomes zero at the voltage antinode or maximum (for $l = l_{max}$).

If l is further increased, ϕ becomes – 180° at the first voltage minimum corresponding to $l = l_{min}$.

Therefore, $$l_{\max} = \frac{48.5605^\circ}{720^\circ} \lambda_g = 0.06745\ \lambda_g,$$

and the next (first) voltage minimum is located at

$$l_{\min} = 0.25\ \lambda_g + 0.06745\ \lambda_g = 0.3174\ \lambda_g.$$

Example 7.2 When an unknown load impedance is connected to a 50 Ω lossless line, a *VSWR* of $S = 3$ is observed on the line and the first voltage minimum is found to occur at a distance of $\frac{5}{8}\lambda_g$ from the load-end. Calculate the unknown impedance.

Solution : : At a voltage minimum, the normalised input line impedance is known to be $z_{in} = \frac{1}{S} + j0$, where S is the *VSWR* existing on the line. z_{in}, that is, the normalised input line impedance is, therefore, given at one point (voltage minimum) on the line, all that is required is to find the line impedance at another point (load-end) on the line. [Note that the input line impedance at the load-end is nothing but the load-impedance itself.] The impedances at two points are related through the impedance transformation given by Eqn. (7.19), wherein, it may be recalled that $Z(l)$ is the line impedance at a distance of ***l towards the generator*** from the point where the impedance is Z_L. In terms of the normalised impedances, Eqn. (7.19) becomes

$$z(l) = \frac{z_L + j\tan(\beta_g l)}{1 + jz_L \tan(\beta_g l)}$$

Identifying z_L with z_{in} = known, and noting that to determine the unknown impedance, one must go along the line a distance, l, ***towards the load***, therefore, putting $l = -\frac{5}{8}\lambda_g$, the above formula gives the unknown impedance to be

$$Z_L = \frac{\frac{1}{3} + j\tan\left(-\frac{5}{8}\times 360°\right)}{1 + j\frac{1}{3}\tan\left(-\frac{5}{8}\times 360°\right)}$$

$$= 50 \times \frac{\frac{1}{3} - j\tan 225°}{1 - j\frac{1}{3}\times\tan 225°}\ \Omega$$

$$= 50 \times \frac{1 - j3}{3 - j1}\ \Omega = 50 \times 1\ \underline{/71.565° - 18.435°}\ \Omega$$

$$= 50\ \underline{/53.130°}\ \Omega = (30.0 + j40.0)\ \Omega$$

Example 7.3 When an unknown impedance is connected at the load-end of a transmission line, a *VSWR of 3 is observed on the line, and the distance between two successive* voltage minima is found to be 22 mm. Further, it is observed that if the load is replaced by a short-circuit, the voltage minima shift through a distance of 4.4 mm toward the generator. Find out the normalized load impedance and the voltage reflection coefficient at the load-end.

Solution : Before taking up the solution of this iproblem, if is achisable to discuss the method employed to experimentally determine an maknown land impedance by making use of the *VSWR* produced by the load and the location of the voltage minimum on the time.

Discussion : Experimental determination of unknown impedance.

This is the standard method employed to determine an unknown impedance at microwave frequencies. The important points to be remembered in this connection and on which this method is based, are the following:

1. The distance between two successive voltage minima on a transmission line $= \frac{\lambda_g}{2}$.
2. The normalised input line impedance at a voltage minimum is

$$z_N = \frac{1}{S} + j0 \text{ (pure resistance), where } S = VSWR.$$

3. From a known impedance at some point on the line, for example, from a knowledge of z_N at the point, N (voltage minimum or node), to determine the line impedance, for instance, Z_L at any other point, L, one must move through a distance, l (which is positive if the movement is towards the generator, and is negative if the movement is towards the load) on the line from the point, N, to *either* the point, L or to a point which is an integral number of halfwavelengths away from the point, L. This follows from the periodicity of line impedance with a period of a halfwavelength. Mathematically, this amounts to transforming the impedance, z_N, through a movement, l, towards the generator. Hence

$$z_L = \frac{z_N + j\tan(\beta_g l)}{1 + j z_N \tan(\beta_g l)}$$

4. The normalised input line impedance at a voltage minimum is a pure resistance (for lossless lines) $z_N = \frac{1}{S} + j0$ and that at a voltage maximum is $z_X = S + j0$. In fact, a voltage minimum and a voltage maximum are the only two points on a transmission line at which the normalised line impedance is real and, therefore, the line impedance a pure resistance (for lossless lines). As a corollary to this, if a load impedance is a pure resistance, the first voltage maximum or minimum occurs at the load-end itself. The first voltage minimum or maximum would, therefore, occur at the load terminals, according as the normalised load resistance is less than or more than one. Thereafter, voltage minima and maxima recur with a periodicity of a halfwavelength. In view of this, therefore, when a short-circuit is connected at the load-end of a line, the voltage minima on the line are integral halfwavelengths away from the load-end.

Hence, in view of these points, a practical method of measuring an unknown impedance would warrant determination of the *VSWR* and the distance from the voltage minimum (with the unknown load connected) to the load-end or to any voltage minimum that occurs when a short-circuit (or any resistance less than the characteristic impedance of the line) is connected at the load-end of the line.

The distance between two successive minima = 22 mm (given), therefore, $\lambda_g = 2 \times 22$ mm = 44 mm. Observed *VSWR* = 3 (given), therefore, normalised input impedance at the voltage minimum,

$$z_N = \frac{1}{3} + j0$$

When load-end is shorted, the shift in minimum towards the generator = 4.4 mm (given)

$$= \frac{4.4}{44}\lambda_g = 0.1\,\lambda_g.$$

Therefore, $\tan(\beta_g l) = \tan(0.1 \times 360°) = \tan 36° = 0.7265.$

$$z_L = \frac{z_N + j\tan(\beta_g l)}{1 + jz_N \tan(\beta_g l)}$$

$$= \frac{\left(\frac{1}{3} + j0\right) + j0.7265}{1 + j\left(\frac{1}{3} + j0\right)0.7265}$$

$$= \frac{1 + j2.1796}{3 + j0.7265} = \frac{2.3981}{3.0867}\,\underline{/(65.355 - 13.614)°} = 0.7769\,\underline{/51.741°}$$

EXERCISES

1. (a) For operation at a frequency of 500 MHz, an air-filled, rigid coaxial line is required to be made by using a copper tube of inner diameter of 22.225 mm and a wall thickness of approximately 1.6 mm, as its outer conductor. Calculate the diameter of its inner conductor, if the characteristic impedance of the line is required to be (1) 30 Ω, (2) 60 Ω, (3) 75 Ω.

[**Ans:** Outer diameter of the inner conductor : (1) 13,4801 mm, (2) 8,1761 mm, (3) 6.3676 mm]

(b) Assuming the air to be a lossless dielectric, estimate the attenuation constants of the three transmission lines designed above at the frequency of operation.

[**Ans:** Attenuation constant :(1) 0.0322 dB/m, (2) 0.0226 dB/m, (3) 0.0218 dB/m].

(c) Calculate the maximum power carrying capacities of the above three lines if the dielectric strength or the flash-over field strength for air is 30 kV/cm.

[**Ans: Theoretical** Max. Power carrying capacity: (1) 3.4071 mw, (2) 2.5068 mw (3) 1.9006 mw. Maximum line voltage : (1) 10.1101 kVolts, (2) 12.2642 k Volts, (3) 11.9392 k Volts. (1) 377.004 Amps. (2) 204.403 Amps, (3) 159.189 Amps.]

(d) Now, if the air dielectric is replaced by polyethylene having the same dielectric strength as that of air, but a relative permittivity of $\varepsilon_r = 2.25$, recalculate the diameters of the inner conductors, attenuation constants and power carrying capacities of the three lines.

[**Ans:** Outer diameter of the inner conductor: (1) 10.4983 mm (2) 4.9591 mm, (3) 3.4083 mm. Attenuation constant : (1) 0.0379 dB/m, (2) 0.0333 dB/m, (3) 0.0365 dB/m.

Theoretical: Max. Power Carrying Capacity : (1) 4.6497 mw, (2) 2.0750 mw, (3) 1.2252 mw.

Max. Line voltage : (1) 11.820 k Volts, (2) 11.2579 k Volts, (3) 9.5859 k Volts

Line current : (1) 393.688 Amps., (2) 185.965 Amps. (3) 127.812 Amps.]

Discussion : Design of a coaxial line.

The problem given as Ex. 1 above, illustrates some important facts about coaxial lines. These as well as some comments on practical aspects are summarised below:

1. For a given diameter ($2a$) of the outer conductor, a coaxial cable has maximum power carrying capacity when the diameter of its inner conductor ($2b$) is given by

$$2b = \frac{2a}{\sqrt{e}}$$

This cable, if it uses air as the dielectric, would have a characteristic impedance of 30 Ω. This may be verified from the answers given for Ex. 1(c). Also given in the answer are the maximum line voltages that are permissible and the line currents that these cables have when the power flow is the maximum permissible.

Note that the values of maximum permissible power flow are only ***theoretical*** in the sense that these have been calculated ignoring the fact that a conductor of a given cross-section can carry but a limited amount of current. For example, the 75 Ω cable would carry its maximum power of 1.9 MW, if and only if its conductors can carry a current (dc) of 160 Amperes. However, the maximum current (dc) that a copper wire of 6.3676 mm diameter is permitted to carry is about 135 Amperes. At high frequencies, because of the skin effect, the effective cross-sectional area of the conductor that carries the current, gets significantly reduced. Hence, the maximum power carrying capacity of the cable, in practice, may be determined by the maximum current carrying capacity individual conductors at the frequency of operation. It is for this reason that the adjective, theoretical, has been emphasised in the answers given.

2. Again, for a given diameter ($2a$) of the outer conductor, a coaxial cable can withstand maximum line voltage, if the diameter ($2b$) is given by the relation,

$$2b = \frac{2a}{e}$$

If the cable uses air as the dielectric, its characteristic impedance would be 60 Ω. This may be verified from the answers given for Ex. 1(c).

3. Lastly, for a given diameter ($2a$) of the outer conductor, a coaxial cable would offer minimum attenuation to signals of a given frequency, if the diameter ($2b$) of its inner conductor, is given by

$$2b = \frac{2a}{e^{1.27846}}$$

Again, if the cable uses air as the dielectric, its characteristic impedance would be 76.7079 Ω. This may also be verified from the answers given for Ex. 1(c).

1. [**Ans.** Outer diameter of the Inner conductor: (1) 13.4801 mm, (2) 8.1761 mm, (3) 6.3676 mm.]
2. [**Ans**: Attenuation Constant: (1) 0.0322 dB/m, (2) 0.0226 dB/m, (3) 0.0218 dB/m.]
3. [**Ans.** ***Theoretical*** Max. Power Carrying Capacity: (1) 3.4071 MW, (2) 2.5068 MW, (3) 1.9006 MW.
   Maximum Line Voltage: (1) 10.1101 kVolts, (2) 12.2642 kVolts, (3) 11.9392 kVolts.
   Line Current: (1) 337.004 Amps., (2) 204.403 Amps., (3) 159.189 Amps.]
4. [**Ans.** Outer diameter of the Inner conductor: (1) 10.4983 mm, (2) 4.4.9591 mm, (3) 3.4083 mm.
   Attenuation Constant: (1) 0.0379 dB/m, (2) 0.0333 dB/m, (3) 0.0365 dB/m.
   Theoretical **Max.** Power Carrying Capacity: (1) 4.6497 MW, (2) 2.0750 MW, (3) 1.2252 MW
   Maximum Line Voltage: (1) 11.8106 kVolts, (2) 22.1579 kVolts, (3) 9.5859 kVolts.
   Line Current: (1) 393.688 Amps., (2) 185.965 Amps., (3) 127.812 Amps.]

2. (a) An open-wire line, required to work at a frequency of 100 MHz, is to be made by running two 3.6 mm diameter copper wires, parallel to each other in air. Calculate the spacing required between the wires, if the line is to have a characteristic impedance of

(1) 150 Ω (2) 300 Ω (3) 600 Ω

[**Ans.** Centre-to-Centre Spacing: (1) 6.7983 mm, (2) 22.0762 mm, (3) 267.1558 mm.]

(b) Assuming the air to be a lossless dielectric, estimate the attenuation constants of the three transmission lines designed above at the frequency of operation.

[**Ans.** Attenuation Constant: (1) 0.0354 dB/m, (2) 0.0152 dB/m, (3) 0.0075 dB/m.]

(c) Calculate the maximum power carrying capacities of the above three lines if the dielectric strength or a flash-over field strength for air is 30 kV/cm.

[**Ans.** *Theoretical* Max. Power Carrying Capacity: (1) 2.2996 MW, (2) 8.1798 MW, (3) 19.1789 MW.]

Maximum Line Voltage: (1) 18.5726 kVolts, (2) 49.5373 kVolts (3) 107.2722 kVolts

Line Current: (1) 123.817 Amps., (2) 165.124 Amps., (3) 178.787 Amps.]

(d) Now, if the air dielectric is replaced by polyethylene having the same dielectric strength as that of air but a relative permittivity of $\varepsilon_r = 2.25$, recalculate the characteristic impedances, attenuation constants and power carrying capacities of the three lines.

[**Ans.** Centre-to-Centre Spacing: (1) 12.0135 mm, (2) 76.5803 mm, (3) 3254.4773 mm.

Attenuation Constant: (1) 0.0314 dB/m,(2) 0.0150 dB/m, (3) 0.0075 dB/m.

Theoretical Max. Power Carrying Capacity: (1) 7.7701 MW, (2) 20.8529 MW, (3) 43.6916 MW.

Maximum Line Voltage: (1) 34.1395 kVolts (2) 79.0940 kVolts, (3) 161.9104 kVolts.

Line Current:(1) 227.596 Amps., (2) 263.647 Amps., (3) 269.851 Amps.]

3. Comment on the following statements :

1. If a line is lossless, its characteristics impedance must be a pure resistance.
2. If the characteristic impedance of a line is a pure resistance, it must be lossless.

4. For a lossless $\frac{\lambda_g}{8}$ long line terminated in an arbitrary impedance Z_L, derive the condition that the input impedance is a pure resistance.

5. Derive an expression for the *VSWR* and *S* produced, when an arbitrary load impedance, $Z_L = mR_o e^{j\theta}$, terminates a lossless line of characteristic impedance, R_o. Discuss the variation is *S* with the magnitude *m*, and also with the angle θ, of the load and obtain expressions for the minimum values of *S* in the following cases :

(1) *m* alone is varied

(2) θ alone is varied

(3) both *m* and θ are varied.

6. A 50 Ω lossless line feeds a load of normalised admittance of $0.5333 - j0.2667$ Ω. Calculate the voltage reflection coefficient at the load-end and the *VSWR* product in the line.

At some distance l_1, from the load-end, the real part (conductance) of the normalised input line admittance is measured to be unity. Calculate the normalised input line susceptance at this point and the value of the distance, l_1.

7. An antenna of input impedance $50 + j37.5$ Ω is fed by a 50 Ω coaxial line feeder, 0.0968 λ_g long. Calculate the impedance of the antenna as measured at the input terminals of the feeder and the *VSWR* produced in the feeder. If this antenna-feeder system is connected to the receiver by a 105 Ω coaxial line, calculate the *VSWR* in the 105 Ω coaxial line.

Chapter 8

LINE AS A CIRCUIT ELEMENT

8.1 INTRODUCTION

A conventional lumped circuit element, because of its practical construction, has parasitics associated with it. For example, an inductor always has associated with it some resistance and capacitance, and a capacitor has parasitic inductance and resistance. To obtain a pure lumped circuit element free from parasitics, is extremely difficult, if not impossible, because, if nothing else, the conductor or the lead used for interconnecting lumped circuit elements itself, has distributed resistance, inductance and capacitance. The parasitics may be reduced by improvement in the construction of the elements, or may be allowed for in the design and construction of the elements, if they are required to operate only at a spot frequency. The effects of these parasitics may be negligible at low frequencies but as the frequency of operation is increased, the effects of the parasitics become more and more appreciable. In fact, for instance, a lumped capacitor may as well start behaving like an inductor at high frequencies because the reactance of the parasitic inductance of the capacitor, goes on increasing with frequency and may even swamp the capacitive reactance of the capacitor. In addition to this difficulty posed by parasitics, it may be difficult even to realise, in lumped-element form, the small inductance or capacitance values that are required at high frequencies. For example, the inductance of an isolated straight wire may be much more than that required. Therefore, new methods of realising pure reactances at high frequencies are to be resorted to.

8.2 REALISATION OF PURE REACTANCES

From the discussion given in Secs. 7.5 and 7.6, it is clear that both short-circuited and open-circuited lossless lines, present, at their input ports or terminals, a pure reactance, positive or negative and of any magnitude, depending on the length of the lines. This property of lossless lines may be exploited at high frequencies, to realise or to simulate a positive or negative reactance of any arbitrary magnitude as discussed in what follows.

Suppose $$X_{rq}, \text{ where } X_{rq} \gtrless 0, \tag{8.1}$$

is the reactance desired to be realised at the given frequency,

$$f_o = \frac{\omega_o}{2\pi} \tag{8.2}$$

Then, either a short-circuited line of length, l_{sc}, or an open-circuited line of length, l_{oc}, may be used to realise the reactance, X_{rq}. The length l_{sc} or l_{oc}, may be determined using the analysis given in Sec. 7.5, as explained below:

Short-circuited Line From Eqn. (7.41), the input impedance (pure reactance) of a short-circuited, lossless, (R_o, β_g)-transmission line of length, l_{sc}, is given by

$$Z_{sc} = j R_o \tan(\beta_g l_{sc}) \tag{8.3}$$

The length, l_{sc}, of the short-circuited line, required to realise a given reactance, X_{rq}, is obtained from Eqn. (8.3) by equating Z_{sc} to jX_{rq}, that is, from the equation,

$$Z_{sc} = j X_{rq} = j R_o \tan(\beta_g l_{sc}) \tag{8.4}$$

which gives,

$$l_{sc} = \frac{1}{\beta_g} \tan^{-1}\left(\frac{X_{rq}}{R_o}\right) = \frac{\lambda_g}{2\pi} \tan^{-1}\left(\frac{X_{rq}}{R_o}\right) \tag{8.5}$$

Open-circuited Line Similarly, from Eqn. (7.42), the input impedance of an open-circuited, lossless, (R_o, β_g)-transmission line of length, l_{oc}, is given by

$$Z_{oc} = -j R_o \cot(\beta_g l_{oc}) \tag{8.6}$$

The length, l_{oc}, of the open-circuited line required to realise a given reactance, X_{rq}, is obtained from Eqn. (8.6) by equating Z_{oc} to jX_{rq}, that is, from the equation,

$$\left.\begin{aligned} Z_{oc} &= j X_{rq} = -j R_o \cot(\beta_g l_{oc}) \\ \text{or,} \quad X_{rq} &= R_o \tan\left(\mp\frac{\pi}{2} + \beta_g l_{oc}\right) \end{aligned}\right\} \tag{8.7}$$

which gives,

$$l_{oc} = \frac{1}{\beta_g}\left[\tan^{-1}\left(\frac{X_{rq}}{R_o}\right) \pm \frac{\pi}{2}\right] = \frac{\lambda_g}{2\pi} \tan^{-1}\left(\frac{X_{rq}}{R_o}\right) \pm \frac{\lambda_g}{4} \tag{8.8}$$

Substitution of Eqn. (8.5) into the above equation, gives

$$l_{oc} = l_{sc} \pm \frac{\lambda_g}{4} \tag{8.9}$$

Lengths, $l_{sc}\left(<\frac{\lambda_g}{2}\right)$, of a short-circuited lossless, (R_o, β_g)-transmission line and $l_{oc}\left(<\frac{\lambda_g}{2}\right)$ of the open-circuited line, given by Eqns. (8.5) and (8.9), respectively, required to realise a prescribed reactance, X_{rq}, are plotted as functions of X_{rq} in Fig. 8.1. Here, it may be pointed out that increasing or decreasing l_{sc} and l_{oc} by an integral multiple of $\frac{\lambda_g}{2}$, has no effect on the reactances simulated. This follows from the fact that the input impedance of a line is a periodic function of its length, with a period of $\frac{\lambda_g}{2}$.

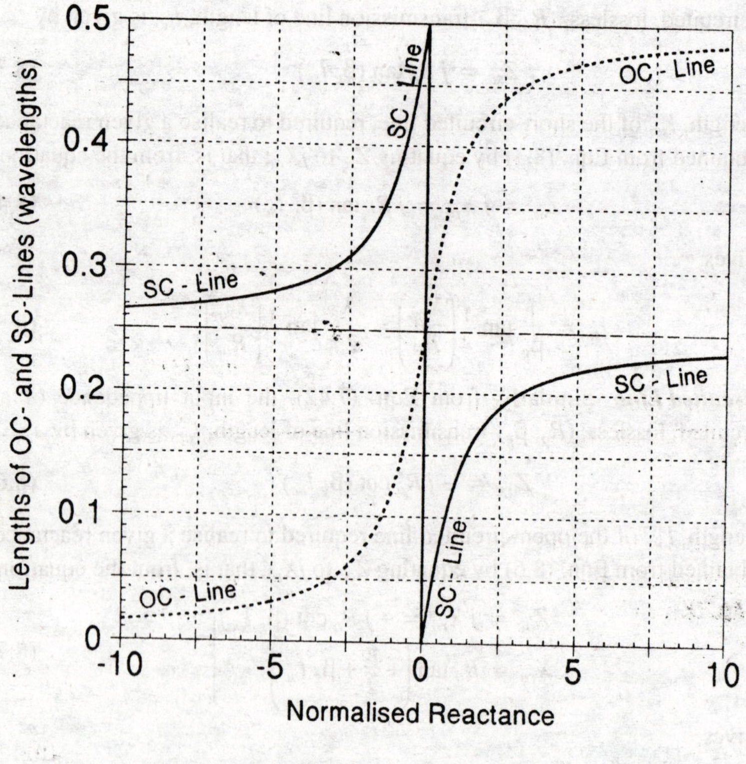

Fig. 8.1 Lengths of Short- and Open-Circuited Lossless Transmission Lines required for Realisation of Pure Reactances.

8.2.1 Positive Reactance

As discussed above, a pure reactance may be realised by either a short-circuited or an open-circuited lossless line of a suitable length. Discussed below is the realisation of positive reactances using a short-circuited and an open-circuited line.

Short-circuited Line If a positive reactance is to be realised at a given frequency, ω_o, that is, if

$$\left.\begin{aligned} &\infty > X_{rq} > 0, \\ \text{so that,}\quad & X_{rq} = \omega_o L_{eff}, \text{ where, } L_{eff} > 0, \\ \text{then,}\quad & \frac{\pi}{2} > \tan^{-1}\left(\frac{X_{rq}}{R_o}\right) > 0. \end{aligned}\right\} \tag{8.10}$$

Accordingly, from Eqn. (8.5), the length of the line required in less than a quarter wavelength, that is,

$$\frac{\lambda_g}{4} > l_{sc} > 0. \tag{8.11}$$

Length of the short-circuited line required as a function of the normalised reactance desired to be realised, is plotted (solid line curve) in Fig. 8.2.

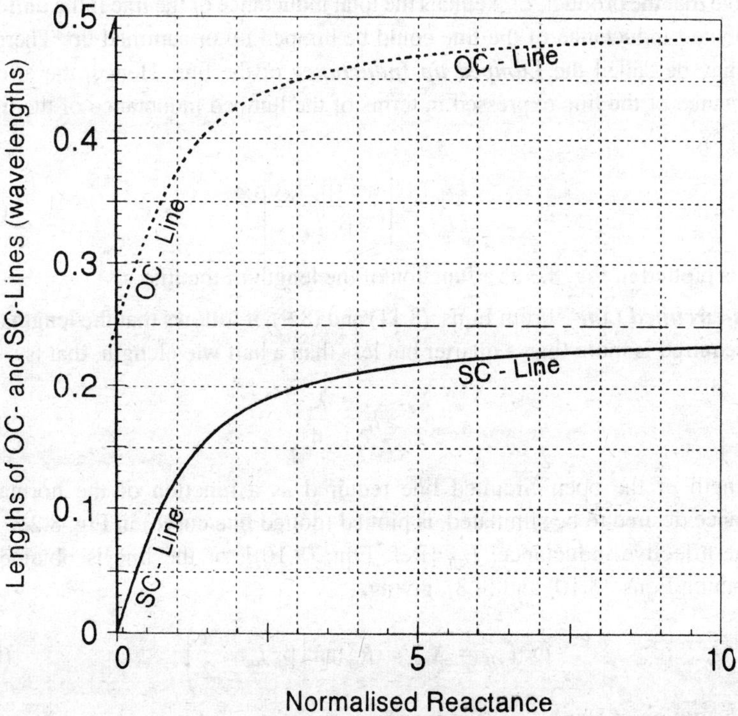

Fig. 8.2 Lengths of Short- & Open-Circuited Lossless Lines required for Realisation of Positive Reactances.

Now, the effective inductance, L_{eff}, defined by the second of Eqns. (8.10), of the line is obtained by combining Eqns. (8.10) and (8.4), giving,

$$\omega_o L_{eff} = R_o \tan(\beta_g l_{sc}). \tag{8.12}$$

Substitution of the following equations,

$$\left.\begin{aligned} R_o &= \sqrt{\frac{L}{C}} \\ \text{and} \quad \omega_o &= \frac{\beta_g}{\sqrt{LC}} \end{aligned}\right\} \tag{8.13}$$

into Eqn. (8.12) gives,

$$L_{eff} = \frac{R_o}{\omega_o} \tan(\beta_g l_{sc}) = \frac{L}{\beta_g} \tan(\beta_g l_{sc}) \tag{8.14}$$

Thus, in so far as the behaviour of the short-circuited line of length, l_{sc}, at the given spot or fixed frequency, f_o, is concerned, it presents a pure reactance equal to that of an inductance, L_{eff}, given by

$$L_{eff} = \left[\frac{\tan(\beta_g l_{sc})}{\beta_g l_{sc}}\right] L l_{sc} \tag{8.15}$$

Note that the product, Ll_{sc}, equals the total inductance of the line if the uniformly distributed inductance of the line could be lumped up or summed up. Therefore, Ll_{sc} may be called the ***Lumped up Inductance*** of the line. Hence, the effective inductance of the line expressed in terms of the lumped inductance of the line, is given by

$$\frac{L_{\text{eff}}}{Ll_{sc}} = \left[\frac{\tan(\beta_g l_{sc})}{\beta_g l_{sc}}\right] \tag{8.16}$$

This is plotted in Fig. 8.3 as a function of the length of the line.

Open-circuited Line From Eqns. (8.11) and (8.9), it follows that the length of the line required is more than a quarter but less than a half wavelength, that is,

$$\frac{\lambda_g}{2} > l_{oc} > \frac{\lambda_g}{4} \tag{8.17}$$

Length of the open-circuited line required as a function of the normalised reactance desired to be simulated, is plotted (dotted line curve) in Fig. 8.2.

The effective inductance, L_{eff} [Ref. Eqn. (8.10)], of the line is obtained by combining Eqns. (8.10) and (8.8), giving,

$$\omega_o L_{eff} = X_{rq} = R_o \tan\left(\beta_g l_{oc} - \frac{\pi}{2}\right) \tag{8.18}$$

Substitution of Eqns. (8.13) in the above equation, gives

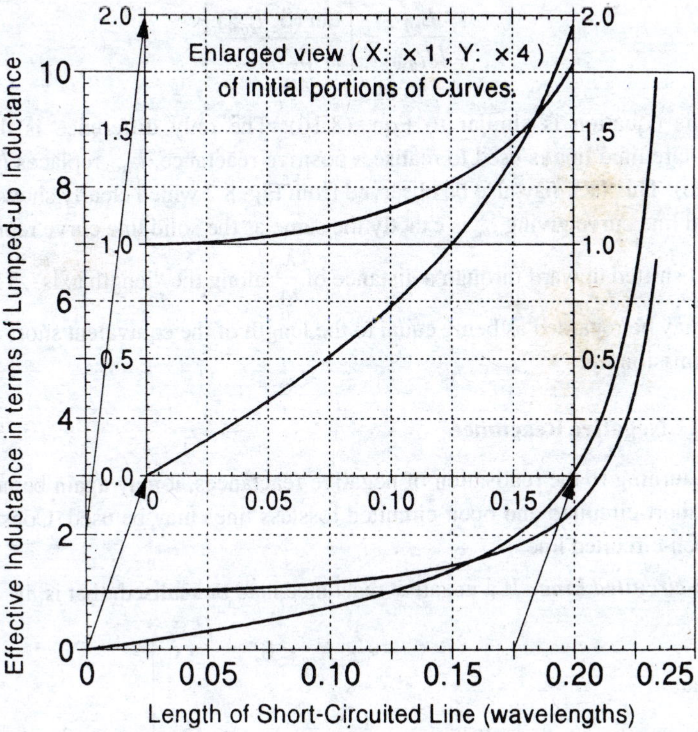

Fig. 8.3 Normalised Input Reactance & Effective Inductance in terms of Lumped-up Inductance of Transmission Line

$$L_{eff} = \frac{R_o}{W_o} \tan\left(\beta_g\, l_{oc} - \frac{\pi}{2}\right) = \frac{L}{\beta_g} \tan\left(\beta_g\, l_{oc} - \frac{\pi}{2}\right) \tag{8.19}$$

$$= \frac{L}{\beta_g} \tan\left\{\beta_g\left(l_{oc} - \frac{\lambda_g}{4}\right)\right\} \tag{8.20}$$

Thus, in so far as the behaviour of the open-circuited line of length, l_{oc}, at the given spot or fixed frequency, f_o, is concerned, it presents a pure reactance equal to that of an inductance, L_{eff}, given by

$$L_{eff} = \left[\frac{\tan\left\{\beta_g\left(l_{oc} - \frac{\lambda_g}{4}\right)\right\}}{\beta_g\left(l_{oc} - \frac{\lambda_g}{4}\right)}\right] L\left(l_{oc} - \frac{\lambda_g}{4}\right) \tag{8.21}$$

Defining

$$l_{eqsc} = l_{oc} - \frac{\lambda_g}{4} = \text{length of the equivalent SC line,} \tag{8.22}$$

Eqn. (8.21) may be rewritten as

$$\frac{L_{eff}}{Ll_{eqsc}} = \left[\frac{\tan(\beta_g\, l_{eqsc})}{\beta_g\, l_{eqsc}}\right] \tag{8.23}$$

This equation is similar to Eqn. (8.16): The only difference is that if an open-circuited line is used to realise a positive reactance, l_{eqsc} replaces l_{sc} in Eqn. (18.16). This fact may also be observed from Fig. 8.2 which clearly shows that the dotted line curve giving l_{oc} is exactly the same as the solid line curve representing l_{sc} but shifted upward through a distance of $\frac{\lambda_g}{4}$ along the "length axis". Therefore, l_{eqsc} may be regarded as being equal to the length of the equivalent short-circuited transmission line.

8.2.2 Negative Reactance

Now, turning to the realisation of negative reactances, it may again be noted that both short-circuited and open-circuited lossless lines may be used. Consider first an open-circuited line.

Open-circuited Line If a negative reactance is to be realised, that is, if

$$\left.\begin{aligned} & -\infty < X_{rq} < 0, \\ \text{so that,}\quad & X_{rq} = -\frac{1}{\omega_o\, C_{eff}},\ \text{where, } C_{eff} > 0, \\ \text{then,}\quad & \frac{\pi}{2} > \tan^{-1}\left(\frac{|X_{rq}|}{R_o}\right) > 0. \end{aligned}\right\} \tag{8.24}$$

then the length, l_{oc}, of the open-circuited transmission line required, is obtained by substituting $-j\,|X_{rq}|$ for Z_{oc} in Eqns. (8.7), giving

$$-j\,|X_{rq}| = -j\,R_o \cot(\beta_g\, l_{oc}) = -j\,R_o \tan\left(\frac{\pi}{2} - \beta_g\, l_{oc}\right) \tag{8.25}$$

This equation yields,

$$l_{oc} = \frac{\lambda_g}{4} - \tan^{-1}\left(\frac{|X_{rq}|}{R_o}\right) \tag{8.26}$$

Using the last of Eqn. (8.24),

$$\frac{\lambda_g}{4} > l_{oc} > 0 \tag{8.27}$$

Length of the open-circuited line required, as a function of the normaiised reactance desired to be simulated, is plotted (dotted line curve) in Fig. 8.4.

The effective capacitance, C_{eff}, defined by the second of Eqns. (8.24), of the line is obtained by combining Eqns. (8.24) and (8.7), giving,

$$-\frac{1}{\omega_o C_{eff}} = -R_o \cot(\beta_g l_{oc}) \tag{8.28}$$

This equation, on using Eqns. (8.13), gives

$$C_{eff} = \frac{C}{\beta_g} \tan(\beta_g l_{oc}) \tag{8.29}$$

Thus, in so far as the behaviour of an open-circuited line of length, l_{oc}, at the given spot or fixed frequency, f_o, is concerned, it presents a pure reactance equal to that of a capacitance, C_{eff}, given by

$$C_{eff} = \left[\frac{\tan(\beta_g l_{oc})}{\beta_g l_{oc}}\right] C l_{oc} \tag{8.30}$$

Note that Cl_{oc} is equal to the ***lumped up capacitance of the line***. Therefore, the equivalent capacitance of the line in terms of the lumped up capacitance of the line, is given by

$$\frac{C_{eff}}{Cl_{oc}} = \left[\frac{\tan(\beta_g l_{oc})}{\beta_g l_{oc}}\right] \tag{8.31}$$

Short-circuited Line Next consider a short-circuited line for realising a negative reactance. It is clear from Eqn. (8.9) that if a negative reactance is to be realised, that is, if

$$-\infty < X_{rq} < 0,$$

so that,

$$X_{rq} = -\frac{1}{\omega_o C_{eff}}, \text{ where } C_{eff} > 0,$$

then,

$$\pi > \tan^{-1}\left(\frac{-|X_{rq}|}{R_o}\right) > \frac{\pi}{2} \tag{8.32}$$

Correspondingly, the length of the line required is more than a quarter wavelength but less than a half wavelength, that is,

$$\frac{\lambda_g}{4} < l_{sc} < \frac{\lambda_g}{2} \tag{8.33}$$

Length of the short-circuited line required, as a function of the normalised reactance desired to be simulated, is plotted (solid line curve) in Fig. 8.4.

The effective capacitance, C_{eff} [Ref. Eqn. (8.32)], of the line is obtained by combining Eqns. (8.32) and (8.4), giving,

$$-\frac{1}{\omega_o C_{eff}} = R_o \tan(\beta_g l_{sc}). \tag{8.34}$$

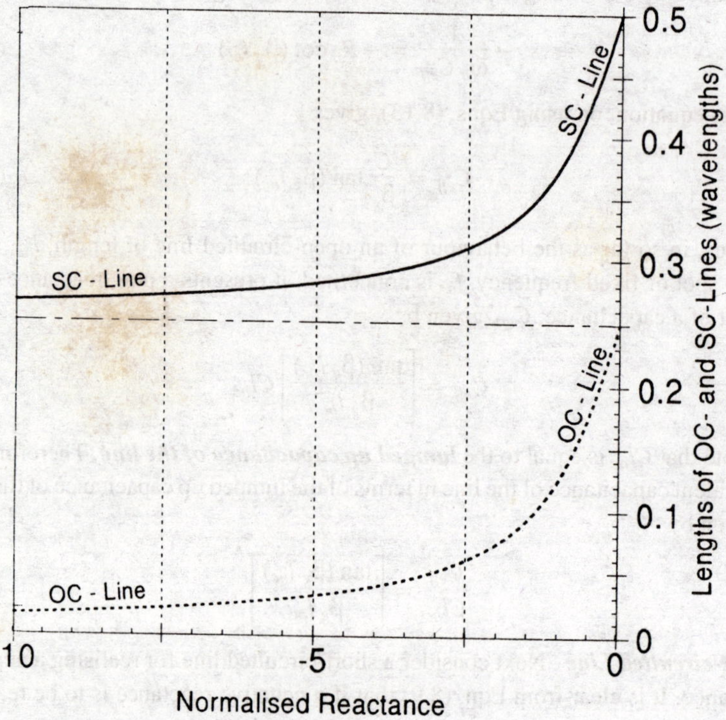

Fig. 8.4 Lengths of Short- and Open-Circuited Lossless Lines required for Realisation of Negative Reactances.

This equation, on using Eqns. (8.13), gives

$$C_{eff} = -\frac{C}{\beta_g}\cot(\beta_g\, l_{sc}) = -\frac{C}{\beta_g}\tan\left(\beta_g\, l_{sc} - \frac{\pi}{2}\right) \tag{8.35}$$

Thus, in so far as the behaviour of the short-circuited line of length, l_{sc}, at the given spot or fixed frequency, f_o, is concerned, it presents a pure reactance equal to that of a capacitance, C_{eff}, given by

$$C_{eff} = \left[\frac{\tan\left\{\beta_g\left(l_{sc} - \frac{\lambda_g}{4}\right)\right\}}{\beta_g\left(l_{sc} - \frac{\lambda_g}{4}\right)}\right] C\left(l_{sc} - \frac{\lambda_g}{4}\right) \tag{8.36}$$

Again, defining

$$l_{eqoc} = l_{sc} - \frac{\lambda_g}{4} = \text{length of the equivalent OC line,} \tag{8.37}$$

Eqn. (8.36) may be rewritten as

$$\frac{C_{eff}}{C\, l_{eqoc}} = \left[\frac{\tan(\beta_g\, l_{eqoc})}{\beta_g\, l_{eqoc}}\right] \tag{8.38}$$

This equation is similar to Eqn. (8.31): The only difference is that if a short-circuited line is used to realise a negative reactance, l_{eqoc} replaces l_{oc} in Eqn. (8.31). This fact may also be observed from Fig. 8.4, which clearly shows that the dotted line curve giving l_{oc} is exactly the same as the solid line curve representing l_{sc} but shifted through a distance of $-\frac{\lambda_g}{4}$ along the "length axis". Therefore, l_{eqoc} may be regarded as being equal to the length of the equivalent open-circuited transmission line.

Discussion The analysis given above in this Section, has established that positive or negative reactance of any magnitude may be realised by both a short-circuited and open-circuited lines. The following conclusions may be drawn based on the analysis given above:

1. If a short-circuited line is used to realise a positive reactance and an open-circuited line for realising a negative reactance, then the line lengths required are less than a quarter wavelength.
2. However, if an open-circuited line is employed for realising a positive and a short-circuited line for negative reactance, then the line lengths required in both cases, are longer than the respective lengths required above, by a quarter wavelength. This is understandable when it is realised that a short-circuit at the end of a quarterwave line appears as an ideal open-circuit at the input of the line; and, an open-circuit at the end of a quarterwave line appears as an ideal short-circuit at the input of the line. Therefore, a short-circuited line of length, l_{sc}, is equivalent to an open-circuited line of length, $l_{sc} - \frac{\lambda_g}{4} = l_{eqoc}$; and, an open-circuited line of length, l_{oc}, is equivalent to a short-circuited line of length, $l_{oc} - \frac{\lambda_g}{4} = l_{eqsc}$.
3. Based on the above, it may be concluded that, theoretically, ***to realise a positive reactance a short-circuited line and to realise a negative reactance an open-circuited line must be used.***
4. However, in practice, whereas an almost ideal short-circuit is not difficult to be simulated, it is almost impossible to simulate an ideal open-circuit. Therefore, in practice, only a short-circuited line is used to realise both a positive and a negative reactance.
5. The equivalent inductance, L_{eff}, and capacitance, C_{eff}, normalised with respect to the lumped up inductance, L and lumped up capacitance, Cl, respectively, vary with the length of the line as given below and as shown in Fig. 8.3.

$$\frac{L_{eff}}{L\, l} \text{ or, } \frac{C_{eff}}{C\, l} = \left[\frac{\tan(\beta_g\, l)}{\beta_g\, l}\right], \tag{8.39}$$

where l is either the length of the line or the equivalent length of the line as discussed above. [Ref. Eqns. (8.16), (8.23), (8.31) and (8.38)].

8.3 SIMULATION OF REACTANCE ELEMENTS

In the analysis given in Sec.8.2, it has been shown that at a specified frequency, f_o, both a positive and a negative reactance, X_{rq}, may, in principle, be realised by using either a given short-circuited, lossless transmission line of length, l_{sc}, or a given open-circuited lossless transmission line of length, l_{oc}. In practice, if lines of the shortest lengths are required for realising a given reactance, that is, if l_{sc} and l_{oc} are required to be as small as possible, then ***it is advisable to use a short-circuited line for realising a positive reactance and an open-circuited line for a negative reactance*** [Ref. item 3 of the Discussion in Sec. 8.2]. This procedure would ensure that for realising any arbitrarily prescribed reactance, the length of the line required would always be less than a quarter of a wavelength. However, as already discussed in Sec. 7.7.6, the use of a short-circuited line becomes imperative for realising even negative reactances because an ideal open-circuit cannot be realised in practice [Ref. item 4 of the Discussion in Sec. 8.2].

At this stage, it is worth noting that the problem of realisation or simulation of a pure reactance by a lossless transmission line, may be of the following two types:

1. One, in which the sole aim of simulation is to realise a given pure reactance at the specified frequency, paying no attention to the nature of variation of the reactance with the frequency of operation. This may be called ***Realisation of Pure Reactance.***

   This type of simulation of arbitrarily prescribed reactance has been discussed in Sec. 8.2 above.
2. The other, in which the aim of simulation is not only to realise a given pure reactance at the specified frequency but also to ensure that it varies with frequency in the same manner as does the reactance of the required circuit element does. This may be called ***Simulation of Reactance Element.*** Simulation of some commonly used circuit elements, is discussed below:

8.3.1 Inductor

Let it be desired to simulate an inductance, L_{eff}. The reactance of this inductor at a frequency, $f_o = \dfrac{\omega_o}{2\pi}$, is

$$X_{rq} = \omega_o L_{eff}, \tag{8.40}$$

and the input reactance, X_{sc}, of a short-circuited lossless line of length, l_{sc}, is given by

$$X_{sc} = R_o \tan(\beta_g l_{sc}). \tag{8.41}$$

The length, l_{sc}, of the short-circuited lossless line required to realise the positive reactance, X_{rq}, is obtained [Ref. Sec. 8.2.1] by setting $X_{sc} = X_{rq}$, at ω_o, giving [Ref. Eqn. (8.5)]

$$\left.\begin{aligned} l_{sc} &= \frac{\lambda_g}{2\pi} \tan^{-1}\left(\frac{X_{rq}}{R_o}\right) \\ \text{where,} \quad & \frac{\lambda_g}{4} > l_{sc} > 0 \\ \text{for} \quad & \infty > X_{rq} > 0 \end{aligned}\right\} \tag{8.42}$$

Using the expression for β_g from Eqn. (8.13), Eqn. (8.41) may be rewritten as

$$X_{sc} = R_o \tan(\beta_g l_{sc}) = R_o \tan(\omega_o \sqrt{LC}\, l_{sc}) \tag{8.43}$$

It may be pointed out that Eqn. (8.42) ensures only that at ω_o, $X_{sc} = X_{rq}$. But, it may be noted from Eqn. (8.41) that X_{sc} does not vary with frequency in the same manner (linearly) as does X_{rq}. However, simulation of the inductor of inductance, L_{eff}, requires that X_{sc} must also vary linearly with frequency. An inspection of Eqn. (8.43) shows that this would, at least approximately, be so,

if $$\omega_o \sqrt{LC}\, l_{sc} = \beta_g l_{sc} << 1, \tag{8.44}$$

that is, if $$l_{sc} << \frac{\lambda_g}{2\pi}. \tag{8.45}$$

Consequently, $$\tan(\beta_g l_{sc}) << 1 \tag{8.46}$$

and, $$X_{rq} = X_{sc} << R_o. \tag{8.47}$$

Under this condition, from Eqns. (8.43) and (8.44), the input reactance of the line is given by

$$X_{rq} = L_{eff}\, \omega_o \approx X_{sc} \approx (R_o \sqrt{LC}\, l_{sc})\, \omega_o. \tag{8.48}$$

giving, $$L_{eff} \approx R_o \sqrt{LC}\, l_{sc} \tag{8.49}$$

$$= L\, l_{sc} \tag{8.50}$$

$$= \text{lumped up inductance of the line.} \tag{8.51}$$

Hence, it may be concluded that in so far as realisation of a positive reactance, X_{rq}, at a spot frequency, ω_o, is concerned, any arbitrary value of X_{rq} may be realised by a length, $l_{sc} \le \frac{\lambda_g}{4}$, of a short-circuited lossless line. But, an inductor of inductance, L_{eff}, can be simulated, only if it is so small that its reactance at ω_o is very much less than R_o. [Ref. Eqn. (8.47)]

In most practical cases, however, the above condition is somewhat relaxed and the simulation of an inductor is carried out as long as $\beta_g l_{sc} \le \frac{\pi}{4}$ which amounts to restricting $X_{rq} \le Z_o$, or to restricting $L_{eff} \le L \frac{\lambda_g}{8}$.

8.3.2 Capacitor

Next, consider simulation of a capacitance, C_{eff}. It is convenient to discuss this case in terms of susceptances. The susceptance of the capacitor at a frequency, f_o is given by

$$j B_{rq} = \frac{1}{j X_{rq}} = j \omega_o C_{eff}, \tag{8.52}$$

and the input susceptance, B_{oc}, of an open-circuited lossless line of length, l_{oc}, is given by

$$B_{oc} = \frac{1}{R_o} \tan (\beta_g l_{oc}). \tag{8.53}$$

The length, l_{oc}, of the open-circuited lossless line required to realise the positive susceptance, $B_{rq} = -\frac{1}{X_{rq}}$, is obtained [Ref. Sec. 8.2.2 by setting $B_{oc} = B_{rq}$, at ω_o giving [Ref. Eqn. (8.6)]

$$\left.\begin{aligned} l_{oc} &= \frac{\lambda_g}{2\pi} \tan^{-1} (R_o B_{rq}) \\ \text{where,} \quad & \frac{\lambda_g}{4} > l_{oc} > 0 \\ \text{for} \quad & \infty > B_{rq} > 0 \end{aligned}\right\} \tag{8.54}$$

Using the expression for β_g from Eqn. (8.13), Eqn. (8.53) may be rewritten as

$$B_{oc} = \frac{1}{R_o} \tan (\omega_o \sqrt{L C}\, l_{oc}) \tag{8.55}$$

It may be pointed out that Eqn. (8.54) ensures only that at ω_o, $B_{oc} = B_{rq}$. But, it may be noted from Eqn. (8.53) that B_{oc} does not vary with frequency in the same manner (linearly) as does B_{rq}. However, simulation of the capacitor requires that B_{oc} must also vary linearly with frequency. An inspection of Eqn. (8.55) shows that this would, at least approximately, be so,

if $$\omega_o \sqrt{LC}\, l_{oc} = \beta_g l_{oc} << 1, \tag{8.56}$$

that is, if $$l_{oc} << \frac{\lambda_g}{2\pi}. \tag{8.57}$$

Consequently, $$\tan (\beta_g l_{oc}) << 1 \tag{8.58}$$

and, $$B_{rq} = B_{sc} << \frac{1}{R_o} \tag{8.59}$$

Under this condition, from Eqns. (8.55) and (8.56), the input susceptance of the line is given by

$$B_{rq} = C_{eff}\,\omega_o \approx B_{oc} \approx (R_o\sqrt{LC}\,l_{oc})\,\omega_o. \tag{8.60}$$

giving,
$$C_{eff} \approx \frac{\sqrt{LC}}{R_o}\,l_{oc} \tag{8.61}$$

$$= C\,l_{oc} \tag{8.62}$$

$$= \text{lumped up capacitance of the line.} \tag{8.63}$$

Hence, it may be concluded that in so far as realisation of a positive susceptance, B_{rq}, at a spot frequency, ω_o, is concerned, any arbitrary value of B_{rq} may be realised by a length, $l_{oc} = \frac{\lambda_g}{4}$, of an open-circuited lossless line. But, a capacitance, C_{eff}, can be simulated, only if it is so small that its susceptance at ω_o is very much less than $\frac{1}{R_o}$.

Again, in most practical cases, however, the above condition is somewhat relaxed at the simulation of a capacitor is carried out as long as $\beta_g\, l_{oc} \le \frac{\pi}{4}$ which amounts to restricting $B_{rq} \le Z_o$, or to restricting $C_{eff} \le C\frac{\lambda_g}{8}$.

8.3.3 Series Resonant Circuit (SRC)

Consider now the simulation of a series resonant circuit (SRC). At the frequency of resonance, f_o, the total reactance, X_{rq}, of this circuit is zero. The input reactance, X_{oc}, of an open-circuited lossless line of length, $l_{oc} = \frac{\lambda_g}{4}$, is also zero. Therefore, the open-circuited quarterwave line would simulate a series resonant circuit [Ref. Fig. 8.5.], if the frequency rate of variation of the line reactance can be made equal to that of the SRC reactance, at least in the neighbourhood of the frequency of resonance.

The total reactance of the series resonant circuit consisting of an inductance, L_{eff}, in series with a capacitance, C_{eff}, is given by

$$X_{rq}(\omega) = \omega L_{eff} - \frac{1}{\omega C_{eff}} \tag{8.64}$$

Application of the condition that at the frequency of resonance, the total reactance of a series resonant circuit (SRC) is zero, to the above equation, gives

$$\omega_o L_{eff} = \frac{1}{\omega_o C_{eff}} \tag{8.65}$$

Also, from Eqn. (8.64)

$$\frac{dX_{rq}}{d\omega} = L_{eff} + \frac{1}{\omega_o^2 C_{eff}} \text{ at } \omega = \omega_o. \tag{8.66}$$

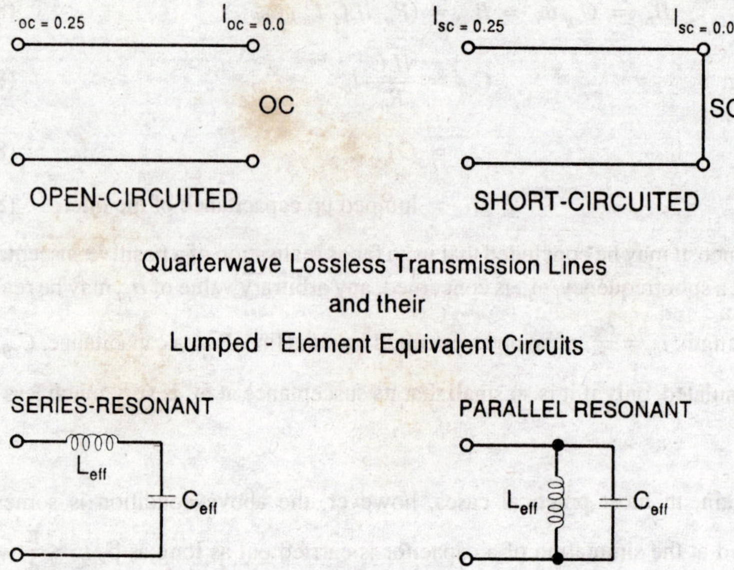

Fig. 8.5 Simulation of Series and Parallel Resonant Circuits using Short- and Open-Circuited Lossless Transmission Lines.

Substitution of Eqn. (8.65) into Eqn. (8.66) gives,

$$\frac{dX_{rq}}{d\,\omega} = 2L_{eff} \text{ at } \omega = \omega_o. \tag{8.67}$$

Now, the input reactance of an open-circuited lossless line of length, l_{oc}, is given by [Ref. Eqn. (8.21)]

$$X_{oc} = -R_o \cot(\beta_g\, l_{oc}) \tag{8.68}$$

Using the expression for β_g from Eqn. (8.13) in this equation gives,

$$X_{oc} = -R_o \cot(\omega \sqrt{LC}\, l_{oc}) \tag{8.69}$$

Differentiating both sides of Eqn. (8.69) with respect to ω gives,

$$\frac{dX_{oc}}{d\,\omega} = R_o \sqrt{LC}\, l_{oc} \operatorname{cosec}^2(\omega \sqrt{LC}\, l_{oc}). \tag{8.70}$$

Therefore, for simulation of the SRC by the above line, the following two conditions are to be satisfied:

At $\omega = \omega_o$,

$$X_{oc} = X_{rq} = 0 \tag{8.71}$$

$$\frac{dX_{oc}}{d\,\omega} = \frac{dX_{rq}}{d\,\omega}. \tag{8.72}$$

To satisfy the condition given by Eqn. (8.71), Eqn. (8.69) gives,

$$\omega_o \sqrt{LC}\, l_{oc} = \frac{\pi}{2} \tag{8.73}$$

which gives,
$$l_{oc} = \frac{\lambda_g}{4} \tag{8.74}$$

Application of Eqn. (8.72) to Eqns. (8.70) and (8.67) and using Eqn. (8.73), give

$$2L_{eff} = R_o \sqrt{LC}\, l_{oc} = L\frac{\lambda_g}{4} \tag{8.75}$$

which gives,
$$L_{eff} = \frac{1}{2} L\, l_{oc} = \frac{1}{8} L\, \lambda_g. \tag{8.76}$$

From Eqns. (8.65), (8.76) and (8.13), the value of the effective capacitance, C_{eff}, is

$$C_{eff} = \frac{1}{\omega_o^2 L_{eff}}$$

$$= \frac{8}{\pi^2} l_{oc} \tag{8.77}$$

$$= \frac{2}{\pi^2} C\, \lambda_g \tag{8.78}$$

8.3.4 Parallel Resonant Circuit (PRC)

Consider now the simulation of a parallel resonant circuit (PRC). It is convenient to discuss this case in terms of susceptances. At the frequency of resonance, f_o, the total susceptance, B_{rq}, of this circuit (PRC) is zero. The input susceptance, B_{sc}, of a short-circuited lossless line of length, $l_{sc} = \frac{\lambda_g}{4}$, is also zero. Therefore, the short-circuited quarterwave line would simulate a parallel resonant circuit [Ref. Fig. 8.5], if the frequency rate of variation of the line susceptance can be made equal to that of the PRC susceptance, at least in the neighbourhood of the frequency of resonance.

The total susceptance of the parallel resonant circuit consisting of a capacitance, C_{eff}, in parallel with an inductance, L_{eff}, is given by

$$B_{rq}(\omega) = \omega\, C_{eff} - \frac{1}{\omega\, L_{eff}}. \tag{8.79}$$

Application of the condition that at the frequency of resonance, the total susceptance of a parallel resonant circuit (PRC) is zero, to the above equation, gives

$$\omega_o\, C_{eff} = \frac{1}{\omega_o\, L_{eff}}. \tag{8.80}$$

Also, from Eqn. (8.79)

$$\frac{dB_{rq}}{d\,\omega} = C_{eff} + \frac{1}{\omega_o^2 L_{eff}} \text{ at } \omega = \omega_o. \tag{8.81}$$

Substitution of Eqn. (8.80) into Eqn. (8.81) gives

$$\frac{dB_{rq}}{d\,\omega} = 2C_{eff} \text{ at } \omega = \omega_o. \tag{8.82}$$

Now, the input susceptance of a short-circuited lossless line of length, l_{sc}, is given by [Ref. Eqn. (8.3)]

$$B_{sc} = -\frac{1}{R_o} \cot(\beta_g\, l_{sc}) \tag{8.83}$$

Using the expression for β_g from Eqn. (8.13) in this equation gives,

$$B_{sc} = -\frac{1}{R_o} \cot(\omega \sqrt{LC}\, l_{sc}) \tag{8.84}$$

Differentiating both sides of Eqn. (8.84) with respect to ω gives,

$$\frac{dB_{sc}}{d\,\omega} = \frac{1}{R_o} \sqrt{LC}\, l_{sc} \operatorname{cosec}^2(\omega \sqrt{LC}\, l_{sc}). \tag{8.85}$$

Therefore, for simulation of the PRC by above line, the following two conditions are to be satisfied.

At $\omega = \omega_o$.

$$B_{sc} = B_{rq} = 0 \tag{8.86}$$

$$\frac{dB_{sc}}{d\,\omega} = \frac{dB_{rq}}{d\,\omega} \tag{8.87}$$

To satisfy the condition given by Eqn. (8.86), Eqn. (8.84) gives,

$$\omega_o \sqrt{LC}\, l_{sc} = \frac{\pi}{2} \tag{8.88}$$

which gives,

$$l_{sc} = \frac{\lambda_g}{4} \tag{8.89}$$

Application of Eqn. (8.87) to Eqns. (8.85) and (8.82) and using Eqn. (8.88), give

$$2C_{eff} = \frac{1}{R_o} \sqrt{LC}\, l_{sc} = C\frac{\lambda_g}{4} \tag{8.90}$$

which gives,

$$C_{eff} = \frac{1}{2} C\, l_{sc} = \frac{1}{8} C\, \lambda_g \tag{8.91}$$

From Eqns. (8.80), (8.91) and (8.13), the value of the effective inductance, L_{eff}, is

$$L_{eff} = \frac{1}{\omega_o^2 C_{eff}}$$

$$= \frac{8}{\pi^2} l_{sc} \quad (8.92)$$

$$= \frac{2}{\pi^2} L \lambda_g \quad (8.93)$$

Based on the discussions given under Secs. 8.2 and 8.3. above, it may be concluded that short-circuited and open-circuited behave as illustrated in Fig. 8.6. For example, a short-circuited line behaves as an inductor or simulates an inductor when its length is very short, say, less than one-eighth of the wavelength. As its length is increased, it presents at its input a positive reactance but does not simulate an inductance. When its length is $\frac{\lambda_g}{4}$, it simulates a parallel resonant circuit. If its length is further increased by an amount not more than about one-eighth of the wavelength, it simulates a capacitor. If its length continues to increase further, it presents a negative reactance at its input which does not behave with frequency in the manner the reactance of a capacitor does. And, it simulates a series resonant circuit when its length equals a half wavelength. This is pictorially depicted in Fig. 8.6.

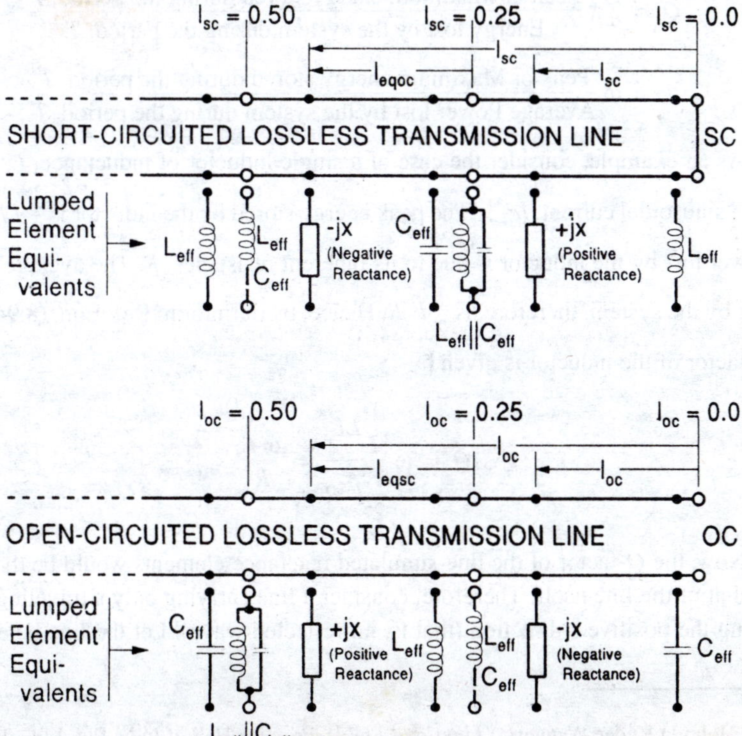

Fig. 8.6 Behaviour of Short- and Open-Circuited Lossless Transmission Lines.

8.4 *Q* OF SIMULATED REACTANCE ELEMENTS

In the foregoing analyses, the line used for the simulation of reactance elements, has been assumed to be lossless, that is, it has been assumed that the line parameters, $R = 0$ and $G = 0$, so that the attenuation constant of the line, $\alpha = 0$. However, in practice, α is not zero although it may be extremely small. This results in a non-infinite Q-factor for the line-simulated reactance elements. A quantitative estimate of the Q-factor of the line-simulated reactance elements is made in this section.

8.4.1 *Q*-factor of a System

The primary function of an energy-storing system, like a reactance element in electrical networks, is to store the energy supplied to it. Physically, the ***figure of merit***, the ***quality-factor***, the ***Q-factor*** or simply, the Q of such a physical system indicates the effectiveness of the system in doing the job expected of it. Quantitatively, the Q of any physical system is defined[1] by the ratio of the ***peak*** or the ***maximum*** energy stored in the system over a period of the excitation, and the total energy lost by the system during the same period. To be exact, the Q may be defined as (for sinusoidal excitation of frequency, $f = \frac{1}{T}$)

$$Q = 2\pi \frac{\text{Peak or Maximum energy stored during the period, } T}{\text{Energy lost by the system during the period, } T}$$

$$= \omega \frac{\text{Peak or Maximum energy stored during the period, } T}{\text{Average Power lost by the system during the period, } T} \quad (8.94)$$

As an example, consider the case of a single inductor of inductance, L, driven by a sinusoidal current, Ie^{jwt}. The peak energy stored by the inductor is $\frac{1}{2}LI^2$. The power lost by the inductor is due to its inherent resistance, R. The average power lost by the system, therefore, is $\frac{1}{2}I^2R$. Hence, by definition [Ref. Eqn. (8.94)], the Q-factor of the inductor is given by

$$Q = \omega \frac{\frac{1}{2}LI^2}{\frac{1}{2}I^2R} = \frac{\omega L}{R}. \quad (8.95)$$

Now, the Q-factor of the line-simulated reactance elements would be the same as that of the line itself. Therefore, consider a line carrying only a traveling wave along the positive z-direction (that is, no reflected wave). Let the line be excited

---

1 John D Ryder: *Networks, Lines and Fields*, Prentice Hall of India Pvt. Ltd., 2nd Ed., 1988, p. 55.

sinusoidally so that the line voltage and current phasors are given by Ve^{jwt} and Ie^{jwt}, respectively. In the case of a line, the energy is stored in both its electric field as the stored energy of its distributed capacitance, and its magnetic field as the stored energy of its distributed inductance. Further, it is to be noted that line voltage and current vary sinusoidally along its length. As the energy depends on the square of the voltage and current, its average with respect to the line length equals a half of its peak value along the length of the line. Therefore, the peak or maximum (with respect to time) energy stored, W, in the length, Δz, of the line is given by

$$W = \frac{1}{2}\left(\frac{1}{2} C \Delta z V^2\right) + \frac{1}{2}\left(\frac{1}{2} L \Delta z I^2\right) \tag{8.96}$$

and, the average power lost, P_l, by the length, Δz, of the line due to dissipation in its series resistance, $R \Delta z$, and in its shunt conductance, $G \Delta z$, is

$$P_l = \frac{1}{2} I^2 R \Delta z + \frac{1}{2} G \Delta z V^2 \tag{8.97}$$

Substitution of Eqns. (8.96) and (8.97) into Eqn. (8.94), gives the Q-factor and of the line segment, Δz, as

$$Q = \omega \frac{\frac{1}{2}\left(\frac{1}{2} C \Delta z V^2\right) + \frac{1}{2}\left(\frac{1}{2} L \Delta z I^2\right)}{\frac{1}{2} I^2 R \Delta z + \frac{1}{2} G \Delta z V^2} \tag{8.98}$$

Dividing both the numerator and the denominator by VI and using the relation

$$\frac{V}{I} = R_o = \sqrt{\frac{L}{C}}, \tag{8.99}$$

and simplification of the resulting equation, give

$$Q = \frac{\omega \sqrt{LC}}{\frac{R}{\sqrt{\frac{L}{C}}} + G\sqrt{\frac{L}{C}}} = \frac{\omega \sqrt{LC}}{2\left[\frac{1}{2}\left\{\frac{R}{Z_{o\infty}} + GZ_{o\infty}\right\}\right]} = \frac{\beta_g}{2\alpha} = \frac{\pi}{\alpha \lambda_g} \tag{8.100}$$

As an illustration, consider the Q-factor of a line segment at 500 MHz. The attenuation constant, α, may typically be above 0.007 Neper/metre. Therefore, from Eqn. (8.100), the value of Q turns out to be of the order of 700. The value is high compared to that normally obtainable with lumped circuit elements at these frequencies.

8.5 SIMULATION OF ARBITRARY IMPEDANCE

A lossless line may also be employed to simulate not only pure reactances, as discussed above, but also any arbitrary impedance. It will now be proved that ***any arbitrarily prescribed impedance, $R + jX$, can be simulated by a suitable length, l, of a lossless transmission line having arbitrarily prescribed line constants, R_o and β_g, terminated by a pure resistance.*** Proving this theorem amounts to solving the problem which may be mathematically formulated thus: Given the input impedance (normalised) $z = \frac{R+jX}{R_o} = r+jx$, of lossless transmission line characterised by R_o and β_g, it is required to find the electrical length, $\theta = \beta_g l$, of the line and the terminating resistance, $r_L = \frac{R_L}{R_o}$ required.

The input impedance, $r+jx$, of a lossless line of electrical length, $\theta = \beta_g l$, terminated in a pure resistance, r_L, is obtained from the Eqn. (7.19) to be

$$r+jx = \frac{r_L+j\tan\theta}{1+jr_L\tan\theta} \tag{8.101}$$

In this equation, $r+jx$ is arbitrarily prescribed, that is,

and
$$\left.\begin{aligned}\infty > r \geq 0,\\ +\infty > x > -\infty\end{aligned}\right\} \tag{8.102}$$

It will now be proved that r_L and θ satisfying Eqn. (8.101) can be uniquely determined for all values of r and x given in (8.102). From Eqn. (8.102), it may be easily shown that

$$r = \frac{(1+\tan^2\theta)\, r_L}{1+r_L^2\tan^2\theta} \tag{8.103}$$

and

$$x = \frac{(1-r_L)^2\tan\theta}{1+r_L^2\tan^2\theta} \tag{8.104}$$

Eqn. (8.103) gives

$$r\,(1+r_L^2\tan^2\theta) = r_L\,(1+\tan^2\theta)$$

or
$$\tan^2\theta = \frac{r_L-r}{r_L\,(r\,r_L-1)}$$

which gives

$$1+r_L^2\tan^2\theta = 1+\frac{r_L^2-r\,r_L}{r\,r_L-1} = \frac{r_L^2-1}{r\,r_L-1} \tag{8.105}$$

Squaring both sides of Eqn. (8.104) and then substitution of (8.105) into it gives the $\tan\theta$ – eliminant of Eqns. (8.103) and (8.104) to be

$$r\,r_L^2-(1+r^2+x^2)\,r_L+r = 0 \tag{8.106}$$

This is a quadratic equation in r_L and, therefore, gives, in general, two possible values for r_L which are

$$\begin{matrix}r_{L1}\\ r_{L2}\end{matrix} = \frac{r^2+x^2+1\pm\sqrt{(r^2+x^2+1)^2-4r^2}}{2\,r} \tag{8.107}$$

Let the expression under the square-root sign in the above equation be denoted by E. It may be noted that the expression, E, is quadratic in r^2. It may be rewritten in the form,

$$E = r^4 + 2(x^2 - 1)\, r^2 + (x^2 + 1)^2 \tag{8.108}$$

Further, note that the discriminant of the expression, E, which is equal to

$$4(x^2 - 1)^2 - 4(x^2 + 1)^2 = -16x^2,$$

is negative or equal to 0 for all real values of x. Hence, the expression, E does not change its sign for any real values of r. Now, as x is real, the expression, E, is positive for $r = 0$ and hence E is positive for all real values of r.

As E has been proved to be positive for all real values of r and x, it follows from Eqn. (8.107) that r_L has two distinct values, r_{L1} and r_{L2} possible, for all real values of r and x.

It is clear from Eqn. (8.106) that the sum and product of the two roots, r_{L1} and r_{L2}, of Eqn. (8.106) are given by

$$\left.\begin{aligned} r_{L1} + r_{L2} &= \frac{r^2 + x^2 + 1}{4} \geq 0 \text{ for } r > 0 \\ \text{and} \qquad r_{L1}\, r_{L2} &= +1 \end{aligned}\right\} \tag{8.109}$$

Eqns. (8.109) show that both r_{L1} and r_{L2} are positive and hence physically realisable.

Now, from Eqn. (8.104),

$$1 + x \tan\theta = \frac{1 + \tan^2\theta}{1 + r_L^2 \tan^2\theta}$$

$$= \frac{r}{r_L} \quad \text{[from Eqn. (8.103)]}$$

or

$$\tan\theta = \frac{1}{x}\left(\frac{r}{r_L} - 1\right) \tag{8.110}$$

From Eqns. (8.107) and (8.110), values of $\tan\theta$ corresponding to the values of r_L are obtained to be

$$\left.\begin{aligned} \tan\theta_1 &= \frac{1}{x}\left(\frac{r}{r_{L1}} - 1\right) \\ &= \frac{1}{x}(r\, r_{L2} - 1) \quad \text{[from Eqn. (8.109)]} \\ \text{and} \qquad \tan\theta_2 &= \frac{1}{x}(r\, r_{L1} - 1) \end{aligned}\right\} \tag{8.111}$$

Substitution of Eqn. (8.107) into Eqn. (8.111) gives

$$\tan\theta_1 = \frac{1}{x}\left[\frac{r^2+x^2+1 \mp \sqrt{(r^2+x^2+1)^2-4r^2}}{2} - 1\right]$$

$$\begin{matrix}\tan\theta_1 \\ \tan\theta_2\end{matrix} = \frac{1}{x}\left[r^2+x^2+1 \mp \frac{\sqrt{(r^2+x^2+1)^2-4r^2}}{2x}\right] \qquad (8.112)$$

Further, this may be noted from Eqn. (8.111) that

$$\tan\theta_1 \tan\theta_2 = \frac{1}{x^2}\,[r^2\,(r_{L1}\,r_{L2})] - r\,(r_{L1}+r_{L2}+1)$$

$$= -1 \quad \text{[from Eqn. (8.109)]} \qquad (8.113)$$

From this it may be inferred that $\theta_1 \sim \theta_2 = \dfrac{\pi}{2}$

or $$l_1 \sim l_2 = \frac{\lambda_g}{4} \qquad (8.114)$$

Now from Eqn. (7.19) it may be easily verified that a resistance $r_{L2} = \dfrac{1}{r_{L1}}$ connected at the end of a $\dfrac{\lambda_g}{4}$ line appears as a resistance r_{L1} at the input of the quarterwave line implying that the two solutions namely, r_{L1} with a line length θ_1 and r_{L2} with a line length $\theta_2 = \theta_1 + \dfrac{\pi}{2}$, are mutually consistent. Hence, the solution is, in effect, unique.

Thus it has been proved that any arbitrarily prescribed impedance, $R + jX$, can be simulated by [Ref. Fig. 8.7]

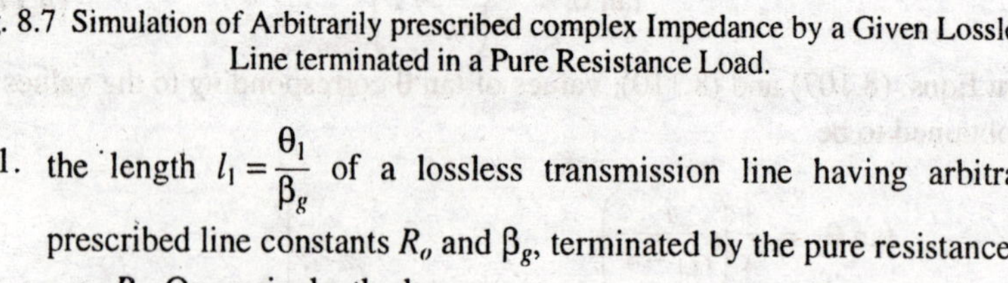

Fig. 8.7 Simulation of Arbitrarily prescribed complex Impedance by a Given Lossless Line terminated in a Pure Resistance Load.

1. the length $l_1 = \dfrac{\theta_1}{\beta_g}$ of a lossless transmission line having arbitrarily prescribed line constants R_o and β_g, terminated by the pure resistance $R_{L1} = r_{L1}R_o$. Or, equivalently, by
2. the length $l_2 = \dfrac{\theta_2}{\beta_g} = l_1 \pm \dfrac{\lambda_g}{4}$ of a lossless transmission line having arbitrarily prescribed line constants R_o and β_g, terminated by the pure resistance $R_{L2} = r_{L2}\,R_o = \dfrac{1}{r_{L1}}\,R_o$.

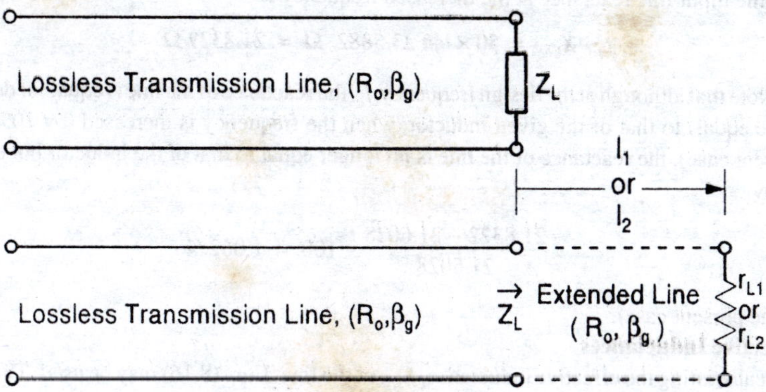

Fig. 8.8 Replacement of a Complex Load Impedance by an Extension of the Given Line and Its Termination in a Pure Resistance.

This theorem is of great practical importance in that a lossless line of length, l, terminated by a complex impedance, Z_L, may be regarded as the same line but of length $l + l_1$ or $l + l_2$ terminated in a pure resistance r_{L1} or r_{L2}, respectively, as illustrated in Fig. 8.8. In the light of this, therefore, ***lossless lines terminated only in pure resistance loads need be considered for analyses.***

SOLVED EXAMPLES

Example 8.1 (a) Simulate an indicator of 5.001 $m\,\mu\,H$ at 625 MHz using a 50 Ω short-circuited, coaxial line. Compare the reactance of the line with that of the lumped inductor at a frequency that is 10% higher than the design frequency.

(b) Determine the effective inductance of the line, in terms of its lumped up inductance, at the design and the increased frequencies. Comment on the simulation provided by the line.

Solution : (a) Given: $L = 5.001 \times 10^{-9}$ H, $f_o = 6.25 \times 10^8$ Hz. $R_o = 50\ \Omega$

Therefore, $X_{rq} = 2\,\pi f_o L = 2\,\pi \times 6.25 \times 10^8 \times 5.001 \times 10^{-9}\ \Omega = 19.6389\ \Omega$

Hence, the length, l_{sc}, of the short-circuited line required, is [Ref. Eqn. (8.42)]

$$l_{sc} = \frac{\tan^{-1}\left(\dfrac{X_{rq}}{R_o}\right)}{2\,\pi}\lambda_g = \frac{\tan^{-1}\left(\dfrac{19.6389}{50}\right)}{2\,\pi}\lambda_g = \frac{21.5538^\circ}{360^\circ}\lambda_g = 0.05957\,\lambda_g.$$

Reactances

When the frequency of operation is increased by 10%, the reactance of the inductor, being linearly dependent on the frequency, increases by 10%, giving

$$X_{inductor} = 1.1 \times X_{rq} = 1.1. \times 19.6389\ \Omega = 21.6028\ \Omega.$$

However, the reactance of the line increases as the tangent of the angle, $\beta_g\, l_{sc}$, which, of course, increases linearly with frequency. Thus, at the increased frequency, the argument of the tan function

$$= 1.1 \times 21.4438^\circ = 23.5882^\circ.$$

and the input line reactance at the increased frequency is

$$X_{line} = 50 \times \tan 23.5882° \ \Omega = 21.8322 \ \Omega.$$

Note that although at the design frequency, f_o, the reactance of the line is equal (or designed to be equal) to that of the given inductor, when the frequency is increased (by 10% in the present case), the reactance of the line is no longer equal to that of the inductor but exceeds it (by

$$\frac{21.8322 - 21.6028}{21.6028} * 100 = 1.062 \ \%$$

in the present case).

Effective Inductances

For calculating the effective inductance, L_{eff}, of the line, Eqn. (8.16) may be used. Therefore,

$$L_{eff}(f_o) = \left[\frac{\tan(\beta_g l_{sc})}{\beta_g l_{sc}}\right](Ll_{sc}) = \left[\frac{\tan(21.4438°)}{21.4438° \times \dfrac{\pi}{180}}\right](Ll_{sc}) = 1.04946\ (Ll_{sc}).$$

Now, if the frequency of operation is increased by 10%, the argument of the tan function also increases by 10%, so that the effective inductance of the line at the increased frequency is

$$L_{eff}(f) = \left[\frac{\tan(1.1 \times 21.4438°)}{1.1 \times 21.4438° \times \dfrac{\pi}{180}}\right](Ll_{sc}) = 1.06061\ (Ll_{sc}).$$

It is, thus, numerically verified that as the frequency of operation is increased, the effective inductance of the line also increases. It may be pointed out here that for ideal simulation, the effective inductance of the line must remain constant with respect to the frequency of operation. In the present example, when the frequency increases by 10%, the effective inductance of the line also increases by

$$\frac{1.06061 - 1.04946}{1.04946} * 100 = 1.0625 \ \%.$$

which is quite a small change and, hence, the simulation may be regarded as a *good* one.

EXERCISES

1. Calculate the lengths of short-circuited lossless 50 Ω line required to realise the following pure susceptances at their inputs at a frequency of 9375 MHz:
   (1) – 80 m mhos, (2) – 80 m mhos (3) – 40 m mhos (4) – 20 m mhos (5) – 20 m mhos (6) – 4 m mhos (7) 4 m mhos (8) 10 m mhos (9) 20 m mhos (10) 40 m mhos (11) 80 m mhos (12) 200 m mhos
   If now the frequency of operation is increased by 10%, calculate the susceptances presented at the inputs of each of the above lines and hence calculate the percentages by which the input susceptances change.
   [**Ans.** (1) 0.016 λ_g, – 181.691 m mhos 9.2 %, (2) 0.039 λ_g, – 72.413 m mhos, 9.5 %,
   (3) 0.074 λ_g, – 35.754 m mhos, 10.6 %,
   (4) 0.125 λ_g, – 17.082 m mhos, 14.6 %,

(5) 0.176 λ_g, – 7.367 m mhos, 26.3 %,
(6) 0.219 λ_g, – 1.203 m mhos, 69.9 %,
(7) 0.281 λ_g, 7.854 m mhos, 96.4 %,
(8) 0.324 λ_g, 15.751 m mhos, 57.5 %,
(9) 0.375 λ_g, 32.637 m mhos, 63.2 %,
(10) 0.426 λ_g, 100.808 m mhos, ...,
(11) 0.461 λ_g, – 447.302 m mhos, ..., (12) 0.484 λ_g, – 96.421 m mhos, ...,]

2. Calculate the lengths of open-circuited lossless 50 Ω line required to realise the following pure susceptances at their inputs:
(1) 4 m mhos (2) 10 m mhos (3) 20 m mhos (4) 40 m mhos (5) 80 m mhos (6) 200 m mhos (7) – 200 m mhos, (8) – 80 m mhos (9) – 40 m mhos (10) – 20 m mhos (11) – 20 m mhos (12) –4 m mhos
If now the frequency of operation is increased by 10%, calculate the susceptances presented at the inputs of each of the above lines and hence calculate the percentages by which the input susceptants change.
[**Ans.** (1) 0.031 λ_g, 4.412 m mhos, 10.3 % (2) 0.074 λ_g, 11.187 m mhos, 11.9 %
(3) 0.125 λ_g, 23.417 m mhos, 17.1 % (4) 0.176 λ_g, 54.295 m mhos, 35.7 %,
(5) 0.211 λ_g, 177.191 m mhos, ..., (6) 0.234 λ_g, – 421.232 m mhos, ...,
(7) 0.266 λ_g, – 73.200 m mhos, 63.4 %,
(8) 0.289 λ_g, –44.008 m mhos, 45.0 %,
(9) 0.324 λ_g, –25.396 m mhos, 36.5 %,
(10) 0.375 λ_g, –12.256 m mhos, 38.7 %,
(11) 0.426 λ_g, – 3.968 m mhos, 60.3 %,
(12) 0.469 λ_g, 1.947 m mhos, ...,]

3. (a) Simulate a capacitance of 0.5306 pF at 3000 MHz using a 50 Ω open-circuited, coaxial line. If the frequency increases by 10%, what would be the susceptances of the line and the lumped capacitor?
[**Ans.** l_{OC} = 0.07380 λ_g At the increased frequency, $B_{\text{capacitor}}$ = 0.01100 mho and B_{line} = 0.01119 mho.]

(b) Determine the effective capacitance of the line, in terms of its lumped up capacitance, at the design and the increased frequencies of operation. Comment on the simulation provided by the line.
[**Ans.** $C_{eff}(f_o) = 1.07843\ (Cl_{oc})$, & $C_{eff}\ (f) = 1.09682\ (Cl_{oc}) = 1.017\ C_{eff}\ (f_o)$ Simulation = Good]

4. Prove that the percentage variation in the effective inductance of a line, may be expressed in terms of the percentage variation in the frequency of operation by the equation,

$$\frac{dL_{eff}}{L_{eff}} \times 100 = \left[\left\{\tan(\beta_g\, l_{sc}) + \frac{1}{\tan(\beta_g\, l_{sc})}\right\}(\beta_g\, L_{eff}) - 1\right]\left(\frac{df}{f} \times 100\right)$$

5. Calculate the length of a short-circuited lossless 50 Ω line if it is required to have an input impedance of $Z_{in} = 0 + j69.587$ Ω at a frequency of 2450 MHz. If the frequency of operation changes by 10%, by what percentage would its effective inductance change? Does the formula given in 2 above yield the same result as obtained following the method given in the solved example above ? If it does not, explain why doesn't it yield the same result.

[**Ans.** l_{sc} = 0.1508 λ_g, $L_{eff}(f_o)$ = 1.4685(Ll_{sc}) L_{eff} (1.1 f_o) = 1.6436 (Ll_{sc}) = 1.119260 × L_{eff} (f_o). But formula (2 above) gives %age increase in L_{eff} = 13.9477 % compared to 11.9260% calculated above.]

6. By terminating a suitable length, l, of a 50 Ω coaxial cable by a pure resistance R at a frequency of 9375 MHz, an admittance (0.04 – j0.02) mho is to be simulated. Determine the length l cm. and the resistance R ohms required. What will be the $VSWR$ in this line?

[**Ans.** R = 19.0985 Ω, l = 0.0369 λ_g = 0.1181 cm, $VSWR$ = 2.618.]

7. If, now, the length of the 50 Ω line obtained in the above question is extended by 0.1631 λ_g and then connected to a 0.3356 λ_g long 75 Ω line, find out the input impedance at the end of the 75 Ω line. What is the $VSWR$ in the 75 Ω line?

[**Ans.** Z_{in} = (37.5946 + j0.0009) Ω, $VSWR$ = 1.995.]

8. A low-loss coaxial line is found to have an attenuation constant of 0.9096×10^{-3} dB/cm. If a resonant circuit is made using this line at 500 MHz, estimate the Q-factor of the circuit. [**Ans.** Q = 500.]

9. A 50 Ω lossless cable feeds a load of impedance, (150 + j0) Ω at a frequency of 2450 MHz. Calculate the distance from the load-end at which the real part of the input line impedance would be 85 Ω. What would be the input reactance at this point?

[**Ans.** l = 0.0500 λ_g, X_{in} = – 66.646 Ω]

Chapter 9

SMITH CHART

Calculation of the impedance, admittance, *VSWR* and transmission coefficient on a lossless line terminated in an arbitrary impedance can, in principle, be carried out using the relevant formulae given in Chapter 7. But these calculations are quite lengthy and involved. Moreover, the accuracy with which these quantities need be known, in practice, may not often be so great as to warrant such calculations. In practice, therefore, use in made of graphical means in the form of transmission line charts. One such chart is called the ***Bicircular Impedance Chart.*** Another chart which is very versatile and is, perhaps, the most commonly used one, is the ***Smith Chart.*** However, before taking up the Smith Chart for an in-depth study, the Bicircular impedance chart is discussed below.

9.1 BICIRCULAR IMPEDANCE CHART

As discussed in Sec. 8.5, a lossless line of length, l, terminated by a complex impedance, Z_L, may be regarded as the same line but of length $l + l_1$ or $l + l_2$ terminated in a pure resistance r_{L1} or r_{L2}, respectively. Therefore, **lossless lines terminated only in pure resistance loads need be considered for analyses without any loss of generality.**

Consider a lossless (R_o, β_g) line terminated in a pure resistance, $R_L = r_L R_o$. The input impedance of the line at a distance, l, from the load-end towards the generator is [Ref. Eqn. (7.19), taking $Z_L = r_L R_0$]

$$R + jX = R_o \frac{(r_L R_o) + jR_o \tan\theta}{R_o + j(r_L R_o)\tan\theta}, \text{ where, } \theta = \beta_g l. \tag{9.1}$$

In terms of the normalised impedances, Eqn. (9.1) may be rewritten as

$$r + jx = \frac{r_L + j \tan \theta}{1 + jr_L \tan \theta}. \tag{9.2}$$

9.1.1 Constant-r_L Circles

Subtracting $\frac{1}{2}\left(r_L + \frac{1}{r_L}\right)$ from both sides of Eqn. (9.2) gives

$$\left[r - \frac{1}{2}\left(r_L + \frac{1}{r_L}\right)\right] + jx = \frac{1}{2}\left(r_L - \frac{1}{r_L}\right)\left[\frac{1 - jr_L \tan \theta}{1 + jr_L \tan \theta}\right]. \tag{9.3}$$

Taking the magnitudes on both sides of this equation and squaring gives the θ_L-eliminant of Eqn. (9.2) to be

$$\left[r - \frac{1}{2}\left(r_L + \frac{1}{r_L}\right)\right]^2 + x^2 = \left[\frac{1}{2}\left(r_L - \frac{1}{r_L}\right)\right]^2 \tag{9.4}$$

In the complex (r, x)-plane (or the z-plane), for a fixed r_L, Eqn. (9.4) gives the locus of the point, $(r + jx)$, as θ is varied, to be a circle which may be called the ***constant-r_L*** circle, having [Ref. Fig. 9.1]

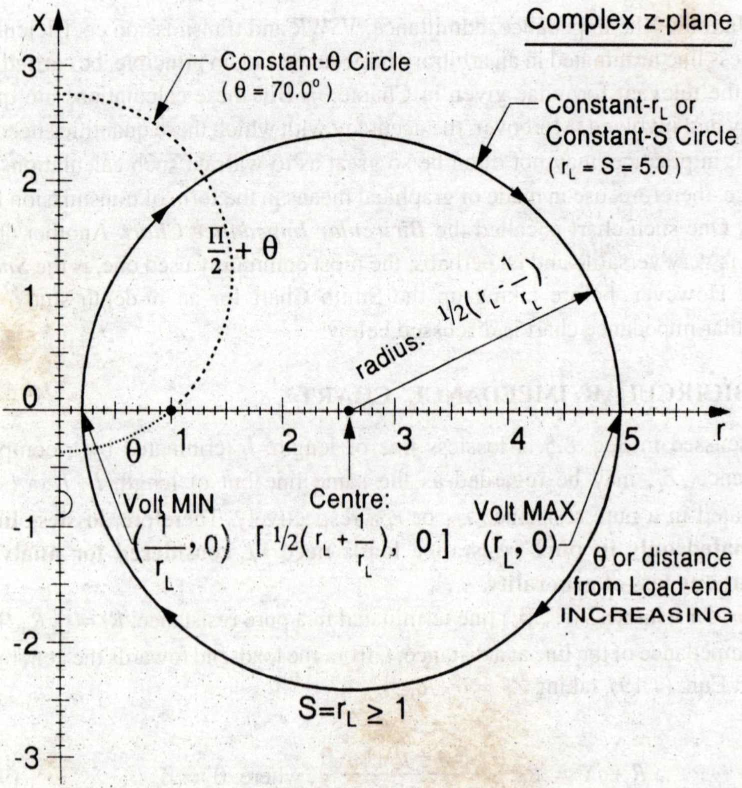

Fig. 9.1 A typical Constant-r_L Circle.

$$\left.\begin{aligned}&\text{its centre at : } \left[\frac{1}{2}\left(r_L+\frac{1}{r_L}\right), 0\right] \text{ on the } r\text{-axis,}\\&\text{and its radius } = \frac{1}{2}\left|r_L-\frac{1}{r_L}\right|, \text{ and,}\\&\left(\frac{1}{r_L}, 0\right) \text{ and } (r_L, 0) \text{ as the points of}\\&\qquad\text{its intersection with the } r\text{-axis.}\end{aligned}\right\} \tag{9.5}$$

Drawn in Fig. 9.1 is a typical constant-r_L circle. Now, as $\infty > r_L \geq 0$, r_L may be $\gtreqless 1$. Therefore, for all permissible values of r_L, one of the points of intersection with the r-axis, lies on the segment, $0 \leq r \leq 1$, and the other, in the region, $1 \leq r < \infty$.

For every value of r_L, Eqn. (9.4) gives a different circle, defined by Eqn. (9.5). On any such circle, r_L is constant and, therefore, it may be labelled by the value of r_L to which it corresponds.

Note that, in Eqn. (9.4), if r_L is replaced by $\frac{1}{r_L}$, it remains unchanged. This means that both r_L and $\frac{1}{r_L}$ give the same circle or correspond to the same circle. Therefore, a constant-r_L which is labelled by the the value of r_L, may equally well be labelled by the value of $\frac{1}{r_L}$. To remove this ambiguity in labelling, let a constant-r circle be labelled by r_L, if $r_L \geq 1$ and by $\frac{1}{r_L}$, if $r_L \leq 1$. That is, a constant-r_L is labelled by a quantity ≥ 1. Now, from the discussion given in Sec. 7.4, this quantity is nothing but the *VSWR*, S, on the line. Hence, a physical line terminated in a pure resistance is represented by the corresponding S-circle. By giving different values to r_L or S, the family of constant-r_L circles [Ref. Eqns. (9.4) and (9.5)], may be drawn, as shown in Fig. 9.2. Note that the constant-r_L corresponding to $r_L = S = 1$ degenerates into the point, $1 + j0$.

The point of the intersection of this circle with the r-axis lying to the right of the point, (1, 0), gives the voltage maxima on the line and the intersection to the left of the point, (1, 0), corresponds to the minima on the line. The line impedances at the two points of intersection are, $z = S + j0$ and $\frac{1}{S} + j0$, respectively.

Further, movement on the line towards the generator corresponds to a clockwise movement on the corresponding S-circle. This follows from the fact that the angle of the input impedance, $\tan^{-1}\left(\frac{x}{r}\right)$, at any point, θ or l, on the line, obtained from Eqn. (9.2), is

$$\tan^{-1}\left(\frac{x}{r}\right) = \tan^{-1}\left[-\frac{1}{2}\left(r_L - \frac{1}{r_L}\right)\sin(2\theta)\right], \tag{9.6}$$

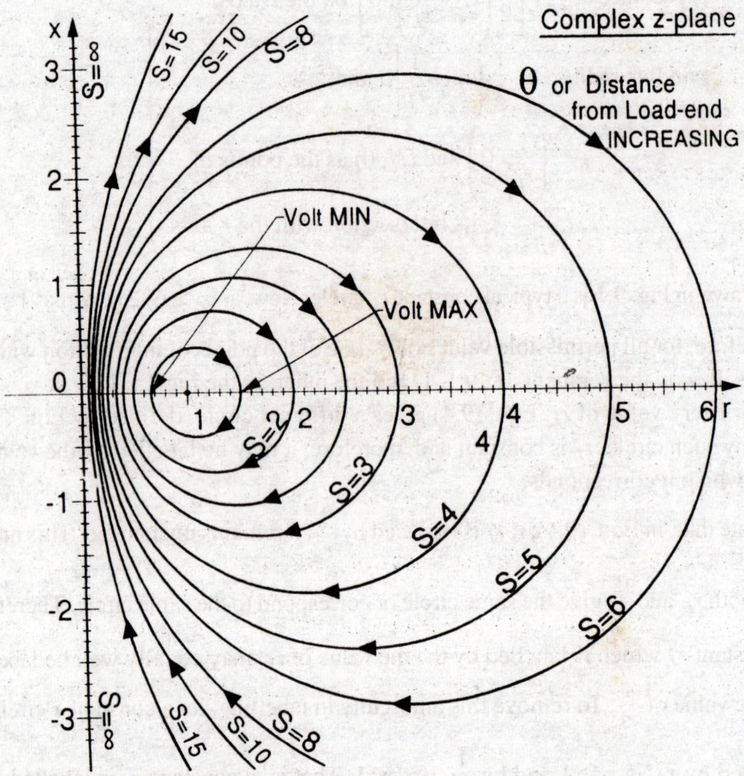

Fig. 9.2 Family of Constant-r_L Circles.

The family of circles given by Eqns. (9.4) and (9.5) can be plotted on the complex z-plane for various values of S or r_L, as shown in Fig. 9.2. A particular S-circle is the locus of the input impedance of the lossless line terminated in the resistance r_L, as the length of the line is varied from 0 to $\frac{\lambda_g}{4}$ to $\frac{\lambda_g}{2}$. If the length is increased further, the same circle is traced again and again in the same (clockwise) sense, implying that the impedance of the line is a periodic function of the length of the line with a period of $\frac{\lambda_g}{2}$

9.1.2 Constant-θ Circles

Again, subtracting $j\frac{1}{2}\left[\tan\theta - \frac{1}{\tan\theta}\right]$ from both sides of Eqn. (9.2) gives

$$r + j\left[x - \frac{1}{2}\left(\tan\theta - \frac{1}{\tan\theta}\right)\right] = \frac{1}{2}\left(\tan\theta - \frac{1}{\tan\theta}\right)\left[\frac{j\,(1 - jr_L\tan\theta)}{(1 + jr_L\tan\theta)}\right]. \quad (9.7)$$

Taking the magnitudes of both sides of this equation and squaring gives the r_L-eliminant of Eqn. (9.2) to be

$$\left.\begin{aligned} r^2 + \left[x - \frac{1}{2}\left(\tan\theta - \frac{1}{\tan\theta}\right)\right]^2 &= \left[\frac{1}{2}\left(\tan\theta + \frac{1}{\tan\theta}\right)\right]^2 \\ \text{or,}\quad r^2 + \left[x + \frac{1}{\tan(2\theta)}\right]^2 &= \frac{1}{\sin^2(2\theta)} \end{aligned}\right\} \tag{9.8}$$

This represents a family of ***constant-θ*** circles in the complex z-plane. For a constant value of θ or l, the circle which may be called the ***constant-θ*** or ***constant-l*** circle, has [Ref. Fig. 9.3]

$$\left.\begin{aligned} &\text{its centre at : } \left[0, \frac{1}{2}\left(\tan\theta - \frac{1}{\tan\theta}\right)\right] = \left[0, -\frac{1}{\tan(2\theta)}\right] \text{ on the } x\text{-axis} \\ &\text{its radius} = \left|\frac{1}{2}\left(\tan\theta + \frac{1}{\tan\theta}\right)\right| = \left|\frac{1}{\sin(2\theta)}\right|, \text{ and} \\ &\text{points of its interaction with } x\text{-axis : } (0, \tan\theta) \text{ and } \left(0, -\frac{1}{\tan\theta}\right) \end{aligned}\right\} \tag{9.9}$$

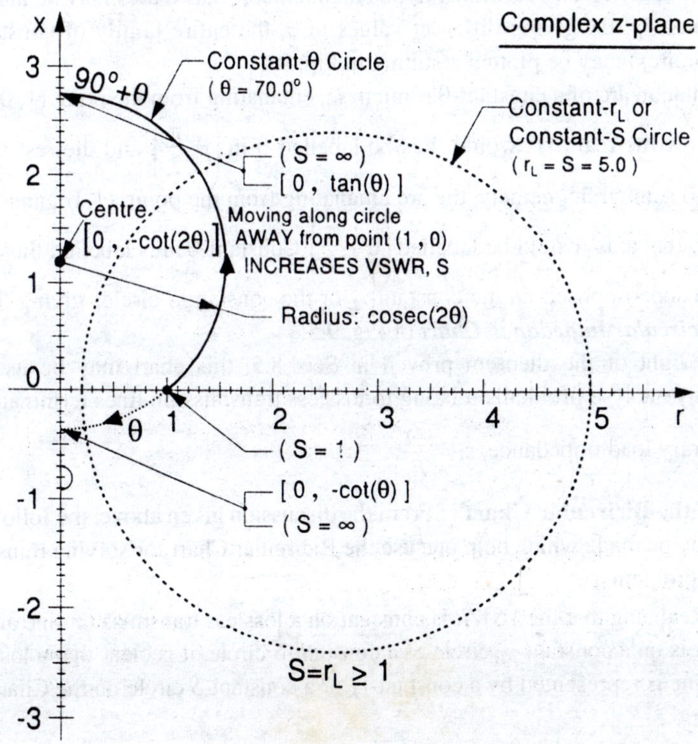

Fig. 9.3 A typical Constant-θ Circle.

Physically, a constant-θ circle is the locus of the input impedance of the lossless line of a constant electrical length θ, when it is terminated in various load resistances r_L. It may be observed from Eqn. (9.8) that all constant-θ circles pass through the point, (1, 0), on the real axis (r-axis).

Further, if only physically realisable impedances ($r \geq 0$) are considered, only the semicircles [Eqn. (9.8)] lying in the right half of the z-plane, need be considered. A typical constant-θ circle is shown in Fig. 9.3.

Note that Eqn. (9.8) remains unaltered if θ is replaced by $\left(\frac{\pi}{2}+\theta\right)$. This gives rise to the ambiguity namely, a particular constant-θ semicircle may equally well be labelled by the value of $\left(\frac{\pi}{2}+\theta\right)$. The ambiguity is removed with the help of Eqn. (9.6) in the following way:

If r_L is assumed to be greater than unity, that is, for $\theta = 0$, $r_L > 1$, then the constant-θ arcs lying below the real axis ($x < 0$), drawn in dotted lines in Fig. 9.3, correspond to values of θ lying between 0 on the real axis $r \geq 1$, and increasing in the clockwise direction to the value, $\theta = \frac{\pi}{2}$ radians corresponding to the segment, $1 \geq r \geq 0$; of the real axis. Constant-θ arcs above the real axis ($x > 0$), drawn in solid lines in Fig. 9.3, correspond to values of θ lying between $\frac{\pi}{2}$ radians and π radians, increasing in the clockwise sense. In this manner, constant-θ arcs may be labelled unambiguously. By giving different values to θ, the entire family of constant-θ circles (or arcs) may be plotted resulting in Fig. 9.4.

Note that an arc of a constant-θ semicircle, emanating from the point (1, 0) and lying below the real axis would thus be labelled θ in $\left(\theta, \frac{\pi}{2}\right)$, and the rest of the constant-θ semicircle, namely, the arc emanating from the point, (1, 0), and lying above the real axis, would be labelled $\left(\theta+\frac{\pi}{2}\right)$. Constant-θ arcs labelled thus may be drawn superimposed on the constant-r_L or the constant-S circles giving the so called ***Bicircular Impedance Chart*** of Fig. 9.5.

In the light of the theorem proved in Sec. 8.5, this chart may be used to graphically analyse problems relating to lossless transmission lines terminated in any arbitrary load impedance, $z_L = \frac{Z_L}{R_o}$.

Usage of the Bicircular Chart : From the discussion given above, the following points may be made which help one use the Bicircular Chart for solving transmission line problems:

1. Realising that the *VSWR* is constant on a lossless transmission line and so it is on a constant-r_L circle or a constant-S circle, it is clear that a lossless line is represented by a constant-r_L or a constant-S circle on the Chart.

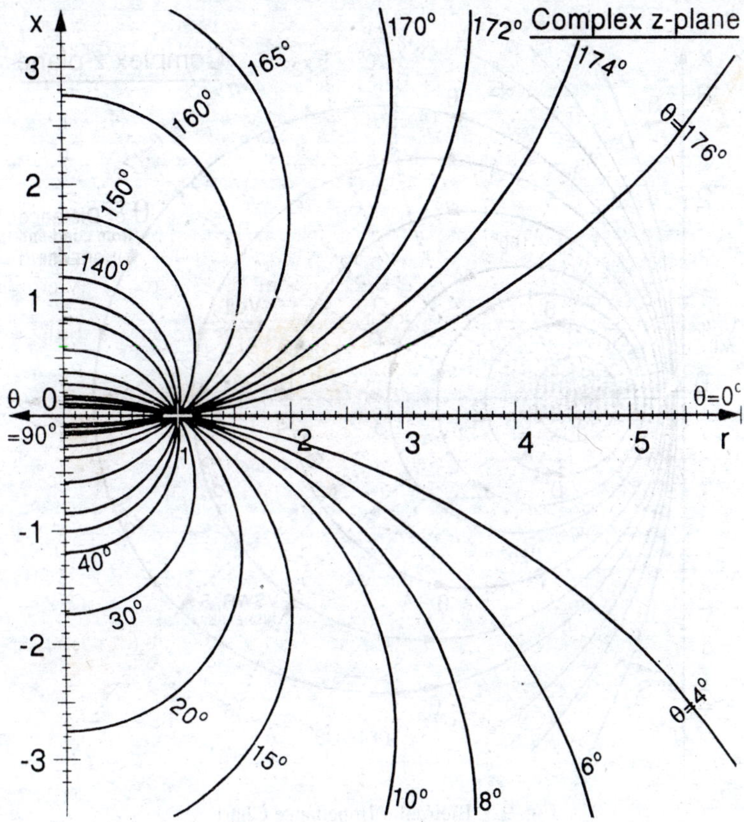

Fig. 9.4 Family of Constant-θ Circles

2. Movement along the physical line ***towards the Generator*** corresponds to a ***clock-wise movement*** on the corresponding constant-*S* circle on the Chart. Constant-θ arcs play the role of graduating the constant-*S* circle so that movement along a constant-*S* circle can be measured.
3. If the Chart is used as an impedance chart, the voltage maxima lie on the part, $r \geq 1$, of the real axis; and the voltage minima lie on the segment, $0 \geq r \geq 1$.
4. Lastly, it must be realised that the given load impedance, z_L, is nothing but an input line impedance (at the load-end). Therefore, only that constant-*S* circle, which passes through the point representing the given load impedance, represents the given line plus the given load impedance system. Interpolation would, in general, be required to identify the constant-*S* circle that represents the given line plus load system on the Chart.
5. Once the constant-*S* circle representing the given line plus load system is identified, most of the problems of practical interest can be solved.

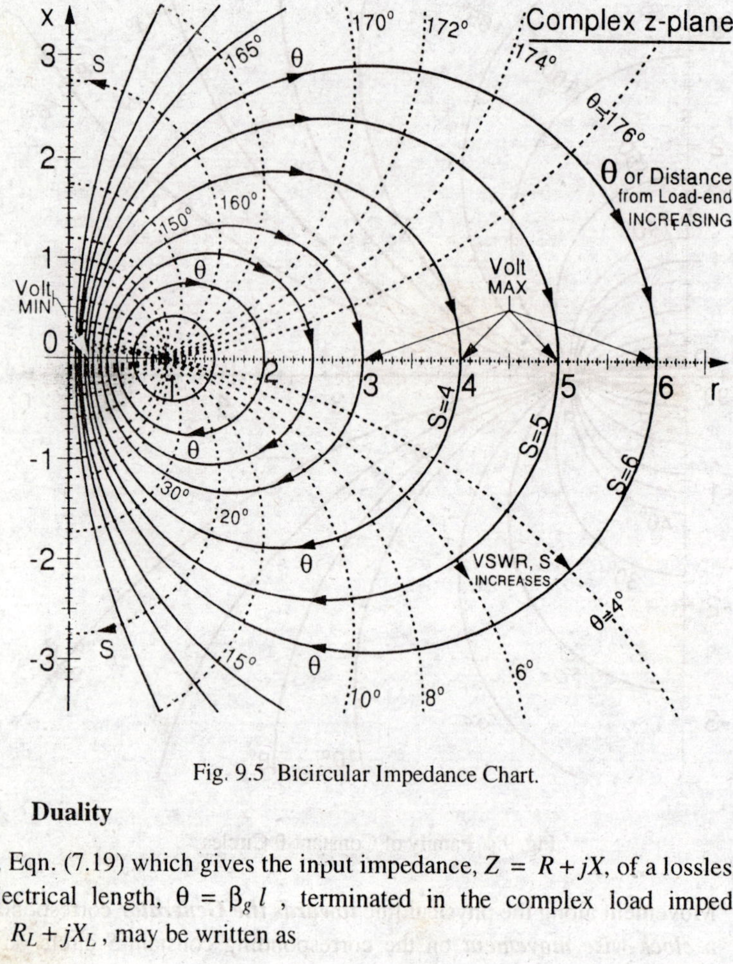

Fig. 9.5 Bicircular Impedance Chart.

9.1.3 Duality

Now, Eqn. (7.19) which gives the input impedance, $Z = R + jX$, of a lossless line of electrical length, $\theta = \beta_g l$, terminated in the complex load impedance, $Z_L = R_L + jX_L$, may be written as

$$Z = R + jX = R_o \frac{Z_L + jR_o \tan\theta}{R_o + jZ_L \tan\theta}$$

or, in terms of the normalised quantities, the input impedance is,

$$z = r + jx = \frac{z_L + j\tan\theta}{1 + jz_L \tan\theta}, \tag{9.10}$$

where,

$$z_L = \frac{Z_L}{R_o} = \frac{R_L}{R_o} + j\frac{X_L}{R_o} = r_L + jx_L .$$

Note that, mathematically, Eqn. (9.10) gives nothing but the transformation or mapping of the point $z_L = r_L + jx_L$, to the point $z = r + jx$ and the Bicircular

Impedance Chart developed and described above, is a graphical representation of this mathematical transformation. From Eqn. (9.10), it is easy to show that the input admittance is given by

$$y = \frac{Y}{G_o} = \frac{G}{G_o} + j\frac{B}{G_o} = g + jb = \frac{y_L + j\tan\theta}{1 + jy_L\tan\theta}, \tag{9.11}$$

where,

$$y_L = \frac{Y_L}{G_o} = \frac{G_L}{G_o} + j\frac{B_L}{G_o} = g_L + jb_L,$$

with $G_o = \dfrac{1}{R_o}$ = characteristic admittance of the line.

This shows that the transformation given by Eqn. (9.11) which transforms the point $y_L = g_L + jb_L$ to the point $y = g + jb$ is mathematically the same as that given in Eqn. (9.10) transforming the point, $z_L = r_L + jx_L$, to the point, $z = r + jx$. Therefore, the Bicircular Impedance Charts may be used equally well as an admittance chart representing the transformation (9.11) with the appropriate changes, namely

	DUALITY TRANSFORMATIONS	
$r \rightarrow g$		Voltage Maxima → Voltage Minima or Current Maxima
$x \rightarrow b$		Voltage Minima → Voltage Maxima or Current Minima
$z \rightarrow y$		

Hence, the Bicircular Chart may be used either as an Impedance Chart or as an Admittance Chart. However, the Bicircular (Cartesian) Chart has the following drawbacks:

1. Constant-S circles are not concentric, and the constant-θ arcs are not spaced uniformly over a constant-S circle. These make interpolation very difficult and inaccurate.
2. Only a limited range of impedance (or admittance) values can possibly be encompassed on a Chart of reasonable size.

These render the Bicircular Chart of little practical usage.

9.2 SMITH IMPEDANCE CHART

A very marked improvement of the Chart results if, in the place of working in the z-plane, z being the line impedance, and using the Cartesian system of coordinates for r and x, ρ-plane, ρ being the voltage reflection coefficient on the line, is employed giving the ***Smith Chart***. However, in the ρ-plane, a curvilinear system of coordinates has to be employed for r and x. The Smith Chart is of great practical value and is used universally for the treatment of transmission line problems. The theory of this chart is given below:

As discussed in Sec. 5.1 the voltage reflection coefficient at any point on a line is defined as the ratio of the instantaneous amplitude of the reflected voltage wave to that of the forward wave at that point. Therefore, for a lossless line terminated in a pure resistance r_L, the voltage reflection coefficient at a distance, l, from the load-end (towards the generator) is given by [Ref. Eqns. (5.19) and (5.28)]

$$\rho(\theta) = u(\theta) + j\,v(\theta) = \rho_L e^{-2j\theta} \tag{9.12}$$

where, $\theta = \beta_g l$ and u and v are the real and the imaginary parts, respectively, of the voltage reflection coefficient, ρ.

The voltage reflection coefficient may be expressed in terms of the line impedance at that point from the Eqn. (5.19) giving

$$\rho = u + j\,v = \frac{z-1}{z+1} = \frac{(r-1)+jx}{(r+1)+jx} \tag{9.13}$$

Equating the real and imaginary parts of the two sides of this equation, gives

$$u = \frac{r^2 + x^2 - 1}{(r+1)^2 + x^2} \tag{9.14}$$

and,

$$v = \frac{2x}{(r+1)^2 + x^2}. \tag{9.15}$$

It may readily be seen from Eqns.(9.14) and (9.15) that

$$\left[u - \left(\frac{r}{r+1}\right)\right]^2 + v^2 = \left[\frac{1}{r+1}\right]^2 \tag{9.16}$$

$$[u-1]^2 + \left[v - \frac{1}{x}\right]^2 = \left[\frac{1}{2}\right]^2, \tag{9.17}$$

and

$$u^2 + v^2 = 1 - \frac{4r}{(r+1)^2 + x^2} \tag{9.18}$$

9.2.1 Constant-r Circles

Eqn. (9.16) is the x-eliminant of Eqns. (9.14) and (9.15) and, therefore, it is the locus of the voltage reflection coefficient, $\rho = u + j\,v$, for various values of x. Note that Eqn. (9.16) represents a family of circles in the complex (u, v)-plane or the ρ-plane with r as the running parameter [Ref. Fig. 9.6). These circles may be called the ***constant-r circles*** and a constant-r circle may be labelled by the value of r for it.

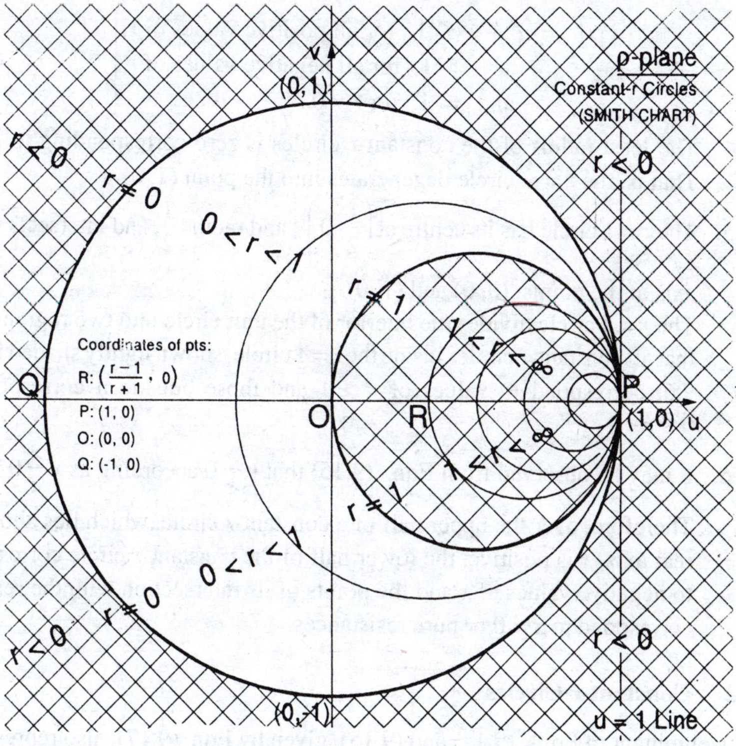

Fig. 9.6 Family of Constant-*r* Circles.

A typical member of the family of the ***constant-r circles*** [Ref. Fig. 9.6] has

its centre at : $\left(\dfrac{r}{r+1}, 0\right)$ on the u-axis, a $\left(\dfrac{r-1}{r+1}, 0\right)$ and (1, 0)

and its radius $= \left(\dfrac{1}{r+1}\right)$, and as the points of its intersection with the u-axis.

Other properties of the constant–*r* circles are [Ref. Fig. 9.6]

1. All constant-*r* circles ($0 \le r < \infty$) pass through the point, (1, 0).
2. The largest radius of constant-*r* circles is unity which corresponds to the $r = 0$ circle whose centre is at the origin (0, 0).
3. All other constant-*r* circles corresponding to positive values of *r* lie inside the unit circle.

   Thus, the unit circle (corresponding to $r = 0$) divides the entire (u, v)-plane into two regions – the inside of the unit circle corresponding to positive values of r, $0 \le r < \infty$, and the outside of the unit-circle (shown shaded in

Fig. 9.6) corresponding to negative values of r. This is borne out by Eqn. (9.18) which shows that

$$\left.\begin{aligned} |\rho|^2 = u^2 + v^2 &\leq 1, \text{ for all positive values or } r, \\ &> 1, \text{ for all negative values of } r, \\ \text{and, } &= 1, \text{ for } r = 0. \end{aligned}\right\} \quad (9.20)$$

4. The least radius of the constant-r circles is zero corresponding to $r = \infty$. That is, the $r = \infty$ circle degenerates into the point (1, 0)
5. The $r = 1$ circle has its centre at $\left(\frac{1}{2}, 0\right)$, and radius $\frac{1}{2}$, and intersects the u–axis at the points, (0, 0) and (1, 0).
   The $r = 1$ circle divides the interior of the unit circle into two regions such that all constant-r circles inside the $r = 1$ circle, shown lightly shaded in Fig. 9.6, correspond to values of $r > 1$ and those outside it correspond to $0 \leq r < 1$.
6. It may be observed from Eqn. (9.15) that $v \gtreqless 0$ according as $x \gtreqless 0$.
   Therefore, over the upper half of a constant-r circle, which lies above the real axis, x is positive, the lower half of the constant-r circle corresponds to negative values of x and the points of its intersection with the real axis correspond to $x = 0$ or pure resistances.

9.2.2 Constant-*x* Circles

The r-eliminant of Eqns. (9.14) and (9.15), given by Eqn. (9.17), also represents a family of circles with x as the running parameter. A typical member of the family of the ***constant-x circles*** [Ref. Fig. 9.7] has

$$\left.\begin{aligned} &\text{its centre at} \;:\; \left(1, \frac{1}{x}\right) \text{ on the strength line, } u = +1, \\ &\text{and its radius } = \frac{1}{|x|}, \text{ and,} \\ &\left(1, \frac{2}{x}\right) \text{ and } (1, 0) \text{ as the points of its intersection with the straight line, } u = +1. \end{aligned}\right\} \quad (9.21)$$

Other properties of the constant-x circles are: [Ref. Fig. 9.7]

1. All constant-x circles ($-\infty < x < \infty$) also, pass through the point, (1, 0), and have their centres on the straight line, $u = +1$.
2. A constant-x circle lies entirely above or entirely below the real axis depending on whether x corresponding to it is positive or negative.
3. The $x = 0$ circle degenerates into the real axis itself and the circles corresponding to $x = \pm\infty$ degenerate into the point (1, 0).
4. The constant-x circles corresponding to $x = \pm 1$ meet the v-axis tangentially at the points $(0, \pm 1)$.

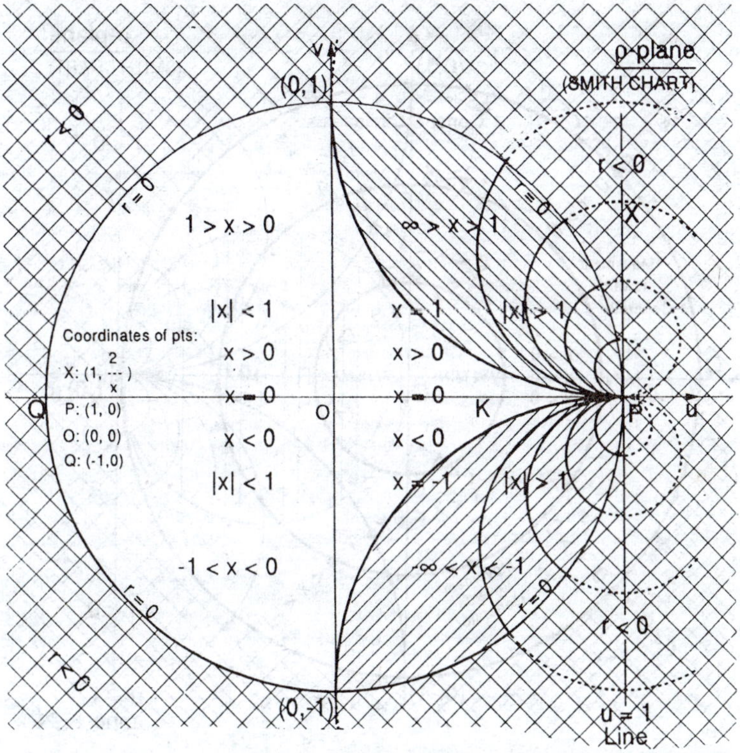

Fig. 9.7 Family of Constaint-*x* Circles

9.2.3 Constant-S Circles

The voltage standing wave ratio on the line is given by [Ref. Eqns. (7.27) and (7.28)]

$$S = \frac{1 + |\rho|}{1 - |\rho|} \geq 1, \tag{9.22}$$

and

$$|\rho| = \frac{S - 1}{S + 1} \leq 1 \tag{9.23}$$

On a lossless transmission line terminated in some load, the *VSWR* and $|\rho|$ remain constant on the line. Therefore, on the Smith Chart, a transmission line is represented by the locus of $|\rho|$ = a constant, or,

$$\sqrt{u^2 + v^2} = |\rho| = \text{a constant} \tag{9.24}$$

which is also a circle of radius equal to $|\rho|$ and its centre located at the origin. [Ref. Fig. 9.8]

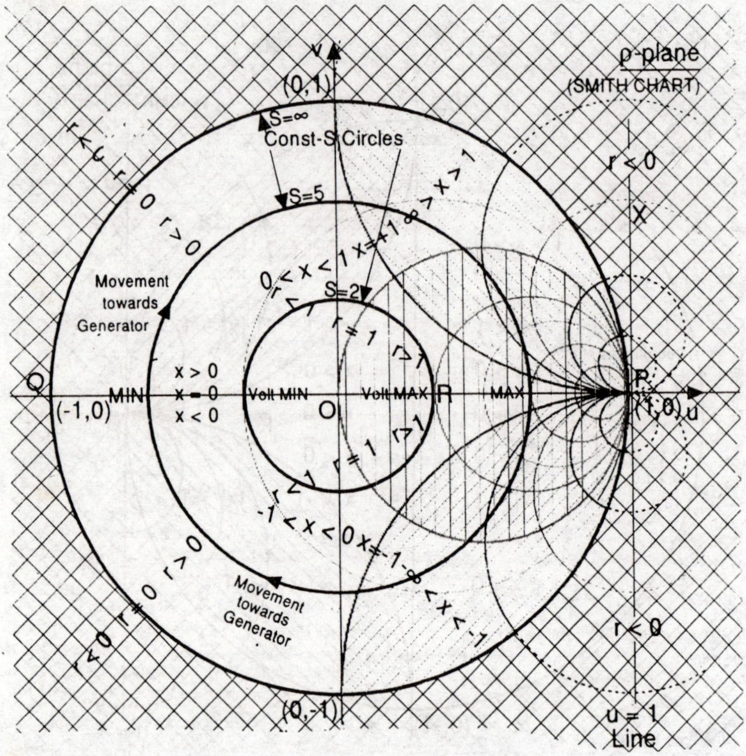

Fig. 9.8 Family of Constant-*S* Circles.

Hence, on the Smith Chart, that is, in the (u, v)-plane, a transmission line is represented by a circle with origin as its centre and its radius equal to the magnitude of the voltage reflection coefficient. The concentric circles may be called the ***constant-S*** circles and a member of this family of circles [Ref. Fig. 9.8] may be labelled by the value of S to which it corresponds. The phase angle of the voltage reflection coefficient at a distance, l, towards the generator, from the load-end, is given by

$$\phi = \phi_L - 2\beta_g\, l \tag{9.25}$$

This shows that movement towards the generator (l increasing) on the physical line, corresponds to decreasing phase angle of the voltage reflection coefficient or, to a movement on the constant-S circle in the clockwise direction. At the points of intersection of a constant-S circle with the u-axis, the line impedances are pure resistances, indicating that these intersection points correspond to voltage maxima and minima on the line. Now, the segment, $0 \le u \le 1$, of the u-axis corresponds to $1 \le r \le \infty$ and the segment, $-1 \le u \le 0$ corresponds to $0 \le r \le 1$. Therefore, the intersection point on the segment $0 \le u \le 1$ corresponds to the voltage maximum

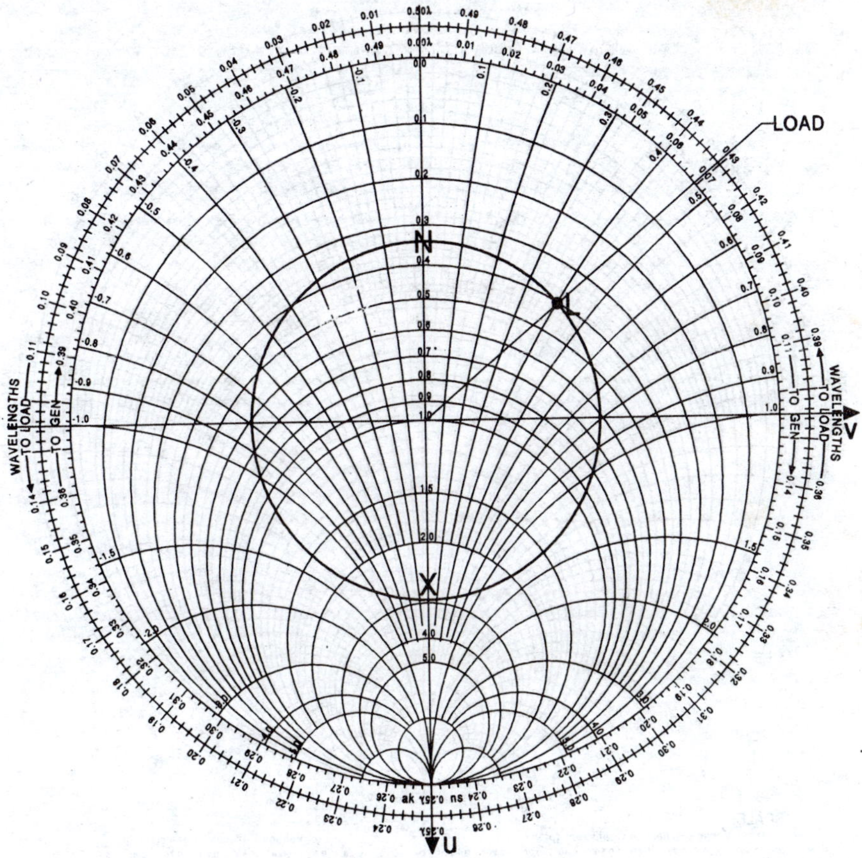

Fig. 9.9 The Smith Chart

and the value of r at the intersection is S itself [Ref. Eqn. (7.30)]. The intersection point on the segment $-1 \le r \le 0$ corresponds to the voltage minimum and the value of r at the intersection is $\frac{1}{S}$ [Ref. Eqn. (7.31)].

Based on the analysis presented above, some typical members of the constant-r and constant-x families of circles may be drawn superimposed on each other, as shown in Fig. 9.9, on a sheet of paper to yield the **Smith Impedance Chart.** It may be pointed out that, as only the interior of the unit circle is of interest ($r \ge 0$), only the portions (arcs) of the constant-x circles which lie inside the unit circle need be drawn resulting the chart shown in Fig. 9.9. For aesthetic reasons, the chart has been shown rotated through –90° so that the v-axis runs from the left to the right and the u-axis from the top to the bottom. A typical detailed Smith Chart together with the *VSWR* and Voltage Reflection Coefficient (magnitude) scales, which may be used for practical line calculations, is shown in Fig. 9.10. Constant-S circles, representing the transmission line, and radial lines marking positions on the line,

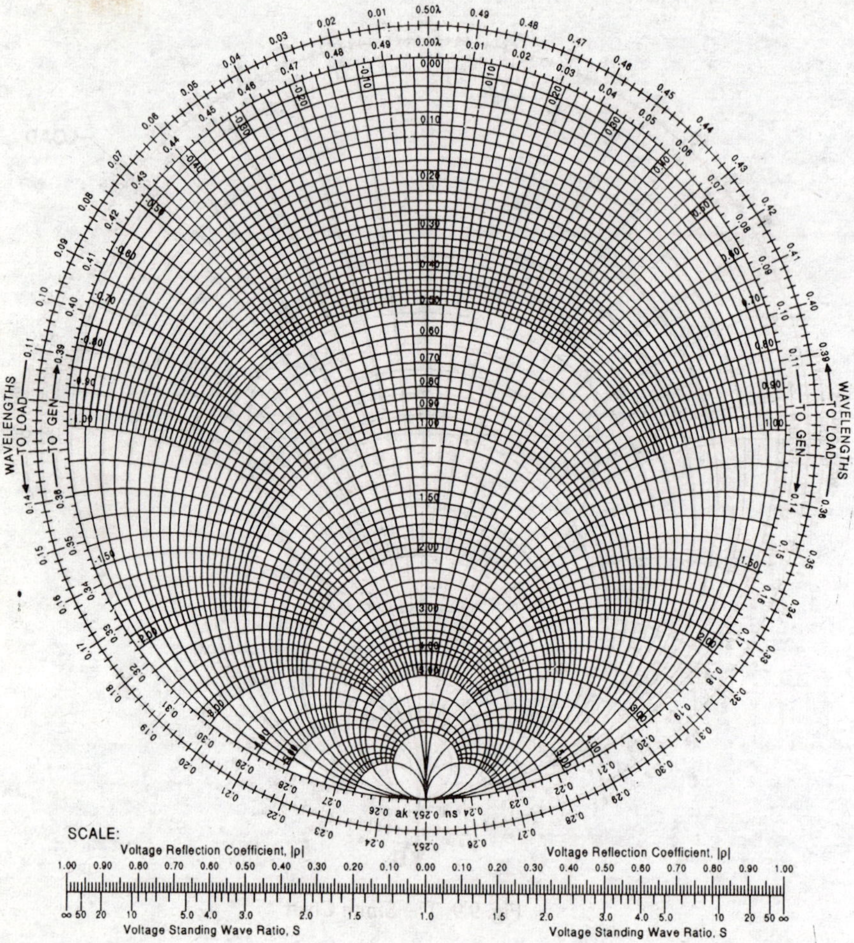

Fig. 9.10 Smith Chart

may be drawn on it as and when required. Usage of the Chart for carrying out transmission line calculations, is discussed in the next Section and illustrated through worked out examples at the end of the Chapter.

Further, as discussed in Sec. 9.1.3, the Smith Chart can also be used as an admittance chart provided the following changes required by the duality are made:

1. r is replaced by g: constant-r circles become constant-g circles
2. x is replaced by b: constant-x circles become constant-b circles
3. voltage maxima and minima become current maxima and minima
4. voltage reflection coefficient, ρ, becomes current reflection coefficient.

Note that a point on the Smith Impedance Chart represents an impedance $r + jx$ which can be read off the curvilinear coordinate system constituted by the constant-r circles and the constant-x circles. It also represents the voltage reflection coefficient at the point on the line where the input impedance is $r + jx$, the real and imaginary parts of the voltage reflection coefficient may be read off the (u, v) (Cartesian) coordinate system. Now, when the Smith Chart is used as an admittance chart, a point on it represents an admittance $g + jb$ and also the current reflection coefficient. It may be recall ›d that the voltage reflection coefficient and the current reflection coefficient at a point have equal magnitudes but their phase angles differ by 180° from each other.

Usage of the Smith Chart

In contrast to the usage of Bicircular Chart, the usage of the Smith Chart is extremely easy, simple and accurate because of the following reasons:

1. All constant-S circles are concentric, having their common centre at the centre of the Chart or the origin of the (u, v)-coordinates system. To find out the constant-S circle representing a given physical transmission line plus load impedance system, all that is required to be done is to draw a circle with the centre of the Chart as its centre and passing through the point representing, on the curvilinear (r, x)-coordinates system, the given load impedance. As pointed out in Sec. 9.1.2, once this is done, most of the practical problems connected with the given line load system may be solved. The magnitude of the reflection coefficient or the corresponding *VSWR* may be read off the linear scale drawn below the Smith Chart in Fig. 9.10.
2. As the phase angle of the reflection coefficient [Ref. Eqn. (9.28)] decreases linearly with the distance from the load-end, l, measurement of a movement along a line is directly proportional to the angle through which one moves on the constant-S circle. Here, a word of caution may be sounded for the benefit of users of the Smith Chart. As may be observed from Eqn. (9.28), the change in the angle of the reflection coefficient is at double the rate at which the electrical length of the line changes. For a physical movement of Δl towards the generator, the phase angle of ρ decreases by $2\beta_g \Delta l$ whereas the corresponding electrical length changes by $\beta_g \Delta l$. To avoid a possible confusion that this fact may create, the user, particularly the beginner, is advised not to use electrical lengths but normalised lengths. Movements like $\frac{l}{\lambda_g}$ and $\frac{\Delta l}{\lambda_g}$, may be read off the outermost circular scales on the Chart. A movement through 180° on the Smith Chart corresponds to a change of 90° in the electrical length or to a movement through $\frac{\lambda_g}{4}$ on the physical line.
3. As in the case of like the Bicircular Chart, in the Smith Chart also a movement along the physical line ***towards the Generator*** corresponds to a ***clockwise movement*** on the corresponding constant-S circle on the Chart.
4. If the Chart is used as an impedance chart, the voltage maxima lie on the lower vertical radius; and the voltage minima lie on the upper vertical radius of the Chart.

9.3 PROPERTIES OF THE SMITH CHART

The Smith Chart has, indeed, many interesting properties. Some of these properties are discussed in this Section to gain deeper understanding and for better utilisation of the Smith Chart.

9.3.1 Locus of Constant Magnitude of Impedance

Every point on the Smith Chart represents a normalised impedance, $z = r + jx$, where r and x are given by the curvilinear coordinates system constituted by the constant-r and the constant-x circles (which are mutually orthogonal). It would be interesting to find out the locus of the point, z, which moves on the Smith Chart such that its magnitude is a constant. This amounts to finding out the relation between u and v– the real and imaginary parts of the voltage reflection coefficient– such that

$$r^2 + x^2 = a^2 = \text{a constant.} \tag{9.26}$$

Substitution of Eqn. (9.26) into Eqns. (9.14) and (9.15) gives,

$$u = \frac{a^2 - 1}{a^2 + 1 + 2r} \tag{9.27}$$

and

$$v = \frac{2x}{a^2 + 1 + 2r} \tag{9.28}$$

From Eqn. (9.27), $r = \dfrac{(a^2 - 1) - (a^2 + 1)\,u}{2\,u}$. (9.29)

Dividing Eqn. (9.28) by Eqn. (9.27) gives,

$$x = \frac{(a^2 - 1)\,v}{2u} \tag{9.30}$$

Substitution of Eqns. (9.29) and (9.30) into Eqn. (9.26) gives the r, x-eliminant of Eqns. (9.27) and (9.28) to be

$$r^2 + x^2 = a^2 = \frac{[(a^2 - 1) - (a^2 + 1)\,u]^2 + (a^2 - 1)^2 v^2}{4u^2},$$

or,

$$\left[u - \frac{a^2 + 1}{a^2 - 1}\right]^2 + v^2 = \left[\frac{2a}{a^2 - 1}\right]^2 \tag{9.31}$$

Eqn. (9.31) represents a family of circles with a as the running parameter. A typical constant-a circle has [Ref. Fig. 9.11]

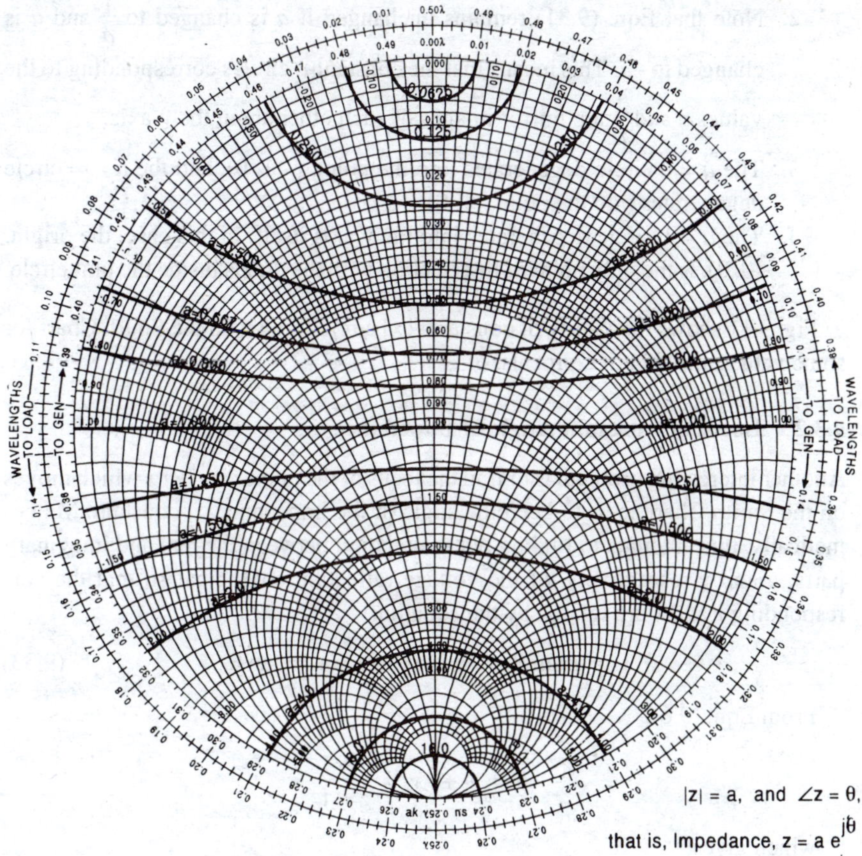

Fig. 9.11 Loci (Circles) of $|z| = a$ = constant.

$$\left.\begin{aligned}&\text{its centre at} \quad : \quad \left(\frac{a^2+1}{a^2-1},\ 0\right) \text{ on the } u\text{-axis,}\\ &\text{and its radius} \quad = \frac{2}{\left|a-\dfrac{1}{a}\right|}, \text{ and,}\\ &\left(\frac{a-1}{a+1},\ 0\right) \text{ and } \left(\frac{a+1}{a-1},\ 0\right) \text{ as the points of its intersection with the } u\text{-axis.}\end{aligned}\right\} \quad (9.32)$$

Other properties of the constant-*a* circles are: [Ref. Fig. 9.11]

1. As $a = |z| \geq 0$, the second point of intersection for a constant-*a* circle lies outside the unit circle. Thus, all the constant-*a* circles intersect the *u*-axis only once inside the unit circle. If $a > 1$, the constant-*a* circle intersects the *u*-axis in the segment, (0, 1); and if $a < 1$, the constant-*a* circle intersects the *u*-axis in the segment, (0, – 1).

2. Note that Eqn. (9.31) remains unchanged if a is changed to $\frac{1}{a}$ and u is changed to $-u$. This means that the constant-a circles corresponding to the values a and $\frac{1}{a}$ are mirror images of each other about the v-axis.
3. The $a = 0$ circle degenerates into the point, $(-1, 0)$, and the $a = \infty$ circle degenerates into the point, $(1, 0)$.
4. The $a = 1$ circle degenerates into the v-axis itself. In this case, the origin, $(0, 0)$, is the only point of intersection with the u-axis inside the unit circle.

Fig. 9.11 shows the family of constant-$|z|$ circles drawn on the Smith Chart for some values and their reciprocals, of $|z|$.

9.3.2 Locus of Constant Angle of Impedance

Another interesting problem is to find the locus of the point $z = r + jx$ which moves on the Smith Chart so as to keep the angle of the impedance, z, constant. Mathematically, this amounts to finding out the relation between the real and imaginary parts, u and v respectively, of the voltage reflection coefficient so that the corresponding r and x are related by the equation,

$$x = r \tan \theta_i = mr, \text{ where, } m = \tan \theta_i. \tag{9.33}$$

From Eqn. (9.6),

$$\rho = u + jv = \frac{z-1}{z+1}$$

which gives,

$$z = r + jx = \frac{(1+u) + jv}{(1-u) - jv}$$

$$= \frac{(1-u^2) - v^2 + 2jv}{(1-u)^2 + v^2} \tag{9.34}$$

Application of Eqn. (9.33) to Eqn. (9.34) gives

$$2v = m\,[1 - u^2 - v^2] \tag{9.35}$$

This equation may be rewritten as

$$u^2 + \left(v + \frac{1}{m}\right)^2 = \left(1 + \frac{1}{m^2}\right) \tag{9.36}$$

Eqn. (9.36) again represents a family of circles with m or θ_i, as the running parameter.

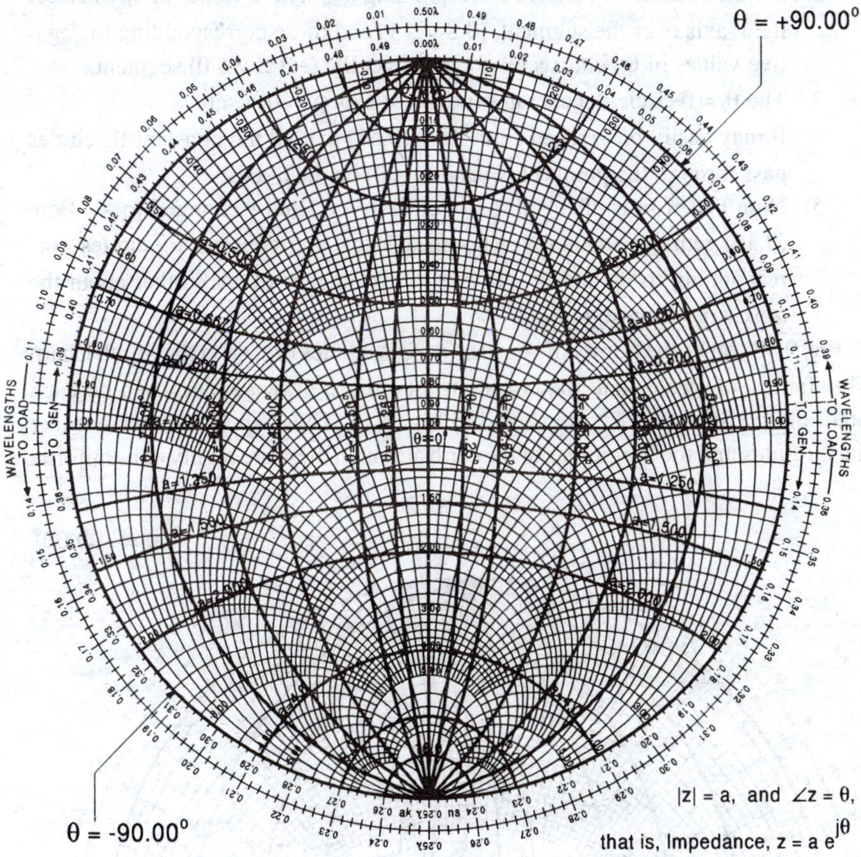

Fig. 9.12 Loci (Circles) of Ang $[z] = \theta$ = constant.

A typical constant-m or constant-θ_i, circle has [Ref. Fig. 9.12]

$$\left.\begin{aligned} &\text{its centre at} \quad : \left(0, -\frac{1}{m}\right) \text{ on the } v\text{-axis,} \\ &\text{and its radius} \;=\; \sqrt{1+\frac{1}{m^2}} = \frac{1}{|\sin\theta|} \text{ and,} \\ &\left(0, -\cot\frac{\theta_i}{2}\right) \text{and} \left(0, \tan\frac{\theta_i}{2}\right) \text{ as the points of its} \\ &\qquad\qquad\qquad \text{intersection with the } v\text{-axis.} \end{aligned}\right\} \tag{9.37}$$

Other properties of the constant-m or θ_i, circles are: [Ref. Fig. 9.12]

1. If $0 \le \theta_i \le \dfrac{\pi}{2}$, one point of intersection of the constant-θ_i circle with the v-axis, is inside but the other is outside the unit circle.

2. All the constant-θ_i, circles corresponding to positive values of θ_i, intersect the v-axis over the segment, $(0 \leq v \leq 1)$ and those corresponding to negative values of θ_i, intersect the v-axis over the $(-1 \leq v \leq 0)$ segment.
3. The $\theta_i = 0$ circle degenerates into the u-axis $(v = 0)$ itself.
4. It may easily be verified from Eqn. (9.36) that all the constant-θ_i, circles pass through the points (1, 0) and (–1, 0) on the u-axis.
5. Note that if m is changed to $-m$ or θ_i to $-\theta_i$, and v changed to $-v$, Eqn. (9.36) remains unaltered. This means that the constant-θ_i, circles corresponding to θ_i, and to $-\theta_i$, are mirror images of each other about the u-axis ($v = 0$ straight line).

Fig. 9.12 shows the family of constant-m or θ_i, circles drawn on the Smith Chart. This family of circles and the constant-$|z|$ circles, drawn superposed on each other on the Smith Chart, are shown in Fig. 9.13. These appear somewhat like the latitudes and the longitudes of the earth although these are highly non-uniform.

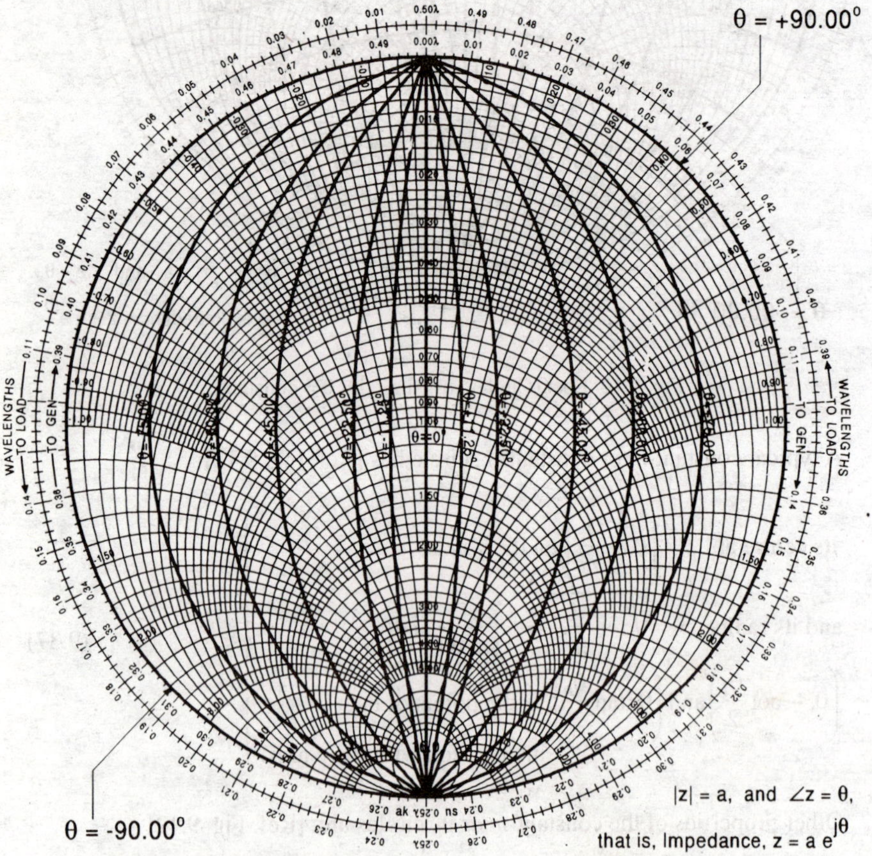

Fig. 9.13 Constant-a Circles superposed on Constant-θ Circles.

9.3.3 Rotation of the $g = 1$ Circle

In the graphical design of a Double-stub Impedance Matching System, discussed later in this book, one encounters the following problem:

In order that it may be possible to match a given load admittance, y_L, to a transmission line, the input admittance, y_2, of the line at the point, T_2, [Ref. Fig. 9.14], is required to be of the form,

$$y_2 = 1 + j\,b, \tag{9.38}$$

where, b may be positive, negative or zero. That is, the admittance, y_2, is required to lie somewhere on the $g = 1$ circle, C_2.

Now, if the locus of the input line admittance, y_2, is required to be the $g = 1$ circle, the question arises: what would be locus of the input line admittance, y_{in1}, at the point, T_1, which is at a distance of d from the point, T_2, towards the load [Ref. Fig. 9.14]? Note that the admittance, y_{in1}, is obtained from y_2, by moving along the transmission line a distance, d, from the point, T_2, towards the load. Remembering that a movement along the line towards the load, is equivalent, on the Smith Chart, to a movement along the corresponding constant-S circle in the counter clockwise direction, the input line admittance, y_{in1}, is obtained from the admittance, y_2, by moving through a distance, d, along the constant-S circle passing through the point, y_2, towards the load [Ref. Fig. 9.15]. For brevity, this operation may be described by stating that the point, y_2 is ***rotated*** counter-clockwise through the distance, d, or, symbolically,

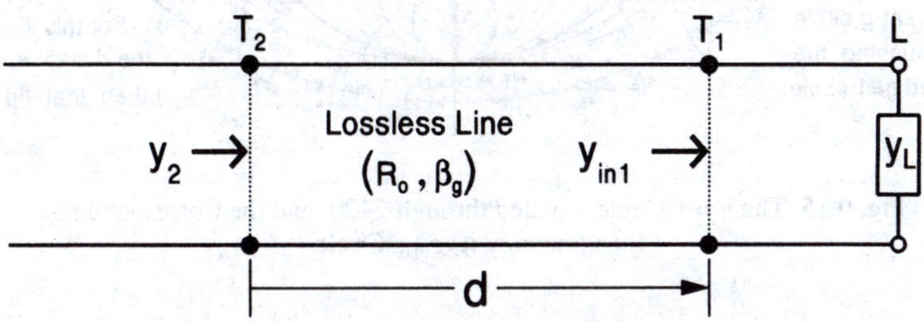

Fig. 9.14 Transformation of Admittance along a Lossless Transmission Line.

$$y_{in\,1} = \overset{d}{\curvearrowleft} y_2 \tag{9.39}$$

Clearly, therefore, the locus of y_{in1} is obtained by rotating counter-clockwise, every point, y_2, on the circle, C_2 (which is the locus of the admittance, y_2), through a distance, d, along the constant-S circle passing through that point. From geometry, it follows that the locus of y_{in1} would be the circle, C_1 [Ref. Fig. 9.15], which is none other than the $g = 1$ circle rotated through the distance, d, towards the load. In fact, not only would the circle C_2 map onto the circle C_1 but also the entire interior of the circle C_2 would map onto the interior of the circle C_1. Consequently, in the (u, v)-plane or on the Smith Chart, the centre and the radius of the circle C_1, are given by

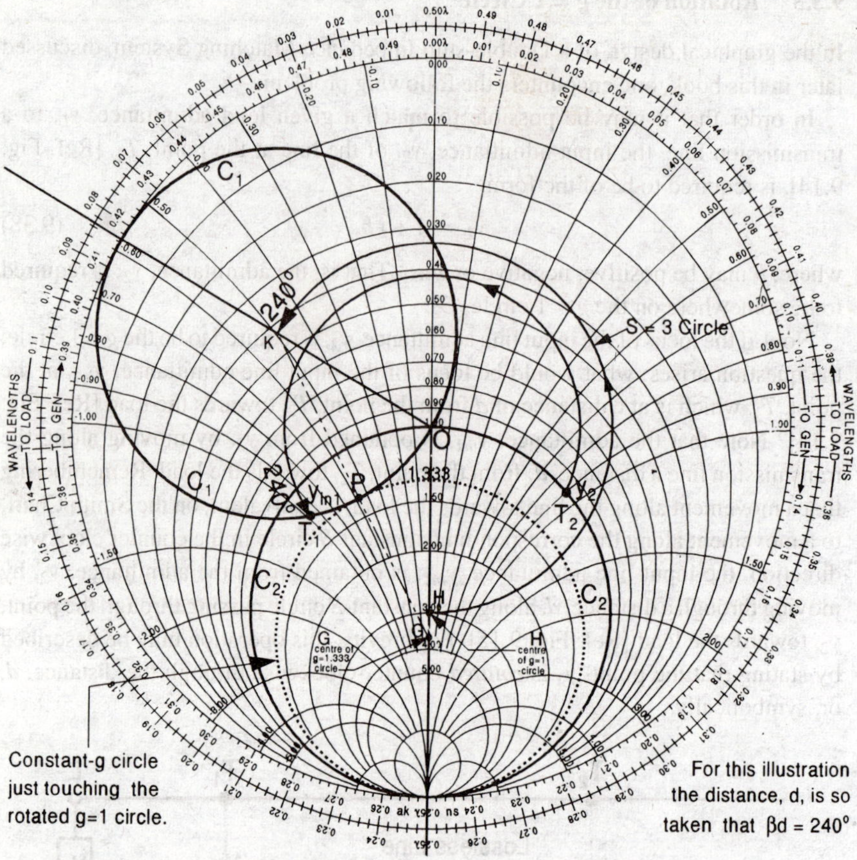

Fig. 9.15 The $g = 1$ Circle roatated through 2400° and the Corresponding (just touching it) g_{max}-Circle.

$$\left.\begin{aligned} \text{centre :} \quad & \left[\frac{1}{2}\cos\theta_o,\ \frac{1}{2}\sin\theta_o\right] \\ \text{and, radius} \quad & = \frac{1}{2} \\ \text{where,} \quad & \theta_o = 2\,\beta_g\, d \end{aligned}\right\} \tag{9.40}$$

Hence, the equation of the locus of the admittance, $y_{in\,1}$, is

$$\left.\begin{aligned} & \left(u - \frac{1}{2}\cos\theta_o\right)^2 + \left(v - \frac{1}{2}\sin\theta_o\right)^2 = \left(\frac{1}{2}\right)^2 \\ \text{or,} \quad & u^2 + v^2 = u\cos\theta_o + v\sin\theta_o \end{aligned}\right\} \tag{9.41}$$

In addition to the above, it is of interest to know the maximum value of the conductance (real part of y_{in1}) that lies on the rotated $g = 1$ circle. This is determined in what follows:

It may be recalled that a constant-g circle, given by [Ref. Eqns. (9.16) and (9.19)]

$$\left(u - \frac{g}{g+1}\right)^2 + v^2 = \left(\frac{1}{g+1}\right)^2, \tag{9.42}$$

has its centre at $\left[\frac{g}{g+1}, 0\right]$ and a radius of $\frac{1}{g+1}$.

A constant-g circle may, in general, intersect the rotated $g = 1$ circle in two points. An inspection of Fig. 9.15 shows that as g is increased the two points of intersection come closer to each other and, become coincident for $g = g_{max}$. If g is increased further, the corresponding constant-g circle will not intersect or meet the rotated $g = 1$ circle. The value of g_{max}, which would be a function of the angle θ_o, may be obtained by imposing the condition that the two circles just meet each other tangentially. This amounts to the condition that the distance between the centres of the two circles (the rotated $g = 1$ circle and the constant-g circle) equals the sum of their radii. Hence, the constant-g circle and the rotated $g = 1$ circle intersect each other in two points, meet in one point or do not intersect each other at all, according as the distance between the centres $\lesseqgtr$ the sum of the radii.

Now, the centres and radii of the two circles are

centres: $\left[\frac{g}{g+1}, 0\right]$ and $\left[\frac{1}{2}\cos\theta_o, \frac{1}{2}\sin\theta_o\right]$

and radii: $\frac{1}{g+1}$ and $\frac{1}{2}$

Application of the above condition, gives

$$\left(\frac{g}{g+1} - \frac{1}{2}\cos\theta_o\right)^2 + \left(\frac{1}{2}\sin\theta_o\right)^2 \lesseqgtr \left(\frac{1}{g+1} + \frac{1}{2}\right)^2$$

or,

$$\frac{g^2}{(g+1)^2} - \frac{g\cos\theta_o}{(g+1)} + \frac{1}{4} \lesseqgtr \frac{1}{(g+1)^2} + \frac{1}{(g+1)} + \frac{1}{4}$$

$$g \lesseqgtr \operatorname{cosec}^2\left(\frac{\theta_o}{2}\right) \tag{9.43}$$

Therefore, $g_{\max} = \operatorname{cosec}^2\left(\frac{\theta_o}{2}\right) = \operatorname{cosec}^2(\beta_g d)$ (9.44)

Hence, all constant-g circles corresponding to g value greater than g_{max} will not intersect the rotated $g = 1$ circle. Therefore, the maximum value of the conductance on the rotated $g = 1$ circle would be $\operatorname{cosec}^2\left(\frac{\theta_o}{2}\right) = \operatorname{cosec}^2(\beta_g d)$.

9.3.4 Renormalisation of the Rotated g = 1 Circle

As stated in the previous section, the $g = 1$ circle plays a pivotal role in the graphical design of a two-stub system for matching an arbitrary load admittance, y_L, to a lossless transmission line. Using the notation of the previous section, an interesting problem is described below :

Consider a typical input line admittance, y_{in1}, represented by the point, P_1, [Ref. Figs. 9.16 and 9.17], lying on the circle C_1, which is the $g = 1$ circle rotated through an angle, θ_o, towards the load or in the counter-clockwise direction, on the Smith Chart. Suppose the admittance, y_{in1}, is re-normalised with respect to the characteristic admittance of another lossless transmission line [Ref. Fig. 9.17], to yield the admittance, y_{ino}, given by

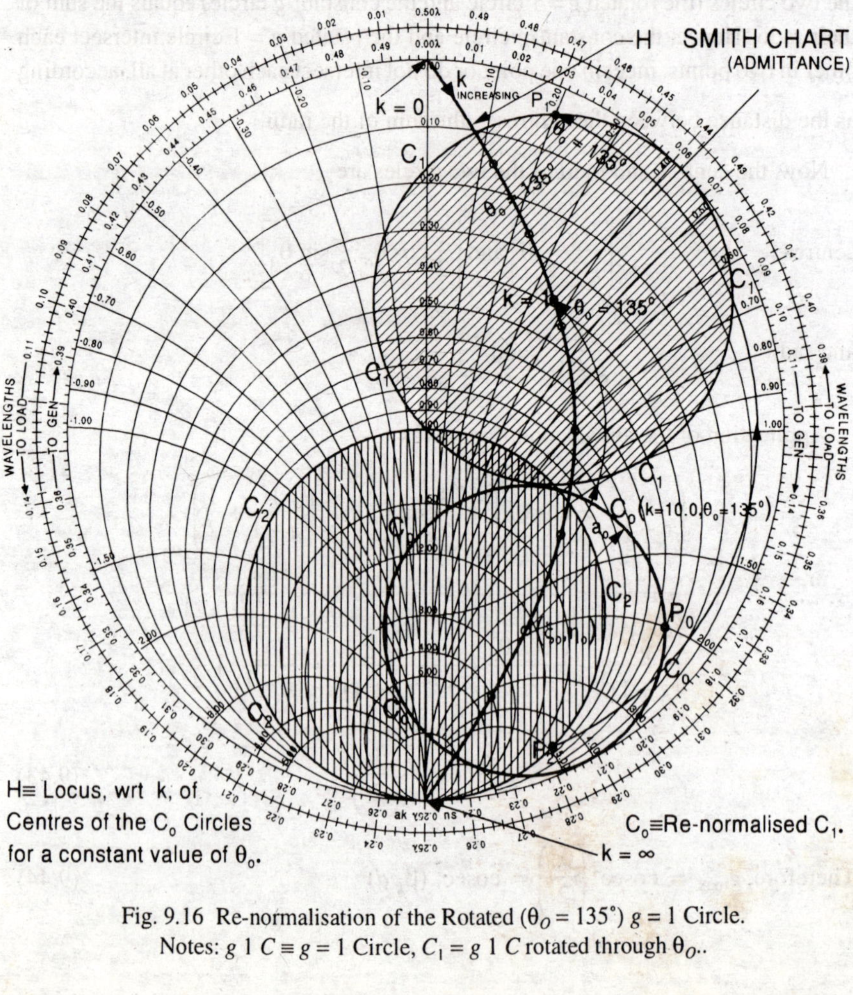

Fig. 9.16 Re-normalisation of the Rotated ($\theta_o = 135°$) $g = 1$ Circle.
Notes: $g\ 1\ C \equiv g = 1$ Circle, $C_1 = g\ 1\ C$ rotated through θ_o..

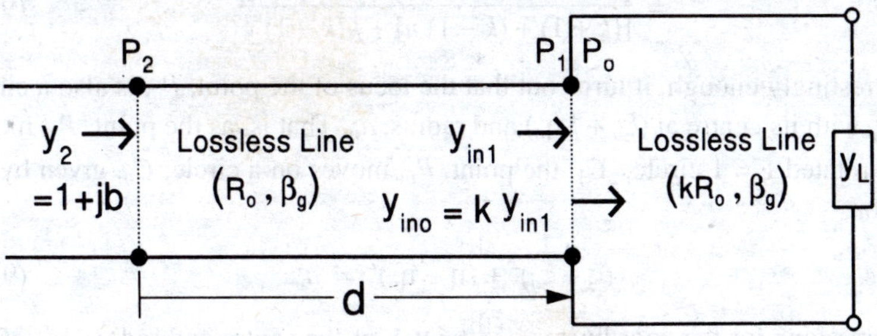

Fig. 9.17 Renormalisation of Admittance at the junction of two Lossless Lines.

$$y_{ino} = ky_{in1}, \text{ where, } k = \text{a real number} \geq 0. \tag{9.45}$$

Let the admittance, y_{ino}, be represented by the point P_o, on the Smith Chart. Then, the process of re-normalisation amounts to transforming the point, P_1, on the circle C_1, to the point, P_o, located somewhere on the Smith Chart, through the transformation given by Eqn. (9.45).

Now, if the re-normalisation of the admittance, y_{in1}, is carried out for all positions of the point P_1, on the circle, C_1, then corresponding to every position of the point P_1, the point, P_o, is obtained. The question that arises now is : What would be the locus of the point, P_o, on the Smith Chart ?

Let the current reflection coefficient corresponding to the admittance, y_{in1}, or the point, P_1, and that corresponding to the admittance, y_{ino}, or the point, P_o, be denoted by $(u + jv)$, and $(\xi + j\eta)$, respectively. Then, the problem may be restated in terms of the current reflection coefficients as follows : If the locus of $(u + jv)$, or of the point, P_1, is the circle, C_1 given by [Ref. Eqn. (9.41)]

$$u^2 + v^2 = u \cos \theta_o + v \sin \theta_o, \tag{9.46}$$

then, what is the locus of the point, $(\xi + j\eta)$, or of the point, P_o ? In what follows, the locus of the point, P_o, is derived.

Note that

$$u + jv = \frac{(y_{in1} - 1)}{(y_{in1} + 1)} \tag{9.47}$$

Or,

$$y_{in1} = \frac{(1 + u) + jv}{(1 - u) - jv}. \tag{9.48}$$

Now,

$$\xi + j\eta = \frac{(y_{ino} - 1)}{(y_{ino} + 1)}. \tag{9.49}$$

Substitution of Eqn. (9.45) into Eqn. (9.49), gives

$$\xi + j\eta = \frac{(ky_{in1} - 1)}{(ky_{in1} + 1)}$$

$$= \frac{[(k-1)+(k+1)\,u]+j\,(k+1)\,v}{[(k+1)+(k-1)\,u]+j\,(k-1)\,v} \tag{9.50}$$

Interestingly enough, it turns out that the locus of the point, P_o, is also a circle, say C_o, with its centre at $(\xi_o + j\,\eta_o)$ and radius, a_o. That is, as the point, P_1, moves on the rotated $g = 1$ circle, C_1, the point, P_o, moves on a circle, C_o, given by the equation,

$$(\xi-\xi_o)^2+(\eta-\eta_o)^2 = a_o^2 \tag{9.51}$$

Expressions for the coordinates, $(\xi_o + j\,\eta_o)$, of the centre and radius, a_o, of the circle, C_O, are derived below:

Subtraction of $(\xi_o + j\,\eta_o)$ from both sides of Eqn. (9.50), gives

$$(\xi-\xi_o)+j\,(\eta-\eta_o)$$

$$= \frac{\{[(k-1)-(k+1)\,\xi_o]+[(k+1)-(k-1)\,\xi_o]\,u+[(k-1)\,\eta_o]\,v\}}{\{[(k+1)]+[(k-1)]\,u+j\,[(k-1)]\,v\}}$$

$$-\,j\,\frac{\{[(k+1)\,\eta_o]+[(k-1)\,\eta_o]\,u-[(k+1)-(k-1)\,\xi_o]\,v\}}{\{[(k+1)]+[(k-1)]\,u+j\,[(k-1)]\,u\}} \tag{9.52}$$

Taking the squares of the magnitudes of both sides of the above equation, gives

$$(\xi-\xi_o)^2+(\eta-\eta_o)^2 = \frac{|\,\mathrm{Num}\,|^2}{|\,\mathrm{Den}\,|^2}, \tag{9.53}$$

where,

$$\begin{aligned}
|\,\mathrm{Num}\,|^2 &= \{[(k-1)-(k+1)\,\xi_o]+[(k+1)-(k-1)\,\xi_o]\,u+[(k-1)\,\eta_o]\,v\}^2 \\
&\quad+\{[(k+1)\,\eta_o]+[(k-1)\,\eta_o]\,u-[(k+1)-(k-1)\,\xi_o]\,v\}^2 \\
&= \{(k-1)^2-2\,(k^2-1)\,\xi_o+(k+1)^2\,(\xi_o^2+\eta_o^2)\} \\
&\quad+2\{(k^2-1)-2(k^2+1)\,\xi_o+(k^2-1)\,(\xi_o^2+\eta_o^2)\}\,u-[8\,k\,\eta_o]\,v \\
&\quad+\{(k+1)^2-2\,(k^2-1)\,\xi_o+(k-1)^2\,(\xi_o^2+\eta_o^2)\}\,(u^2+v^2),
\end{aligned} \tag{9.54}$$

and,

$$\begin{aligned}
|\,\mathrm{Den}\,|^2 &= [(k+1)+(k-1)\,u]^2+[(k-1)\,v]^2 \\
&= (k+1)^2+2\,(k^2-1)\,u+(k-1)^2\,(u^2+v^2)
\end{aligned} \tag{9.55}$$

By substituting $(u\cos\theta_o + v\sin\theta_o)$ for (u^2+v^2) from Eqn. (9.46) into Eqns. (9.54) and (9.55) and by simplifying the resulting equations, it may be shown that the expressions for $|\,\mathrm{Num}\,|^2$ and $|\,\mathrm{Den}\,|^2$ may be written as

$$|\,\mathrm{Num}\,|^2 = a_1+a_2\,u+a_3\,v, \quad \text{and}$$

$$|\,\mathrm{Den}\,|^2 = b_1+b_2\,u+b_3\,v, \tag{9.56}$$

where

$$\left.\begin{aligned} a_1 &= (k-1)^2 - 2\,(k^2-1)\,\xi_o + (k+1)^2\,(\xi_o^2+\eta_o^2) \\ a_2 &= [(k+1)^2 - 2\,(k^2-1)\,\xi_o + (k-1)^2\,(\xi_o^2+\eta_o^2)]\cos\theta_o \\ &\quad + 2[(k^2-1) - 2\,(k^2+1)\,\xi_o + (k^2-1)\,(\xi_o^2+\eta_o^2)] \\ a_3 &= [(k+1)^2 - 2\,(k^2-1)\,\xi_o + (k-1)^2\,(\xi_o^2+\eta_o^2)]\sin\theta_o - 8\,k\,\eta_o, \\ \text{and,}& \\ b_1 &= (k+1)^2 \\ b_2 &= (k-1)^2\cos\theta_o + 2\,(k^2-1) \\ b_3 &= (k-1)^2\sin\theta_o \end{aligned}\right\} \tag{9.19}$$

Substitution of Eqns. (9.56) into Eqn. (9.53) gives

$$(\xi-\xi_o)^2 + (\eta-\eta_o)^2 = \frac{a_1 + a_2\,u + a_3\,v}{b_1 + b_2\,u + b_3\,v} = a_o^2 = \text{a constant.} \tag{9.58}$$

if Eqn. (9.58) is required to hold good for all values of u and v, it follows that

$$\frac{a_1}{b_1} = \frac{a_2}{b_2} = \frac{a_3}{b_3} = a_o^2 = \text{a constant.} \tag{9.59}$$

These conditions may equivalently be stated as

$$\left.\begin{aligned} a_3\,b_1 &= a_1\,b_3 \\ a_2\,b_3 &= a_3\,b_2 \\ a_1\,b_2 &= a_2\,b_1 \end{aligned}\right\} \tag{9.60}$$

Substitution of the expressions for a's and b's from the Eqns. (9.57) into the above Eqns. (9.60) gives, on simplification, the relations which the coordinates, (ξ_o, η_o), of the centre of the circle, C_o, satisfy, to be

$$\left.\begin{aligned} [(k^2-1)\sin\theta_o]\,\xi_o + [(k+1)^2\,\eta_0 - [(k^2+1)\sin\theta_o] &= 0, \\ [(k-1)\sin\theta_o]\,\xi_o + [(k-1)\cos\theta_o + 2\,(k+1)\,\eta_o - [(k+1)\sin\theta_o &= 0, \end{aligned}\right\} \tag{9.61}$$

$$\text{and, } [(k^2-1)\cos\theta_o + (k+1)^2]\,\xi_o - [(k^2+1)\cos\theta_o + (k^2-1) = 0. \tag{9.62}$$

which gives

$$\xi_o = \frac{(k^2-1)\cos\theta_o + (k^2-1)}{(k^2-1)\cos\theta_o + (k+1)^2} \tag{9.63}$$

Solving the simultaneous Eqns. (9.62) for the unknowns, ξ_o and η_o, yields

$$\left.\begin{aligned} \xi_o &= \frac{(k^2+1)\cos\theta_o + (k^2-1)}{(k^2-1)\cos\theta_o + (k+1)^2}, \\ \eta_o &= \frac{2\,k\sin\theta_o}{(k^2-1)\cos\theta_o + (k+1)^2}. \end{aligned}\right\} \tag{9.64}$$

Note that the same expression for ξ_o as given in Eqn. (9.64), is obtained from Eqn. (9.62) [Ref. Eqn. (9.63)].

Now, from Eqn. (9.58), the radius, a_o , of the circle, C_o , is given by

$$a_o = \sqrt{\frac{a_1}{b_1}}. \tag{9.65}$$

By substituting the expressions for a_1 and b_1 from Eqn. (9.57), into the above equation and the expressions for ξ_o and η_o from Eqns. (9.64), the expression for the radius of the circle, C_o, is obtained to be

$$a_o = \frac{2k}{(k^2-1)\cos\theta_o + (k+1)^2}. \tag{9.66}$$

Hence, it may be concluded that as the point, P_1, [Ref. Fig. 9.16], moves on the rotated $g = 1$ circle, C_1, and if the parameter, k, is held constant (= 10 in Fig. 9.16), then the corresponding point, P_o, also moves on a circle, C_o, whose centre, (ξ_o , η_o), and radius, a_o, are given by Eqns. (9.64) and (9.66), respectively. Shown in Fig. 9.18 are the C_o circles that are generated by the re-normalisation of the C_1 circle rotated through 135° towards the load, by using 7 different re-normalisation factors, k. Also drawn in Fig. 9.18 is the locus with respect to k for a constant value of the angle, θ_o, Ref. Sec. 9.3.4.1 (2) below, of the centres of the C_o circles. Drawn in Fig. 9.19 are the C_o circles that are generated by the re-normalisation of the C_1 circle rotated towards the load, through various angles θ_o, but using a constant value of the re-normalisation factor, k. Also, drawn in Fig. 9.19 is the locus with respect to the angle, θ_o, but for a constant value of re-normalisation factor, k, Ref. Sec. 9.3.4.1.(3) below, of the centres of the C_o circles.

9.3.4.1 Properties of the Circles, *Co* :

1. All Co-Circles meet the Chart periphery tangentially

This may be proved easily as given below :

Squaring both sides of Eqns. (9.64) and adding, and then taking square-roots of both sides of the resulting equation, give

$$\sqrt{\xi_o^2 + \eta_o^2} = \frac{\sqrt{[(k^2+1)\cos\theta_o + (k^2-1)]^2 + 4k^2\sin^2\theta_o}}{[(k^2-1)\cos\theta_o + (k+1)^2]} \tag{9.67}$$

On simplification, the expression under the radical sign turns out to be a perfect square yielding, therefore,

$$\sqrt{\xi_o^2 + \eta_o^2} = \frac{(k^2-1)\cos\theta_o + (k^2+1)}{(k^2-1)\cos\theta_o \cdot (k+1)^2} \tag{9.68}$$

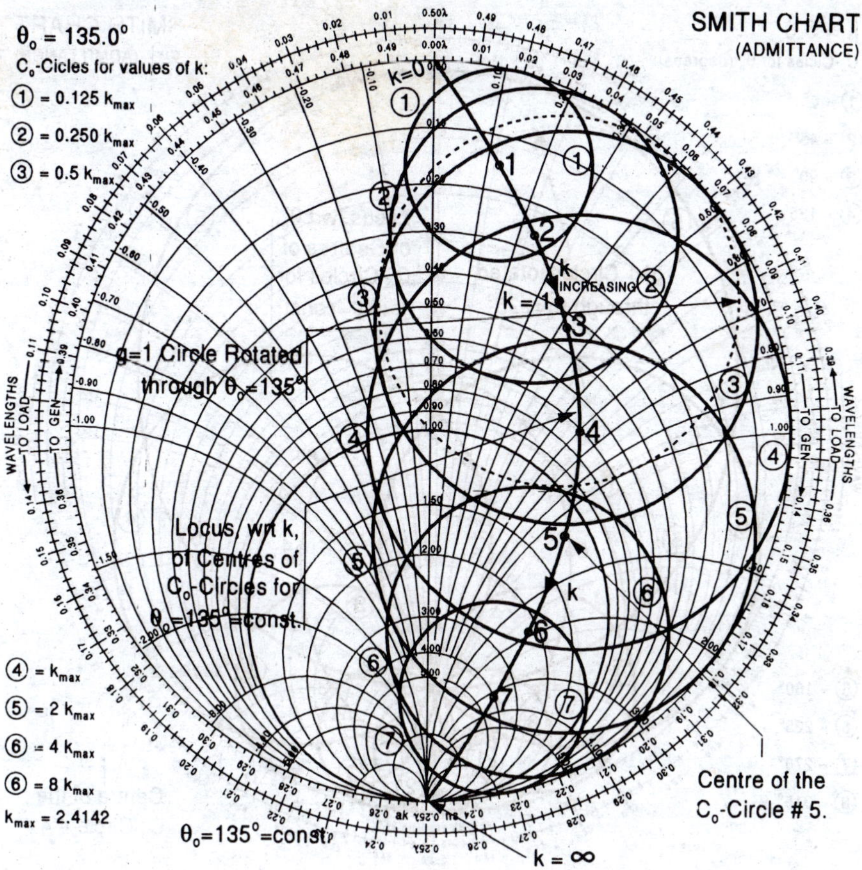

Fig. 9.18 C_0-Circles [Re-normalised rotated (θ_O = 135°) g = 1 circle] for various values of k but for a fixed value of θ_O (= 135°).

Now, addition of Eqns. (9.68) and (9.66) gives

$$\sqrt{\xi_o^2 + \eta_o^2} + a_o \;=\; 1, \text{ for all values of } k \text{ and } \theta_o. \qquad (9.69)$$

Eqn. (9.69) implies that all C_o (k, θ_o)-circles (that is, for all values of k and θ_o) meet the periphery of the Smith Chart [that is, the (g = 0)-circle] tangentially.

2. *Locus of the Centre with respect to k*

It may be noted from Eqns. (9.64) and (9.66) that for all values of the angle, θ_o, if

1. $k = 0$, then $\xi_o = -1, \eta_o = a_o = 0$.

   that is, the circle, C_o, degenerates to the point, (–1, 0),

2. $k = \infty$, then $\xi_o = 1, \eta_o = a_o = 0$.

   that is, the circle, C_o, degenerates to the point, (1, 0).

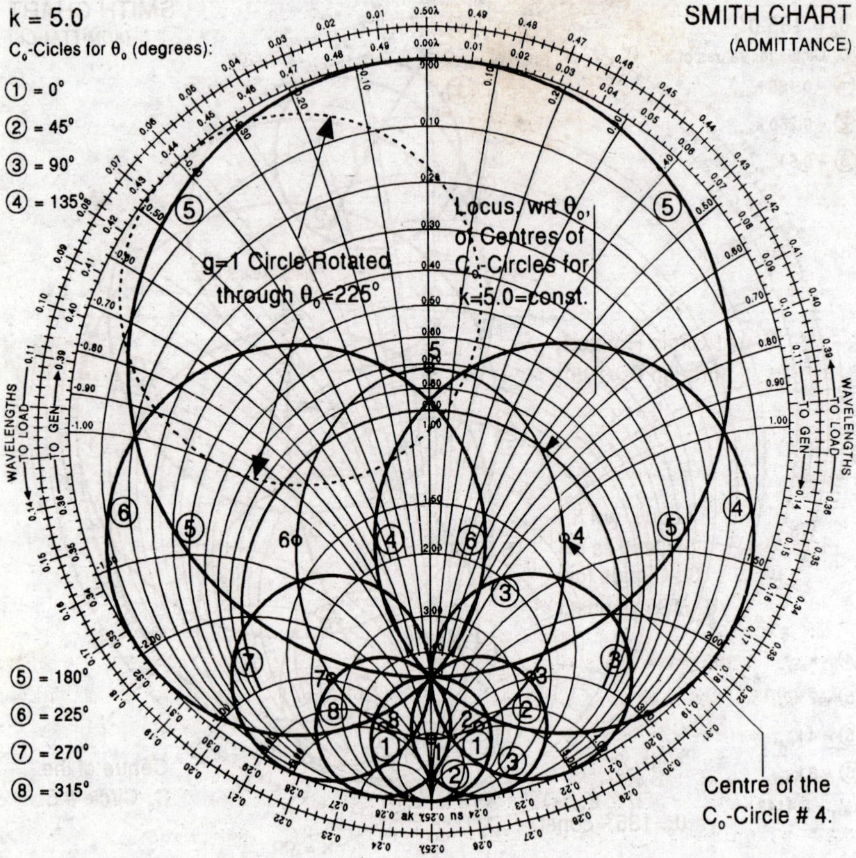

Fig. 9.19 C_0-Circles [Re-normalised (using $k = 5$) rotated $g = 1$ circle] for various values of θ_O but for a fixed value of k (= 5).

3. For a given value of the angle θ_o, of the circle, C_o, attains its maximum value when

$$\left.\begin{aligned} k &= k_{\max} = \sqrt{\frac{1-\cos\theta_o}{1+\cos\theta_o}} \\ \text{Maximum value of} \quad a_o &= a_{o\max} = \frac{1}{1+|\sin\theta_o|} \\ \text{also, for} \quad k = k_{\max,}\ \xi_o(k_{\max}) &= 0, \\ \text{and,} \quad \eta_o(k_{\max}) &= a_{o\,\max}\sin\theta_o \end{aligned}\right\} \tag{9.70}$$

It would be interesting to determine the locus of the centre of the circle, C_o, as k takes up various values in its range, $0 \le k < \infty$. Mathematically, it amounts to obtaining the k-eliminant of Eqns. (9.61) that the coordinates, ξ_o and η_o, satisfy.

Dividing the first and the second of Eqns. (9.61) throughout by $(k+1)^2$ and $(k+1)$, respectively, and substituting

$$x = \frac{k-1}{k+1} \tag{9.71}$$

in the resulting equations, yield the following equations in x :

$$(x \sin \theta_o)\, \xi_o + \eta_o - \frac{1}{2}(x^2+1) \sin \theta_o = 0, \tag{9.72}$$

$$(x \sin \theta_o)\, \xi_o + (x \cos \theta_o + 2)\, \eta_o - \sin \theta_o = 0. \tag{9.73}$$

The x-eliminant of these equations, which is nothing but the locus of the centre with respect to k, may be shown to be

$$\xi_o^2 \sin^2 \theta_o - \eta_o^2 \cos^2 \theta_o + 2\, \eta_o \sin \theta_o - \sin^2 \theta_o = 0, \tag{9.74}$$

or,
$$\xi_o^2 \sin^2 \theta_o - (\eta_o \cos \theta_o - \tan \theta_o)^2 + (\tan^2 \theta_o - \sin^2 \theta_o) = 0. \tag{9.75}$$

This equation may be rewritten as

$$\frac{(\eta_o - \tan \theta_o \sec \theta_o)^2}{\tan^4 \theta_o} - \frac{\xi_o^2}{\tan^2 \theta_o} = 1. \tag{9.76}$$

This equation represents a family of hyperbolas with θ_o as the running parameter, and the η_o-axis as their axis. The following observations may be made :

1. Hyperbolas corresponding to all values of θ_o pass through the point, $(-1, 0)$ for $k = 0$; and through the point, $(1, 0)$ for $k = \infty$.
2. Hyperbolas corresponding to $0° < \theta_o < 180°$ lie in the half-plane, $\eta_o > 0$, and those corresponding to $180° < \theta_o < 360°$ lie in the half-plane, $\eta_o < 0$.
3. For $\theta_o = 0°$ or $180°$, the hyperbola degenerates to the $\xi_o = 0$ or the η_o -axis itself.
4. A typical hyperbola corresponding to θ_o intersects the η_o-axis at the point, or, the vertex of the hyperbola, is given by

$$\eta_v = \left(0, \frac{\sin \theta_o}{1 + |\sin \theta_o|}\right). \tag{9.77}$$

3. Locus of the Centre with respect to θ_o

The locus of the centre, (ξ_o, η_o), of the C_o-circle as θ_o takes up various values in its range, $0 \le \theta_o < 2\pi$, the normalising factor, k, being held constant, is the θ_o-eliminant of Eqns. (9.61) that the coordinates, ξ_o and η_o, satisfy.

From Eqn. (9.63), the expression for $\cos \theta_o$ may be obtained, giving

$$\cos \theta_o = \frac{-(k+1)^2\, \xi_o + (k^2-1)}{(k^2-1)\, \xi_o - (k^2+1)} \tag{9.78}$$

Now, from the second of Eqns. (9.64), the following expression for $\sin\theta_o$ may be obtained by substituting the expression for $\cos\theta_o$ and simplifying :

$$\sin\theta_o = \frac{-(k+1)^2\,\eta_o}{(k^2-1)\,\xi_o - (k^2+1)} \tag{9.79}$$

Substituting the expressions for $\cos\theta_o$ and $\sin\theta_o$ from Eqns. (9.78) and (9.79) into the identity,

$$\cos^2\theta_o + \sin^2\theta_o = 1, \tag{9.80}$$

the following θ_o-eliminant may be obtained after a bit of simplification :

$$\frac{\left[\xi_o + \frac{1}{2}\left(\frac{k-1}{k+1}\right)\right]^2}{\left(\frac{1}{2}\right)^2} + \frac{\eta_o^2}{\left(\frac{\sqrt{k}}{k+1}\right)^2} = 1. \tag{9.81}$$

This represents a family of ellipses having,

$$\left.\begin{array}{ll} \text{Centre at} & \left[-\frac{1}{2}\left(\frac{k-1}{k+1}\right), 0\right] \\ a = \frac{1}{2} & \text{for all } k \text{ along the } \xi_o\text{-axis} \\ b = \frac{\sqrt{k}}{k+1} & \text{along the } \eta_o\text{-axis} \end{array}\right\} \tag{9.82}$$

The following observations on the loci with respect to θ_o for fixed k, may be made : [Ref. Fig. 9.18]

1. If k is changed to $\frac{1}{k}$ and ξ_o to $-\xi_o$, Eqn. (9.81) remains unchanged. This means that the ellipses corespondenting to k and $\frac{1}{k}$ are mirror images of each other with respect to the line, $\xi_o = 0$ (or the η_o-axis).
2. Semi-major axis of all ellipses (for all values of k is $\frac{1}{2}$. But, the semi - minor axis, starting from its value zero for $k = 0$, increases as k increases, reaches its maximum values of $\frac{1}{2}$ for $k = 1$ and then it again decreases to zero as k continues to increase to ∞.
3. The ellipse (9.81) degenerates to the point, $(-1, 0)$ for $k = 0$ and to the point, $(1, 0)$ for $k \to \infty$. When $k = 1$, the ellipse reduces to a circle of radius $\frac{1}{2}$ and having its centre at the origin $(0, 0)$.

4. The centre of the ellipse lies on the upper vertical radius $[-0.5 < \xi_o < 0, \eta_o = 0]$ of the Smith Chart $k < 1$ and on the lower vertical

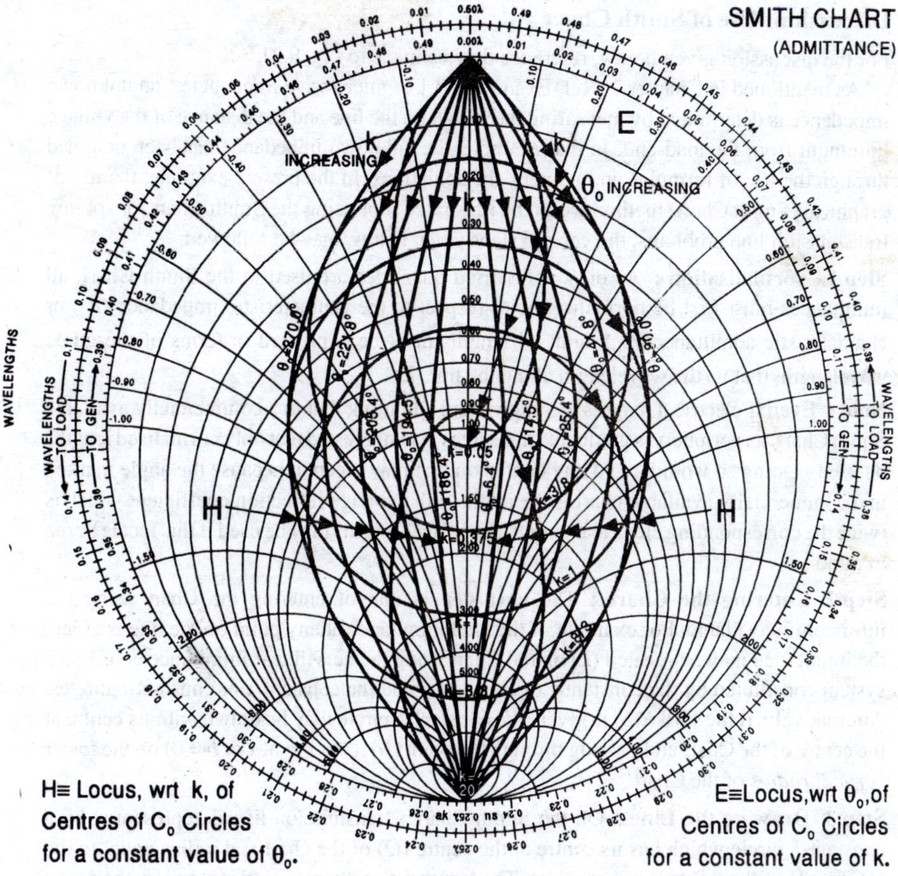

Fig. 9.20 Loci of Centres of C_0-Circles with respect to (1) k for constant θ_O, and (2) θ_O for constant k.

radius $[0 < \xi_o < 0.5, \eta_o = 0]$ and radius equal to $\frac{1}{2}$.

Loci of the centre of the circle, $C_o(k, \theta_o)$, with respect to k for several fixed values of the angle, θ_o, as well as those with respect to θ_o for several fixed values of the re-normalisation constant, k, are plotted on the Smith Chart in Fig. 9.20.

SOLVED EXAMPLES

Example 9.1 The *VSWR* in a 50 Ω coaxial line is found to be 2.6 and the first voltage minimum in the line is found to occur at a distance of 0.1021 λ_g from the load towards the generator. Use Smith chart to find out both the impedance and admittance of the load. Also, determine the *VSWR* on the line and the voltage reflection coefficient at the load-end.

Solution :

Discussion: Use of Smith Chart

For the discussion given below, reference may be made to Fig. 9.21.

As mentioned in Chapter 7 [Ref. Example 7.3.], at microwave frequencies, an unknown impedance is determined by measuring the *VSWR* on the line and the distance of the voltage minimum from the load-end. In that example, the unknown impedance has been obtained through the use of formulae and numerical calculations. In the present example, the use of graphical (Smith Chart) method would be illustrated. For using the Smith Chart for solving transmission line problems, the general steps given below, may be followed:

Step 1. Normalisation : As only normalised quantities are used in the Smith Chart, all immitances must first be normalised with respect to the characteristic impedance (R_o) or characteristic admittance (G_o); and, all lengths must be expressed in terms of the guide wavelength, that is, the wavelength along the line, λ_g .

Note : Even if electrical lengths are known and admitting that in a commercially available Smith Chart, its periphery is graduated in degrees, it is suggested that only normalised lengths be used. Doing so would avoid any confusion that may creep in because the angle marked in a commercially available chart is the angle of the voltage reflection coefficient which is twice the corresponding electrical length. (Electrical lengths may be used if this fact is borne in mind.)

Step 2. Entering the Chart : The next step is that of entering the Chart using the information available. For example, if the load immitance at any point on the line is given, the immitance may be located (*L*) on the Chart using the curvilinear immitance coordinate system constituted by the constant-*r* or constant-*g* and the constant-*x* or constant-*b* circles. Alternatively, if the *VSWR(S)* is given, the constant-*S* circle may be drawn with its centre at the centre of the Chart and passing through the point (*N*), (r or $g = S$, x or $b = 0$) on the *lower vertical radius* of the Chart.

Step 3. Drawing the Line : On the Smith Chart a transmission line is represented by a constant-*S* circle which has its centre at the centre (*O*) of the chart and radius equal to the magnitude of the reflection coefficient. The fact that a constant-*S* circle intersects the *lower vertical radius* of the chart at the point, $r = S$ or $g = S$, helps in drawing the circle in practice. This procedure for drawing the line can be used, as given in Step 2, if the *V S W R* is known. However, if the line immitance at the load-end (load immitance) or at any other point on the line is given, the line may then be drawn by drawing a circle with its centre at the centre (*O*) of the chart and the radius being so adjusted that the circle passes through the known immitance point (*L*) marked in Step 2 above.

Note : Voltage Maxima and Minima : If the Smith Chart is used as an impedance chart, the point of intersection of the constant-*S* circle with the *lower vertical radius* of chart, represents Voltage Maxima on the line and the intersection with the *upper vertical radius* gives the Voltage Minima. And, if the Smith Chart is used as an admittance chart, the point (*N*) of intersection of the constant-*S* circle with the *lower vertical radius* of the chart, represents Voltage Minima on the line and the intersection (*X*) with the *upper vertical radius* gives the Voltage Maxima.

Step 4. Location of Points : The location of a point, say *L*, lying on a constant-*S* circle, may be obtained by drawing the radius vector ($\overline{OLL_o}$) passing through that point (*L*) and reading the distance (0.3521 λ_g, in the present case) off the circular, uniformly graduated distance scale at the periphery of the chart. In the same way, the locations (0.50 λ_g and 0.25 λ_g) of the voltage maximum (X) and minimum (N) may be read off the same circular, uniformly graduated distance scale at the periphery of the chart. Then, the location of the point (L) may be specified in terms of its distance from the voltage maximum or minimum. In the present example, L is located at a distance of $(0.3521 - 0.25)\ \lambda_g = 0.1021\ \lambda_g$ from the voltage minimum *towards the load* or, equivalently, $(0.50 - 0.3521)\ \lambda_g = 0.1478\ \lambda_g$ from the voltage maximum *towards the generator.*

After these general steps, the procedure to be followed would depend on what is required and what data is available. This is illustrated in the solved examples that follow.

Given data: *VSWR*, $S = 2.6$, $R_o = 50\ \Omega$, therefore, $G_o = 20$ m Ω.

First Voltage Minimum located 0.102 λ_g from the load-end *towards the generator.*

Required: Load Admittance, Load Impedance and Voltage Reflection Coefficient.

Draw the transmission line on the Smith Chart by drawing a circle with its centre at O and radius $= ON$, where N is the point ($g = 2.6$, $b = 0$). By doing so, the line gets represented on the chart and the voltage maxima and minima also get located on the chart automatically. As the chart is being used as an admittance chart, the intersection (N) of the constant-S circle with the *lower vertical radius* of the chart gives the minima and the intersection (X) of the constant-S circle with the *upper vertical radius* of the chart gives the maxima.

Now to determine the Load Admittance, the best way to find out what to do, is to ask oneself the question: What would one do on an actual or physical line, to find out the load admittance. The answer would be *go from the voltage minimum along the line to the load-end, that is, go along the line through a distance of* 0.1021 λ_g *towards the load, and then measure the line admittance there.* All that is now required to determine the load admittance using the Smith Chart is to translate or transform the actions shown italicised above, to Smith Chart language, as given below.

Starting from the point, N, move along the constant-S circle through a distance of 0.1021 λ_g *towards the load, and then read the admittance off the curvilinear (g, b) coordinates.* This is done in the following way:

Locate the point, L_o, on the distance scale at a distance of 0.1021 λ_g from N_o, which is the point of intersection of the radius vector, $(\overline{ON})$, through the minimum (N) and the distance scale. Join the centre (O) to the point, L_o. The intersection of the straight line, $\overline{OL_o}$, with the S-circle gives the load point, L. Therefore, the normalised load admittance $= 0.85 + j0.90$, and, the load admittance $(= 0.85 + j\,0.90) \times 20$ m mhos $= (17 + j\,18)$ m mhos.

Note that diametrically opposite points on an S-circle, correspond to two points separated by a distance of one quarter wavelength on the physical line. In view of the immitance inversion property of a quarterwave line, the normalised immitances at these two diametrically opposite points are reciprocals of each other.

Applying it to the present case, the admittance at the point, M, which is diametrically opposite to the load point, L, is equal to the reciprocal of y_L or z_L.

Reading off the coordinates of the point, M, gives the load impedance $z_L = 0.55 - j\,0.59$.

Hence, $Z_L = (0.55 - j\,0.59) \times 50\ \Omega = (27.5 - j\,29.5)\ \Omega$

Voltage Reflection Coefficient at the load-end:

$$|\rho_L| = \frac{\text{the distance, } \overline{OL}}{\text{the radius of the (S } = \infty\text{)-circle}} = 0.444.$$

Angle of the voltage reflection coefficient at the voltage minimum, N, is $-180°$.

Therefore, the angle of the voltage reflection coefficient at the load-end, $\phi_L = -180^\circ + 2\beta_g l_{min}$

or, $\phi = -180^\circ + 720^\circ \times 0.1021 = -180^\circ + 73.512^\circ = -106.488^\circ$.

Example 9.2 When a load is connected at the load-end of a 50 Ω coaxial line, a *VSWR* of 2.5 is observed on the line and the position of the first voltage minimum is noted. If the load is now replaced by a load of $(10 + j0)$ Ω, the first minimum is found to shift through a distance of 0.48 λ_g towards the load. Calculate the load impedance. If an inductance having a reactance of 49.13 Ω is shunted across the load terminals, using the Smith Chart, find out the *VSWR* produced.

Solution : Before proceeding with the solution of this problem, the following comments on the experimental measurement of an unknown impedance may be made.

Discussion: Experimental Measurement of Impedances at Microwave Frequencies : For the discussion given below, reference may be made to Fig. 9.21 and Fig. 9.22.

Experimental measurement of an unknown load impedance at microwave frequencies, that is, by employing transmission line techniques, has already been discussed in Chapter 7 [Ref. Example 7.3.] and under Example 9.1 in this chapter also. However, one more point that remains to be explained, is discussed here.

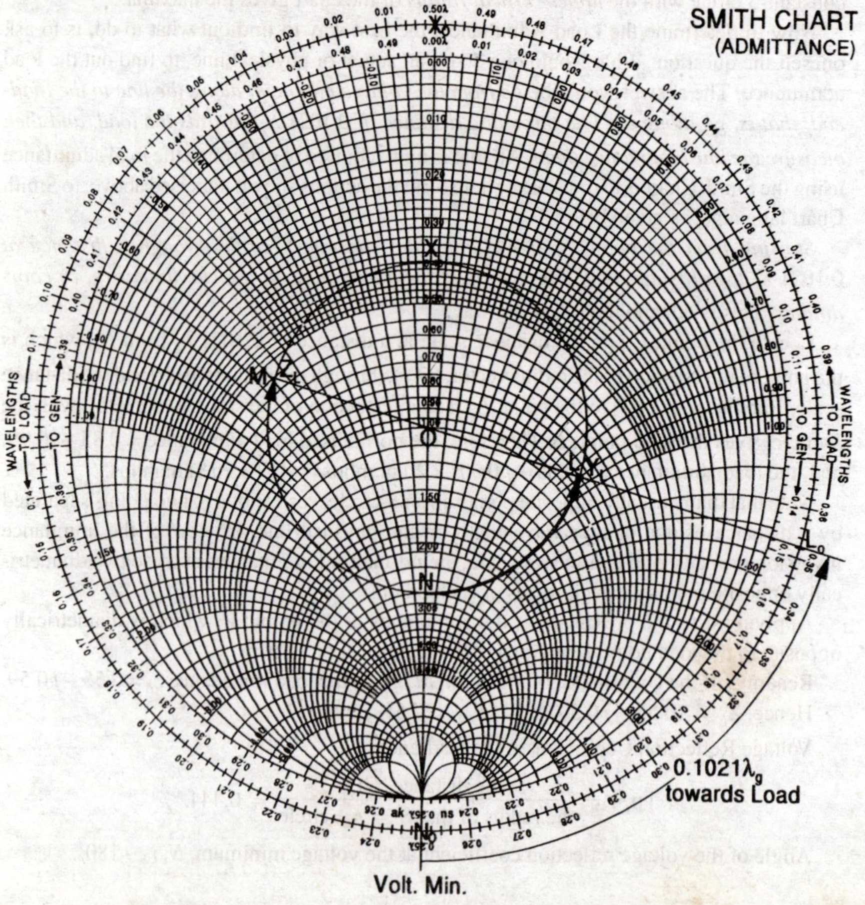

Fig. 9.21 Determination of Z_L. Given: VSWR and l_{min}.

At microwave frequencies, the distance of the first voltage minimum when the unknown load is connected, from the load-end, cannot be measured accurately because the load-end is usually not accessible. Even if the *physical* load-end is accessible, at these frequencies, the *effective or electrical load*-end is not precisely known. To circumvent this problem, a short circuit or a pure resistance of value less than the characteristic Impedance of the line, is connected at the *physical* load-end and the *VSWR* pattern is observed. The first voltage minimum in this case would occur at the *effective or electrical* load-end itself and therefore voltage minima would be observed every half-wavelength distance along the line. In view of the periodicity of the line impedance along the length of the line, with a period of a half wavelength, the unknown impedance may be determined if the distance from the voltage minimum (with the unknown load connected) to the load-end or to **any voltage minimum** that occurs when a short-circuit (or any resistance less than the characteristic impedance of the line) is connected at the *physical* load-end of the line.

In practically all experimental setups, even the *physical load*-end of the line is inaccessible. This situation is depicted in Fig. 9.22, wherein the region between the load-end and the thick dotted line is inaccessible. Summarising, therefore, the experimental procedure for determining an unknown impedance is to connect the unknown load at the load-end of the line and observe the *VSWR* produced in the line and also note down the position of the first **observable** voltage minimum. Then the unknown impedance is replaced by a short-circuit and the *VSWR* pattern is again observed and the position of the first minimum is again noted. Depending on the unknown impedance, the first minimum may shift either towards the load-

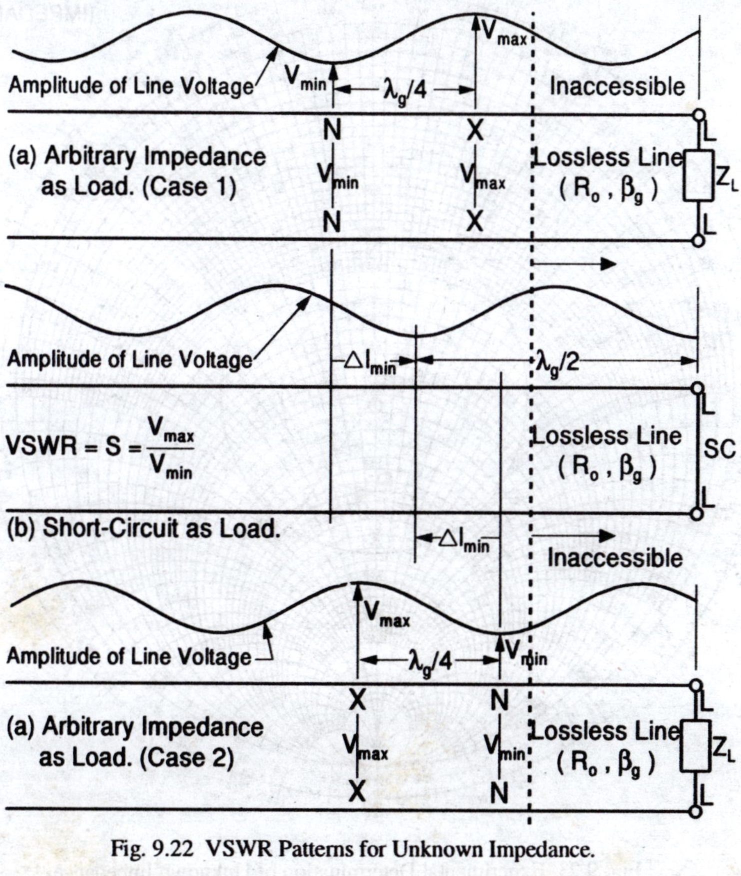

Fig. 9.22 VSWR Patterns for Unknown Impedance.

end or towards the generator-end, as is shown in Fig. 9.22 and Fig. 9.23. Whatever be the direction of movement of the voltage minimum, all that one has to remember to determine the unknown immitance is that one has to go from the position of the voltage minimum with the unknown immitance as the load to the position of the voltage minimum with the short-circuit as the load. **[From the unknown to the known (short-circuit) makes the unknown known** may serve as a mnemonic!]

Given data:

VSWR, $S = 2.5$, $\Delta l_{min} = 0.048\,\lambda_g$ *towards the load.*

Use the Smith Chart as in Impedance Chart.

Step 1. Draw the line, that is, the constant-*S* circle having its centre at *O* and passing through the impedance point, $2.5 + j\,0$.

Step 2. Starting from the voltage minimum, N_o, move along the outermost circular distance scale, through a distance of $0.048\,\lambda_g$ and mark the point, L_o on it. Join the point, L_o, to the centre *O*, by a straight line intersecting the *S*-circle in the point, *L*, representing the load point. Read off its coordinates, (0.43, – 0.26), on the curvilinear (*r, x*) coordinate system.

Therefore, $z_L = 0.43 - j\,0.26$ or $Z_L = (0.43 - j\,0.26) \times 50\,\Omega = (21.5 - j\,13.0)\,\Omega$

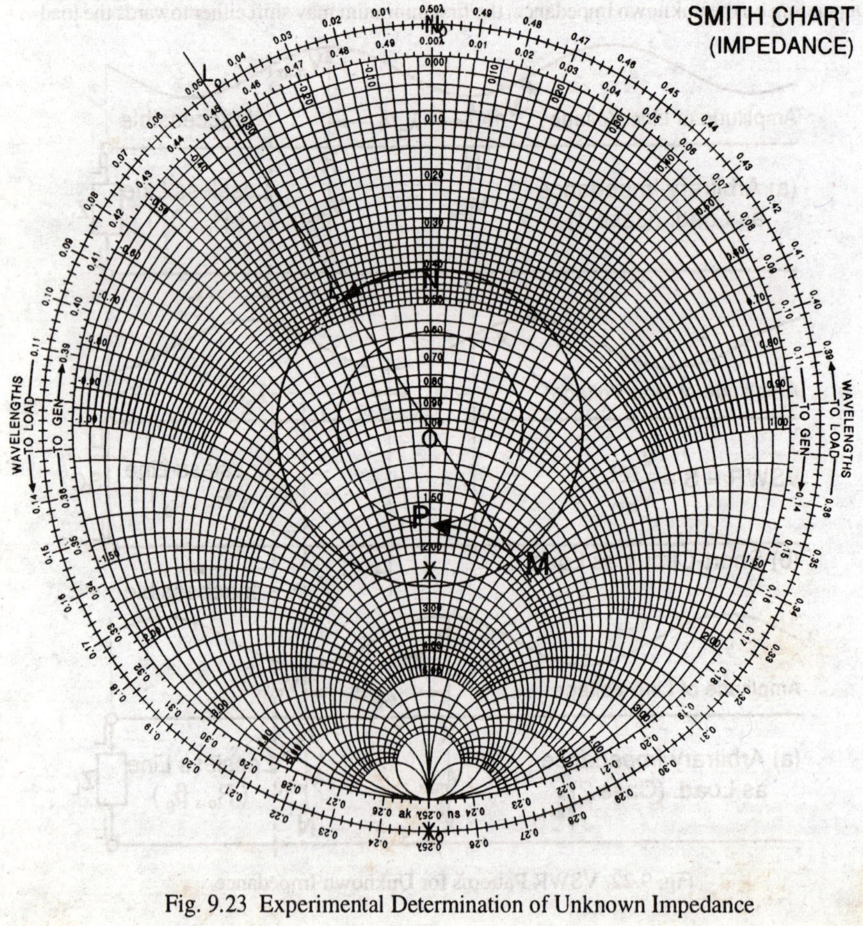

Fig. 9.23 Experimental Determination of Unknown Impedance

Step 3. From the point, L, go through the centre, O, to the diametrically opposite point, M, representing the normalised load admittance. Read off its coordinates on the curvilinear (g, b) coordinate system, giving

$$y_L = 17.1 + j\,1.02 \text{ or } Y_L = (1.71 + j\,1.02) + 20 \text{ m mhos} = (34.2 + j\,20.4) \text{ m mhos}$$

Step 4. To the load admittance, add the susceptance of the inductance (reactance = $j\,49.13\ \Omega$, normalised susceptance $= -\dfrac{50}{49.13} = -j\,1.02$) to get the total admittance at the load-end = $1.71 + j0.00$. Graphically, move along the constant-g circle from the point, M, until a susceptance of -1.02 is added, and thus arrive at the point, P. [Ref. Fig. 9.23]

Step 5. Draw the constant-S circle through the point, P. Read off its interaction with the *lower vertical radius* ($S = 1.71$, in the present problem) to yield the value of the *VSWR*, S on the line.

EXERCISES

1. What are the disadvantages of a Bicircular Impedance Chart ?
2. Every point on the Smith Chart represents a normalised impedance $z = r + jx$. Show that the locus of the point z which moves on the Smith Chart such that $|z| = a$ = constant, is a circle. Derive the equation to the family of such circles. Discuss the variation in the locus as the magnitude of the impedance, a, is varied.
3. Show that the locus of the point $z = r + jx$ which moves on the Smith Chart so as to keep the angle of the impedance, $\theta = \tan^{-1}\left(\dfrac{x}{r}\right)$, constant, is a circle. Show that these constant–θ circles are orthogonal to the constant $-|z|$ circles. Discuss the variation in the locus as the angle of the impedance, θ, is varied over its permitted range.
4. Carry out an analysis to show that on the Smith Chart, if the $g = 1$ circle is rotated through an angle βd towards the load, the maximum conductance (normalized) encountered on the ***rotated* g = 1 *circle*** is $\operatorname{cosec}^2 (2\beta d)$.
5. The input impedance of a lossless line terminated in the impedance $\mathbf{Z}_L$, is $\mathbf{Z}_1$ at a distance l_1 from the load-end. Is it necessary that the input impedance of this line becomes $\mathbf{Z}^*$ at some distance l_2 from the load-end ? Justify your answer. If your answer is *yes*, then establish the relation between l_2 and l_1.
6. (a) Prove both analytically and graphically, making use of the geometry of the Smith Chart, that the real part of the input impedance of any given line, can never lie outside the range, $\left(\dfrac{1}{S}, S\right)$, where S is the *VSWR* on the line.

   (b) Prove both analytically and graphically, making use of the geometry of the Smith Chart, that the magnitude of the imaginary part of the input impedance of any given line, can never exceed the value, $\dfrac{1}{2}\left(S - \dfrac{1}{S}\right)$, where S is the *VSWR* on the line.
7. (a) An admittance (normalised), $y_L = g_L + jb_L$, is connected at the load-end of a lossless (G_o, β_g) line. Another admittance (normalised), $y = g + jb$ is connected in shunt with the line at a point, T, at a distance of l towards the gen-end from the load-end [Ref. Fig. 9.24]. Let the input line admittance (normalised), y_{in}, at the point, T, which would be the sum of y_s and the admittance, y_L as "seen" through the length, l of the line, be represented by the point, P, on the Smith Chart. Prove that, as the distance, l, is varied, the point, P, moves on the circle,

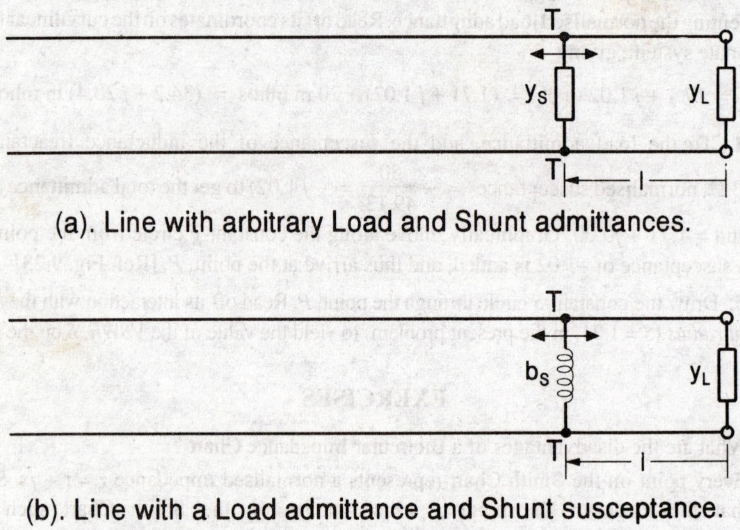

Fig. 9.24 Line with an arbitrary Load Admittance and an arbitrary Shunt Admittance.

$$\left[\xi - \frac{g_S^2 + g_S\left(g_L + \frac{1}{g_L}\right) + b_S^2}{g_S^2 + (g_S+1)\left(\sqrt{g_L} + \frac{1}{\sqrt{g_L}}\right)^2 + b_S^2}\right]^2 + \left[\frac{2\,b_S}{g_S^2 + (g_S+1)\left(\sqrt{g_L} + \frac{1}{\sqrt{g_L}}\right)^2 + b_S^2}\right]^2 = \left[\frac{\left(g_L - \frac{1}{g_L}\right)}{g_S^2 + (g_S+1)\left(\sqrt{g_L} + \frac{1}{\sqrt{g_L}}\right)^2 + b_S^2}\right]^2$$

(b) Discuss a special case of the above probelm, in which the shunt admittance at T is a pure susceptance only; that is, when $g_S = 0$.

Let (ξ_o, η_o) denote the centre of this circle. Then show that the locus of the centre with respect to the susceptance, b_S, that is, the equation of the curve on which the point, (ξ_o, η_o), moves when the susceptance, b_S, takes up all permissible values, is,

$$\frac{\left(\xi_o - \frac{1}{2}\right)^2}{\left[\frac{1}{2}\right]^2} + \frac{\eta_o^2}{\left[\sqrt{g_L} + \frac{1}{\sqrt{g_L}}\right]^2} = 1 - \text{Ellipse.}$$

8. An unknown load impedance when connected to a line produces a *VSWR* of 2.618 on the line and the distance between two successive voltage minima is found to be 16 mm. Further, it is observed that if this load is replaced by a short-circuit, the voltage minima shifts a distance of 3.2 mm towards the generator. Use the Smith Chart to calculate the normalised load impedance and the voltage reflection coefficient at the load terminals.

[**Ans.** $z_L = 0.54 + j\,0.58$ $\rho_L = 0.447 \underline{/108^\circ}$]

9. A 75 Ω lossless line feeds a pure resistance load and the observed *VSWR* on the line is 3. The real part of the input line admittance (input conductance) is found to be unity within a distance of 0.1 λ_g from the load. Calculate the input susceptance at this point and also determine the load resistance.[**Ans.** $R_L = 25\ \Omega$, $B_{in} = -j\,14.2667$ m mhos.]

10. An antenna having an input impedance of $(73 + j\,37)\ \Omega$ is connected to the generator through a $\frac{7}{8}\lambda_g$ long 75 Ω feeder, followed by a $\frac{3}{8}\lambda_g$ length of a 50 Ω feeder. Use the Smith Chart to determine the *VSWRs* in each of the two lines. Also obtain the complex voltage reflection coefficients at both the generator and antenna terminals and the impedance that the generator sees.

[**Ans.** $S_{75} = 1.642$, $S_{50} = 1.135$, $\rho_L = 0.2429 \underline{/79.058^\circ}$,

$\rho_S = \underline{/-140.710^\circ}$, $Z_{in} = 45.3541 \underline{/-4.584^\circ}$.]

11. When the load-end of a 75 Ω line is terminated in a pure resistance of 7.5 Ω, a *VSWR* of 10 is observed and the wavelength is measured to be 4.48 cm. When the resistance (load) is replaced by an unknown impedance, a *VSWR* of 6 is observed and the first voltage minimum is found to shift a distance of 0.842 cm towards the load-end. Calculate the value of the unknown impedance in ohms.

[**Ans.** $Z_L = (49.540 + j\,101.590)\ \Omega$.]

12. A lossless line of $\frac{\lambda_g}{8}$ length is terminated in an arbitrary impedance, Z_L. Derive, first analytically and then by using the Smith Chart, the condition that Z_L must satisfy in order that the input impedance of the line is a pure resistance. [**Ans.** $|z_L| = 1$.]

[Hopefully, you would be able to appreciate the simplification that grapical methods, like using the Smith Chart, offer in solving some transmission line problems]

13. Determine the load impedance having its resistance and reactance equal, which when connected to the load-end of a 50 Ω lossless coaxial line, produces a *VSWR* of 2.618.

[**Ans.** $(50 + j\,50)\ \Omega$ or $(25 + j\,25)\ \Omega$.]

Chapter 10

IMPEDANCE MATCHING

10.1 TYPES OF MATCHING

In a communication network, it is often desirable to adjust its elements, if possible, so that the power delivered to the load is maximum. In order to derive the conditions under which maximum power transfer occurs, consider a load impedance, $Z_L = R_L + j\ X_L = |Z_L|\, e^{j\theta_L}$, connected to a generator of internal impedance, $Z_G = R_G + j\ X_G = |Z_G|\, e^{j\theta_G}$, generating a voltage V_G, as shown in Fig. 10.1(a). The power, P_L, delivered to the load is

$$P_L = \left| \frac{V_G}{Z_G + Z_L} \right|^2 R_L$$

$$= \frac{|V_G|^2}{(R_G + R_L)^2 + (X_G + X_L)^2} R_L \qquad (10.1)$$

Alternatively,

$$P_L = \frac{|V_G|^2}{\left| |Z_G|\, e^{j\theta_G} + |Z_L|\, e^{j\theta_L} \right|^2} |Z_L| \cos\theta_L$$

$$= \frac{|V_G|^2}{\left| Z_L| + |Z_G|\, e^{j(\theta_G - \theta_L)} \right|^2} |Z_L| \cos\theta_L$$

$$= \frac{|V_G|^2}{|Z_G|^2 + |Z_L|^2 + 2|Z_G||Z_L| \cos(\theta_G - \theta_L)} |Z_L| \cos\theta_L \qquad (10.2)$$

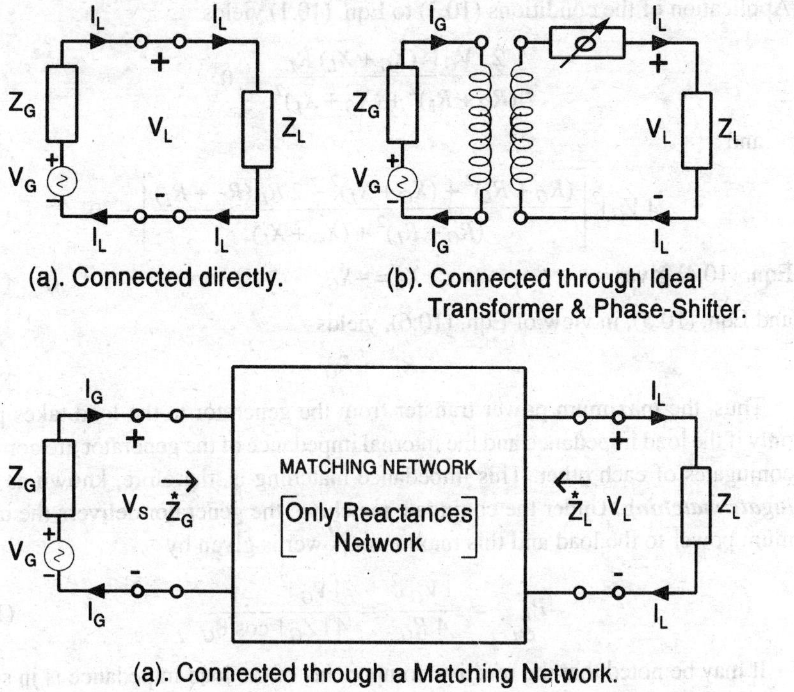

Fig. 10.1 A Load connected to a Generator.

Usually, the internal impedance of the generator is fixed so that the problem of maximising the power, P_L, amounts to finding out the values of R_L and X_L for which the power delivered to the load is maximum. The process of adjusting the load impedance so that it extracts or receives maximum power from the generator, is called ***Matching***. Two cases of practical interest arise: The first case being the one in which both R_L and X_L can be varied independently, that is, both the magnitude and the angle of the load-impedance can be varied independently. In practice, this case is encountered, when a matching network [Ref. Fig. 10.1(b)] is used to effect the matching. The second case arises when only the magnitude of the load impedance, but not its angle, can be varied. This case is encountered in practice when the matching is effected by using an ideal transformer [Ref. Fig. 10.1 (c)]. Conditions of matching in these two cases are derived below.

10.1.1 Conjugate Matching

If both R_L and X_L can be varied independently [Ref. Fig. 10.1 (b)], then the condition that the power delivered to the load is maximum, may mathematically be given by

$$\frac{\partial P_L}{\partial X_L} = 0 \text{ and } \frac{\partial P_L}{\partial R_L} = 0, \qquad (10.3)$$

because, in this case, P_L is a function of the two independent variables, R_L and X_L. Application of the conditions (10.3) to Eqn. (10.1) yields

$$\frac{2\,|V_G|^2 (X_G + X_L)\, R_L}{(R_G + R_L)^2 + (X_G + X_L)^2} = 0 \tag{10.4}$$

and

$$|V_G|^2 \left[\frac{(R_G + R_L)^2 + (X_G + X_L)^2 - 2\,R_L (R_G + R_L)}{(R_G + R_L)^2 + (X_G + X_L)^2} \right] = 0 \tag{10.5}$$

Eqn. (10.4) gives, $$X_L = -X_G, \tag{10.6}$$

and Eqn. (10.5), in view of Eqn. (10.6), yields

$$R_L = R_G \tag{10.7}$$

Thus, the maximum power transfer from the generator to the load takes place only if the load impedance and the internal impedance of the generator are complex conjugates of each other. This impedance matching is, therefore, known as ***conjugate matching***. Under the conjugate matching, the generator delivers the maximum power to the load and this maximum power is given by

$$P_{\substack{L\max \\ CM}} = \frac{|V_G|^2}{4\,R_G} = \frac{|V_G|^2}{4\,|Z_G|}\frac{1}{\cos\theta_G} \tag{10.8}$$

It may be noted that the reactive components of the load impedance is in series resonance with the reactive component of the generator impedance. That is, under the matched condition, the load reactance of the generator is of the correct value required to ***tune out*** the reactance of the generator: For this reason, in microwave engineering, the process of matching is colloquially referred to as ***tuning***. Therefore, the total reactance in the circuit then becomes equal to zero.

10.1.2 Modulus Matching

Now, consider the case when only the magnitude of the load impedance can be varied, but not its angle [Ref. Fig. 10.1(c)]. In this case, therefore, maximum power transfer from the generator to the load occus when

$$\frac{d\,P_L}{d\,|Z_L|} = 0, \tag{10.9}$$

because $|Z_L|$ is the only variable available in this case. Application of this condition to Eqn. (10.2) gives,

$$\cos\theta_L\,[|Z_G|^2 + |Z_L|^2 + 2\,|Z_G|\,|Z_L|\cos(\theta_G - \theta_L)]$$
$$= 2\,|Z_L|\cos\theta_L\,[|Z_L| + |Z_G|\cos(\theta_G - \theta_L)]$$

or

$$|Z_L|^2 = |Z_G|^2$$

or

$$|Z_L| = |Z_G| \tag{10.10}$$

Thus, the maximum power transfer from the generator to the load under the condition that only the magnitude, but not the angle of the load impedance, can be varied, takes place when the magnitude of the load impedance equals that of the generator impedance. The value of the maximum power delivered is given by

$$\begin{aligned}
P_{\substack{L\max \\ MM}} &= \frac{|V_G|^2}{4Z_G} \frac{\cos\theta_L}{\cos^2\left(\dfrac{\theta_G-\theta_L}{2}\right)} \\
&= P_{\substack{L\max \\ CM}} \frac{\cos\theta_G \cos\theta_L}{\cos^2\left(\dfrac{\theta_G-\theta_L}{2}\right)} \\
&= P_{\substack{L\max \\ CM}} \frac{1}{2} \frac{\cos(\theta_G+\theta_L)+\cos(\theta_G-\theta_L)}{\cos^2\left(\dfrac{\theta_G-\theta_L}{2}\right)} \\
&= P_{\substack{L\max \\ CM}} \frac{1}{2} \frac{\left[2\cos^2\left(\dfrac{\theta_G-\theta_L}{2}\right)-1\right]+\left[1-2\sin^2\left(\dfrac{\theta_G+\theta_L}{2}\right)\right]}{\cos^2\left(\dfrac{\theta_G-\theta_L}{2}\right)} \\
&= P_{\substack{L\max \\ CM}} \left[1-\frac{\sin^2\left(\dfrac{\theta_G+\theta_L}{2}\right)}{\cos^2\left(\dfrac{\theta_G-\theta_L}{2}\right)}\right]
\end{aligned} \tag{10.11}$$

Note from Eqn. (10.11) that, in general, the power delivered to the load under modulus matching condition is less than that under conjugate matching conditions. However, if $\theta_G = \theta_L = 0$, that is, if both the generator and the load impedance are pure resistances, the modulus match condition becomes identical with the conjugate match condition.

10.1.3 Reflectionless Matching

A microwave network is composed of a number of component parts which are interconnected by lengths of low-loss transmission lines [Ref. Fig. 10.2] which are likely to be electrically long although physically short. In so far as transfer of power from one component part to the other is concerned, the two component parts may be regarded to form a load connected to a generator through a long transmission line. Although the transformed load impedance as seen from the generator-end of the transmission line can be matched to the generator for the maximum power transfer (conjugate match), a small change in the operating frequency will change the electrical length ($= \beta_g l$) of the interconnecting transmission line by an appreciable

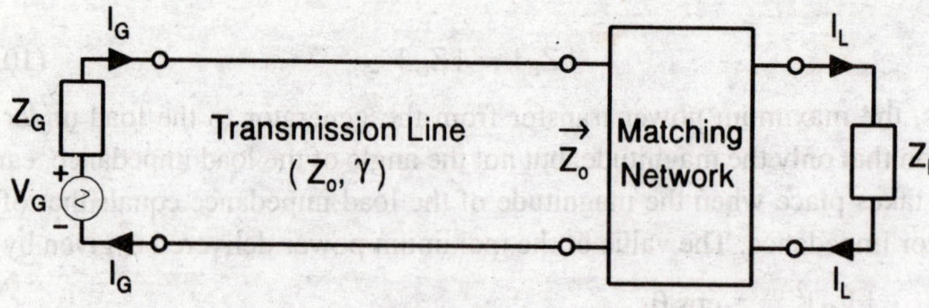

Fig. 10.2 Reflectionless Matching between a Load and a Transmission Line.

fraction of 2π radians, and hence the effective load impedance seen at the generator-end will greatly change, resulting in a large mismatch. To avoid this frequency sensitivity of the matching, the load and the generator should be individually matched to the interconnecting transmission line on the reflectionless match basis, that is, $Z_L = Z_o$ and $Z_G = Z_o$. In this case, only travelling waves exist on the lines, that is, the *VSWR*, $S = 1$. It may be shown that the power delivered to the load is less than that which may be delivered under conjugate matching conditions. However, if the microwave system is required to perform satisfactorily at a number of spot frequencies without re-tuning or over a finite bandwidth, which is most often the case, reflectionless matching must be resorted to, as

1. It results in the best average performance.

   Some of the other advantages of the reflectionless matching are :

2. The transmission loss of a given line is minimum and hence its transmission efficiency maximum, under the reflectionless match conditions [Ref. Sec. 2.9.7].
3. The power carrying capacity of a transmission line is also the largest under the reflectionless match conditions [Ref. Sec. 3.5.1].

In view of these advantages, in microwave engineering, only the reflectionless matching is of practical interest and the term ***matching*** usually means reflectionless matching.

The condition for the reflectionless matching is

$$\left.\begin{aligned} S &= 1 \\ \text{or,} \quad Z_L &= Z_G \text{ (in general)} \end{aligned}\right\} \tag{10.12}$$

The power delivered to the load under these conditions may easily be derived to be

$$P_{L \text{ reflectionless matching}} = \left| \frac{V_G}{2\, Z_G} \right|^2 R_G$$

$$= \frac{|V_G|^2}{4\,|Z_G|} \cos \theta_G \tag{10.13}$$

10.1.4 Mismatch Loss

For an arbitrary load impedance, Z_L, the power delivered to the load by the generator of internal impedance, Z_G, is given by Eqn. (10.2). This is the power delivered to the load under the mismatch conditions. However, if the load is matched on conjugate match basis or the reflectionless match basis, the power that would be delivered to it by the generator are given by Eqns. (10.8) and (10.13). Therefore, a mismatch loss relative to the conjugate match or to the reflectionless match may be defined as:

Mismatch loss relative to conjugate match

$$= 10 \log \left[\frac{P_{L_{\max} CM}}{P_L} \right]$$

$$= 10 \log \left[\frac{|Z_G|^2 + |Z_L|^2 + 2|Z_G||Z_L| \cos(\theta_G - \theta_L)}{4 |Z_G||Z_L| \cos\theta_G \cos\theta_L} \right] \text{dB} \qquad (10.14)$$

and Mismatch loss relative to reflectionless match

$$= 10 \log \left[\frac{P_{L_{\max} RM}}{P_L} \right]$$

$$= 10 \log \left[\frac{|Z_G|^2 + |Z_L|^2 + 2|Z_G||Z_L| \cos(\theta_G - \theta_L)}{4 |Z_G||Z_L| \dfrac{\cos\theta_L}{\cos\theta_L}} \right] \text{dB} \qquad (10.15)$$

10.2 TECHNIQUES OF MATCHING

In microwave networks, as discussed above, the components are matched, on reflectionless matching basis, to the interconnecting transmission lines by means of reactive networks. Components of these matching networks may be short lengths of transmission lines or may take the form of diaphragms, holes or rods in waveguides. These impedance matching devices may be structures in themselves, usually tunable or adjustable, forming two-port networks which may be connected between the transmission lines and the components that are to be matched to the lines. Alternatively and perhaps, more commonly, the elements of the impedance-matching networks are built into the components, sometimes adjustable and sometimes not. Some of the frequently used impedance-matching techniques are discussed below :

10.2.1 Quarterwave Line Transformer

The input impedance, Z_{in}, of a lossless line of length, l, and characteristic impedance, R_{om}, when terminated in a load impedance, Z, is given by [Ref. Eqn. (7.19)], Sec. 7.2,

$$Z_{in} = R_{om}\left[\frac{Z_L + j\,R_{om}\tan(\beta_g\,l)}{R_{om} + j\,Z_L\tan(\beta_g\,l)}\right] \tag{10.16}$$

For a quarterwave line, $l = \frac{\beta_g}{4}$ or $\beta_g\,l = \frac{\pi}{2}$, therefore, Eqn. (10.16) yields.

$$Z_{in} = \frac{R_{om}^2}{Z_L} \tag{10.17}$$

Thus, if $Z_L = R_{o1} =$ a resistance, Z_{in} is also a pure resistance, say R_O. This property of a quarterwave line can be exploited to match two lossless transmission lines of characteristic impedances R_{o1} and R_o to each other by using a quarterwave section of a transmission line whose characteristic impedance R_{om}, is

$$R_{om} = \sqrt{R_o\,R_{o1}} \tag{10.18}$$

A quarterwave section of a lossless transmission line [Ref. Fig. 10.3] may also be used to match a pure resistance load, R_L, to a lossless transmission line of characteristic impedance, R_O, by choosing its characteristic impedance, R_{om}, to be given by

$$R_{om} = \sqrt{R_L\,R_o} \tag{10.19}$$

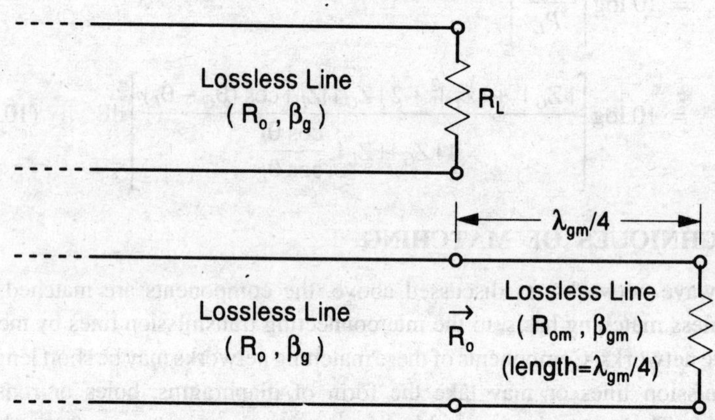

Fig. 10.3 Quarterwave Transformer matching a Pure Resistance Load to Lossless Line.

Next, consider a complex load impedance $Z_L = R_L + j\,X_L$ connected to the load-end of a lossless transmission line of characteristic impedance, R_o, producing a voltage standing wave ratio S [Ref. Fig. 10.4(a)]. The impedance at a voltage minimum looking towards the load is a pure resistance equal to $\frac{R_o}{S}$ and that at a voltage maximum is also a pure resistance but of the magnitude SR_o [Ref. Fig. 10.4 (b)]. This means that the given complex impedance Z_L is transformed to or, is equivalent to a pure resistance $\frac{R_o}{S}$ at the voltage minimum. Therefore, the given

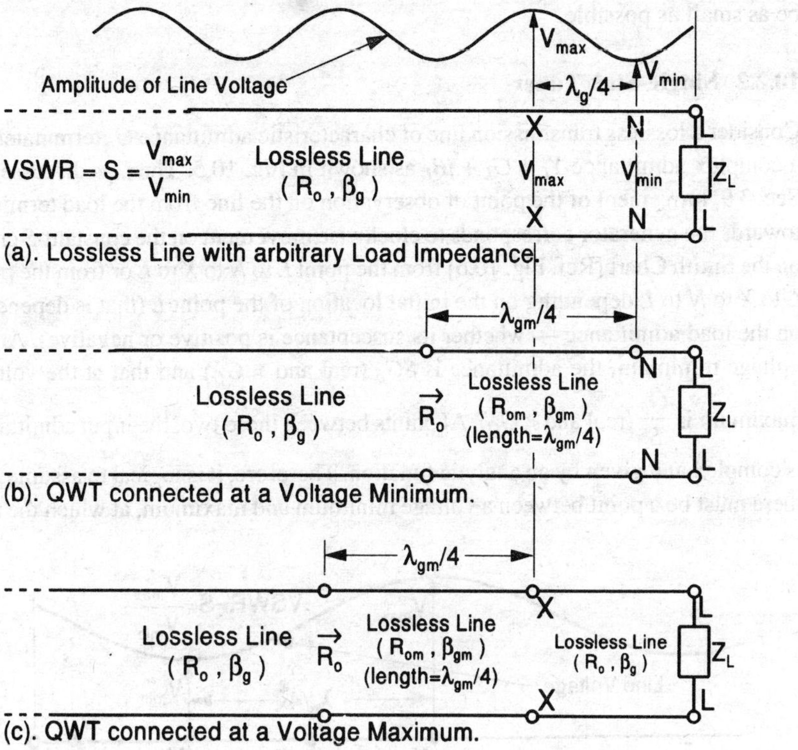

Fig. 10.4 Quarterwave Transformer matching an Arbitrary Complex Load Impedance to a Lossless line.

load impedance can be matched to the line by means of a quarter wave line of characteristic impedance,

$$R_{om} = \sqrt{R_o \frac{R_o}{S}} = \frac{R_o}{\sqrt{S}} \tag{10.20}$$

connected such that its load-end is at a voltage minimum [Ref. Fig. 10.4 (b)].

Alternatively, matching may be effected by using a quarterwave line of characteristic impedance,

$$R_{om} = \sqrt{R_o \, S \, R_o} = \sqrt{S} \, R_o \tag{10.21}$$

connected such that its load-end is at a voltage maximum [Ref. Fig. 10.4 (c)].

Designed and connected in the manner described above, a quarter wave transformer produce unit *VSWR* in the line on the generator side of the transformer. It may be emphasised, however, that in the line between the load-end and the quarter-wave transformer, standing waves do exist having a *VSWR* of *S*. Therefore, to derive the benefits of matching to the maximum, the length of the line between the load and the quarterwave transformer–the matching network or device–must be as small as possible.

10.2.2 Single-Stub Tuner

Consider a lossless transmission line of characteristic admittance G_o terminated by a complex admittance $Y_L = G_L + jB_L$ as shown in Fig. 10.5. Then, as discussed in Sec. 3.9, movement of the point of observation on the line from the load terminals towards the generator corresponds to clockwise movement on the constant-S circle on the Smith Chart [Ref. Fig. 10.6] from the point L to N to X to L or from the point L to X to N to L depending on the initial location of the point L (that is depending on the load admittance — whether its susceptance is positive or negative). At the voltage minimum, the admittance is SG_o (real and $> G_O$) and that at the voltage maximum is $\frac{G_o}{S}$ (real and $< G_O$). At points between these two, the input admittance is complex and given by an analytic function. Therefore, it is logical to assume that there must be a point between a voltage minimum and maximum, at which the real

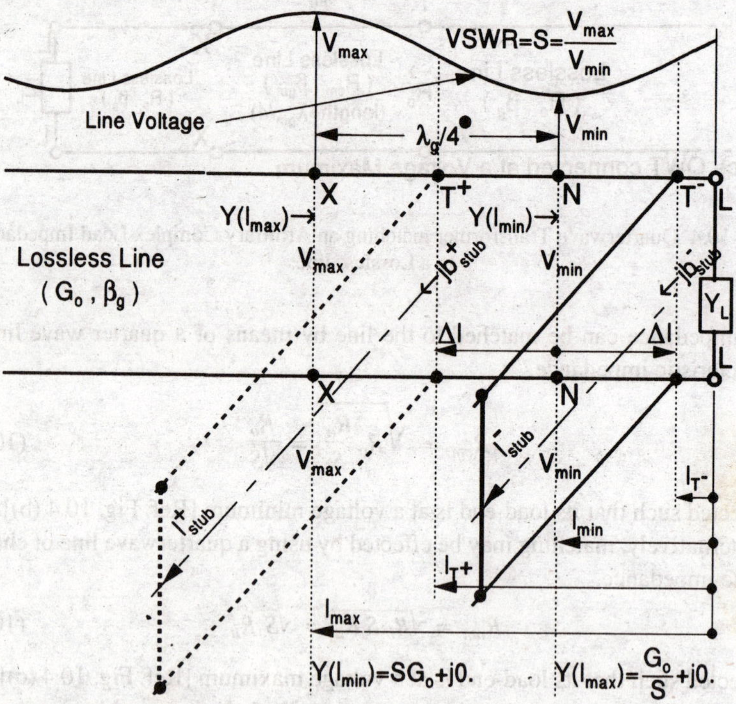

$$\Delta l = \left(\frac{\lambda_g}{4\pi}\right)\cos^{-1}\left(\frac{S-1}{S+1}\right); \quad l_{max} = l_{min} \pm \lambda_g/4, \quad l_{T^-} = l_{min} - \Delta l, \quad l_{T^+} = l_{min} + \Delta l.$$

Fig. 10.5 Single-Stub Matching System (Theory and Configuration).

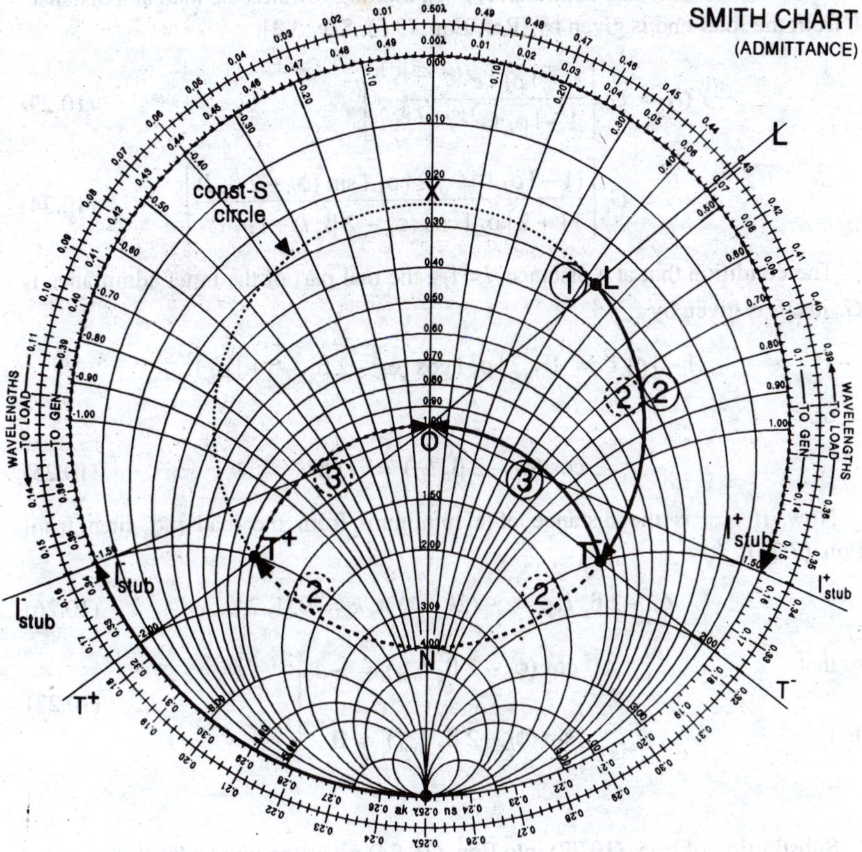

Fig. 10.6 Single Stub Matching (Graphical Design Procedure).

part of the input admittance would be G_o itself. This result is obvious from the graphical representation because the constant S circle cuts the $g = 1$ circle in T^- and T^+.

At the points T^- and T^+, where the real part of the input admittance is G_o, the input admittance is of the form

$$Y_{T^-} = (1 \pm j\, b_{T^-})\, G_o \tag{10.22}$$

The susceptance $B_{in} = \pm\, b_{T^-} G_o$, may be tuned out by connecting at the points T^- or T^+ in parallel with the line, a susceptance, $B_{stub} = -B_{in} = \mp\, b_{T^-} G_o$, thus producing a match condition. The susceptance B_{stub} required to tune out the input susceptance at the point T^- or T^+ may be provided by a short length of a transmission line called the stub line or simply, a stub. This is the principle underlying the operation of a single stub tuner.

10.2.2.1 Analysis

The design equations for the design of a single stub matching system may be derived as follows :

For a lossless line, the admittance, $Y(l)$, looking towards the load at a distance, l, from the load-end is given by [Ref. Eqn. (3.6), Sec. 3.2],

$$Y(l) = G_o\left[\frac{1 - |\rho_L| e^{j(\phi_L - 2\beta_g l)}}{1 + |\rho_L| e^{j(\phi_L - 2\beta_g l)}}\right] \tag{10.23}$$

$$= G_o\left[\frac{(1 - |\rho_L|^2) - j2|\rho_L| \sin(\phi_L - 2\beta_g l)}{1 + 2|\rho_L| \cos(\phi_L - 2\beta_g l) + |\rho_L|^2}\right] \tag{10.24}$$

The condition that at a distance, $l = l_T^{\pm}$, the real part of the input admittance is G_o itself, is given by

$$1 - |\rho_L|^2 = 1 + 2|\rho_L| \cos(\phi_L - 2\beta_g l_T^{\pm}) + |\rho_L|^2$$

or,

$$\cos(\phi_L - 2\beta_g l_T^{\pm}) = -|\rho_L| \tag{10.25}$$

Now, if $l_{\min}$ is the distance of a minimum from the load-end, then from Eqn. (7.25),

$$\phi_L - 2\beta_g l_{\min} = -(2k+1)\pi,\ k = 0, 1, 2, \ldots \tag{10.26}$$

so that

$$\left.\begin{aligned} \cos(\phi_L - 2\beta_g l_{\min}) &= -1 \\ \text{and} \quad \sin(\phi_L - 2\beta_g l_{\min}) &= 0 \end{aligned}\right\} \tag{10.27}$$

Substitution of Eqn. (10.27) into Eqn. (10.24) gives the input admittance at the voltage minimum to be

$$Y(l_{\min}) = G_o\left[\frac{1 + |\rho_L|}{1 - |\rho_L|} - j0\right] = G_o S - j0 \tag{10.28}$$

Now,

$$\cos[(\phi_L - 2\beta_g l_{\min}) - (\phi_L - 2\beta_g l_T^{\pm})] = \cos(\phi_L - 2\beta_g l_{\min}) \cos(\phi_L - 2\beta_g l_T^{\pm})$$
$$+ \sin(\phi_L - 2\beta_g l_{\min}) \sin(\phi_L - 2\beta_g l_T^{\pm}) \tag{10.29}$$

Substituting Eqns. (10.25) and (10.27) into Eqn. (10.29) gives

$$\cos[(\phi_L - 2\beta_g l_{\min}) - (\phi_L - 2\beta_g l_T^{\pm})] = |\rho_L|$$

or,

$$\cos[2\beta_g (l_T^{\pm} - l_{\min})] = |\rho_L| \tag{10.30}$$

or,

$$l_T^{\pm} = l_{\min} \pm \Delta l, \tag{10.31}$$

where, $$\Delta l = \frac{\lambda_g}{4\pi} \cos^{-1} |\rho_L| \tag{10.32}$$

$$= \frac{\lambda_g}{4\pi} \cos^{-1}\left(\frac{S-1}{S+1}\right) \tag{10.33}$$

$$= \frac{\lambda_g}{2\pi} \tan^{-1}\left(\frac{1}{\sqrt{S}}\right) \tag{10.34}$$

$$= \frac{\lambda_g}{4} - \frac{\lambda_g}{2\pi} \tan^{-1} (\sqrt{S}) \tag{10.35}$$

This shows that a distance, Δl, on either side of a voltage minimum, the real part of the normalised input admittance is unity. The value of the susceptance at this point is given by

$$B(l_T^{\mp}) = -\frac{2|\rho_L| \sin(\phi_L - 2\beta_g l_T^{\pm})}{1 - |\rho_L|^2} G_o \tag{10.36}$$

Now,

$$\sin(\phi_L - 2\beta_g l_T^{\pm}) = \sin(\phi_L - 2\beta_g l_{\min} \mp \cos^{-1} |\rho_L|)$$

$$= \pm \sin(\cos^{-1} |\rho_L|)$$

$$= \pm \sqrt{1 - |\rho_L|^2} \tag{10.37}$$

Therefore,

$$B_{T^\pm} = B(l_T^{\pm}) = \mp G_o \frac{2|\rho_L|}{\sqrt{1 - |\rho_L|^2}} = \mp\left(\sqrt{S} - \frac{1}{\sqrt{S}}\right) G_o \tag{10.38}$$

The susceptance, $B_{stub}^{\pm}$ of the stub connected at $T^{\pm}$ required to tune out the susceptance of the lime is, therefore, given by

$$B_{stub}^{\pm} = -B_{T^\pm} = \pm\left(\sqrt{S} - \frac{1}{\sqrt{S}}\right) G_o \tag{10.39}$$

and location of the stub is

$$l_T^{\pm} = l_{\min} \pm \frac{\lambda_g}{8}\left[1 - \frac{2}{\pi} \tan^{-1}\left\{\frac{1}{2}\left(\sqrt{S} - \frac{1}{\sqrt{S}}\right)\right\}\right] \tag{10.40}$$

Let the characteristic admittance of the stub line be the same as that of the main line, namely, G_o. The stub may be either a short-circuited one or an open-circuited one. Consider first a short-circuited stub. Its length, l_{stub} is given by the condition that its input susceptance equals that given by Eqn. (10.40), that is

$$Y_{in}(l) = -j G_o \cot \beta_g l^{\pm}_{\substack{stub \\ SC}} = \pm j \left(\sqrt{S} - \frac{1}{\sqrt{S}}\right) G_o \tag{10.41}$$

Therefore,

$$
\begin{aligned}
l^{\pm}_{\substack{stub\\SC}} &= \frac{\lambda_g}{2\,\pi}\tan^{-1}\left(\frac{\mp 1}{\sqrt{S}-\frac{1}{\sqrt{S}}}\right)\\
&= \frac{\lambda_g}{2\,\pi}\left[\frac{\pi}{2}-\tan^{-1}\left\{\mp\left(\sqrt{S}-\frac{1}{\sqrt{S}}\right)\right\}\right]\\
&= \frac{\lambda_g}{4}\left[1-\frac{2}{\pi}\tan^{-1}\left\{\mp\left(\sqrt{S}-\frac{1}{\sqrt{S}}\right)\right\}\right]\\
&= \frac{\lambda_g}{4}\left[1\pm\frac{2}{\pi}\tan^{-1}\left(\sqrt{S}-\frac{1}{\sqrt{S}}\right)\right] \qquad (10.42)
\end{aligned}
$$

Similarly, it may be shown that the length of open-circuited stubs required are given by

$$
l^{\pm}_{\substack{stub\\SC}} = \begin{cases} \dfrac{\lambda_g}{2\,\pi}\tan^{-1}\left(\sqrt{S}-\dfrac{1}{\sqrt{S}}\right) \\[2ex] \dfrac{\lambda_g}{2}\left[1-\dfrac{1}{\pi}\tan^{-1}\left(\sqrt{S}-\dfrac{1}{\sqrt{S}}\right)\right] \end{cases} \qquad (10.43)
$$

From Eqns. (10.42) and (10.43) it may be noted that

$$
l^{+}_{\substack{stub\\SC}} + l^{\pm}_{\substack{stub\\SC}} = l^{+}_{\substack{stub\\OC}} + l^{-}_{\substack{stub\\OC}} = \frac{\lambda_g}{2} \qquad (10.44)
$$

and

$$
l^{+}_{\substack{stub\\SC}} - l^{+}_{\substack{stub\\SC}} = l^{-}_{\substack{stub\\OC}} - l^{-}_{\substack{stub\\OC}} = \frac{\lambda_g}{4} \qquad (10.45)
$$

These results or alternative solutions are given in Fig. 10.5

$$
\Delta\, l = \frac{\lambda_g}{8}\left[1-\frac{2}{\pi}\tan^{-1}\left\{\frac{1}{2}\left(\sqrt{S}-\frac{1}{\sqrt{S}}\right)\right\}\right] \qquad (10.46)
$$

$$
l_{T^{\pm}} = l_{\min} \pm \Delta\, l \qquad (10.47)
$$

$$
l^{\pm}_{\substack{stub\\SC}} = \frac{\lambda_g}{4}\left[1\pm\frac{2}{\pi}\tan^{-1}\left(\sqrt{S}-\frac{1}{\sqrt{S}}\right)\right] \qquad (10.48)
$$

$$
l^{\pm}_{\substack{stub\\SC}} = \frac{\lambda_g}{4}\left[1\mp\left\{1-\frac{2}{\pi}\tan^{-1}\left(\sqrt{S}-\frac{1}{\sqrt{S}}\right)\right\}\right] \qquad (10.49)
$$

10.2.2.2 Graphical Design Procedure

It may be observed that the centre of the Smith Chart corresponds to $|\rho| = 0$ or $S = 1$. Therefore, the centre of the Smith Chart may be called the ***Match Point*** and the problem of designing a stub matching system for a given load admittance, boils down to arriving at the centre of the Smith Chart starting from the load admittance point by restricting movements to only along the two families of circles–the constant-S circles and the constant-g circles. A movement on an S-circle in the clockwise direction on the Smith Chart corresponds to a movement on the transmission line towards the generator-end. And, a movement on an S-circle is usually only along the clockwise direction. But movements on a g-circle in the clockwise direction is equivalent to the stub presenting positive susceptance to the line. On the other hand, a movement along the g-circle in the anti-clockwise direction is equivalent to the stub presenting negative susceptance to the line.

The design of a single stub system to match a given load to a line using the Smith Chart, therefore, is analogous to the maze puzzle game which children play in which a prince standing at the entrance of a maze has to find his way through the maze to the princess (his *match*) sitting at the centre of the maze. And, while going through the lanes and alleys that constitute the maze, the prince must observe certain rules like jumping over a lane and piercing through the end of an alley are not permitted.

The design of a single stub system to match a given load admittance to a lossless transmission line using the Smith Chart is quite straightforward following the procedure given below :

Step 1 : Draw the constant-S circle passing through the load point, L, [Ref. Fig. 10.6] representing the given transmission line on the Smith admittance chart. This can be done if the *VSWR*, S, is known or the normalised load admittance is known.

The points, T^- and T^+, at which the constant-S circle intersects the $g = 1$ circle, give the locations where the stub may possibly be located because at these points the line admittances are, $1 + jb_{T^-}$ and $1 + jb_{T^+}$, respectively.

Note that T^- and T^+ are located at the same distance on either side of the voltage minimum point, N. At T^- which is towards the load from the voltage minimum, the input susceptance is positive (capacitive) and at T^+ the input susceptance is negative (inductive). Also note that $b_{T^+} = -b_{T^-}$.

Step 2 : For locating the stub at T^- (or T^+), move from the point, T^- (or T^+) on the constant-g circle and arrive at the match point or the centre of the Chart, O. Find out the susceptance, b_{stub} that must be ***added*** to the admittance at the point T^- (or T^+) in order to arrive at the match point, O. In fact, $b_{stub} = -b_{T^-}$ (or $-b_{T^+}$). b_{stub} is the input susceptance of the stub or the susceptance that the stub connected at the point, T^- (or T^+), must present to the line to effect matching. It is clear that at T^-, an inductive (negative susceptance) stub is required and at T^+ a capacitive (positive susceptance) stub is required.

Step 3 : To determine the length of the short-circuited stub line required, mark off the admittance point, $(0 + jb_{stub})$, on the $S = \infty$ circle, (the outermost circle or the periphery of the Smith Chart). The length of the short-circuited stub is then the distance travelled in moving from the $g = \infty$ point on the $S = \infty$ *circle*, to the point, $(0 + jb_{stub})$, in the clockwise direction.

The length of the open-circuited stub is the clockwise distance from the point, $(0 + j0)$, to point, $(0 + jb_{stub})$.

It may be pointed out that the above analysis and the graphical procedure have been developed assuming that the input impedance of a stub comes in shunt with the line. Accordingly, it has been found convenient to use admittances and the Smith admittance Chart. However, if a **series** stub whose input impedance comes in series with the line, the whole theory and the graphical procedure can be easily re-developed on similar lines using impedances and the Smith Impedance Chart.

10.2.3 Double-Stub Tuner

In so far as the practical implementation of the single stub matching technique is concerned, it has a serious drawback in that it requires the stub to be movable along the main line.

An improved system is the double stub tuner which employs two stubs of variable lengths fixed at a distance *d* apart along the main line. The design procedure using Smith admittance chart for a double stub tuner may be developed as given below for the case of shunt stubs (connected in parallel with the line and, therefore, admittances are considered).

Let the two stubs be a distance *d* apart and the first stub be at a distance *l* from the load-end. Using the following notation : [Ref. Fig. 10.7]

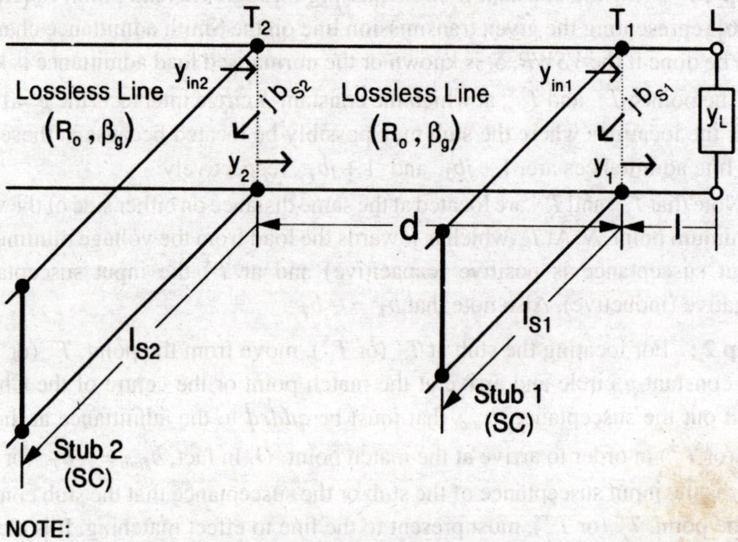

NOTE:
For Matching, $y_{in2} = 1 + j0$: Possible only if $\text{Re}[y_1] = g_1 < \text{cosec}^2(\beta_g d)$.

Fig. 10.7 Double-Stub Matching System.
(Theory and Configuration).

y_1 = the input admittance that is, the admittance looking towards the load, at the point T_1 before the first stub is connected.

y_{in1} = the input admittance at the point T_1 including the effect of the first stub.

y_2 = the input admittance at the point T_2 before the second stub is connected.

y_{in2} = the input admittance at the point T_2 including the effect of the second stub.

b_{s1} = Input susceptance of the first stub.

b_{s2} = Input susceptance of the second stub.

Relationship between various admittances are the following :

The load admittance y_L when transformed by the length l of the transmission line appears as y_1 at T_1 [Ref. Fig. 10.8]. On the Smith admittance chart y_1 is obtained from y_L by moving on the constant-S circle passing through y_L, through a distance l in the clockwise direction. Symbolically, this may be written as

$$y_1 = \overset{l}{\overrightarrow{y_L}} \tag{10.50}$$

$$y_{in1} = y_1 + jb_{S1} \tag{10.51}$$

$$y_2 = \overset{d}{\overrightarrow{y_{in1}}} \tag{10.52}$$

$$y_{in2} = y_2 + jb_{S2} \tag{10.53}$$

In the matching system, b_{S1} and b_{S2} are the only variables which have to be adjusted so that the given y_L is matched to the line. If the system is to effect matching, y_{in2} must be the admittance $1 + j0$, that is, y_{in2} must lie at the centre (match point) of the chart. Therefore, for matching

$$y_{in2} = 1 + j0 \tag{10.54}$$

From Eqns. (10.54) and (10.53), it is clear that if matching is to occur, the admittance y_2 must be given by

$$y_2 = y_{in2} - jb_{S2} = 1 - jb_{S2} \tag{10.55}$$

That is, from Eqn. (10.55), if matching is to occur, y_2 must lie somewhere on the $g = 1$ circle (circle C_2).

From Eqn. (10.52), y_{in1} is obtained by rotating y_2 towards the load through d. That is,

$$y_{in1} = \overset{d}{\overleftarrow{y_2}} \tag{10.56}$$

As the locus of y_2 is the circle C_2 ($g = 1$ circle), the locus of y_{in1} is the circle C_1 which is obtained by rotating the circle C_2 through d in the anticlockwise (towards the load) direction. Hence, matching can be effected, if and only if, y_{in1} lies somewhere on the circle C_1. From Eqn. (10.51), it is clear that y_{in1} is obtained from y_1 by the addition of a susceptance only. That is, y_{in1} is obtained by starting from the point y_1 and moving along a constant-conductance circle. Hence, the condition that the system can match y_L to the line boils down to the following : Starting from the point y_1, it must be possible to arrive somewhere on the circle C_1 by moving along the constant-conductance circle passing through y_1. That is, the constant-conductance passing through y, must intersect the circle C_1 which is the $g = 1$ circle rotated in the anticlockwise direction through the angle $2\,\beta_g\, d = 4\,\pi \dfrac{d}{\lambda_g}$. Referring to Sec. 3.9.3.1., it is clear that the condition that y_L may be matched by the double stub system is that

$$g_1 \leq \operatorname{cosec}^2\left(\frac{2\pi d}{\lambda_g}\right) \tag{10.57}$$

10.2.3.1 Graphical Design Procedure

Using the Smith admittance chart, the given two stub system may be designed [Ref. Fig. 10.8] to match a given load y_L to the line, in the following way :

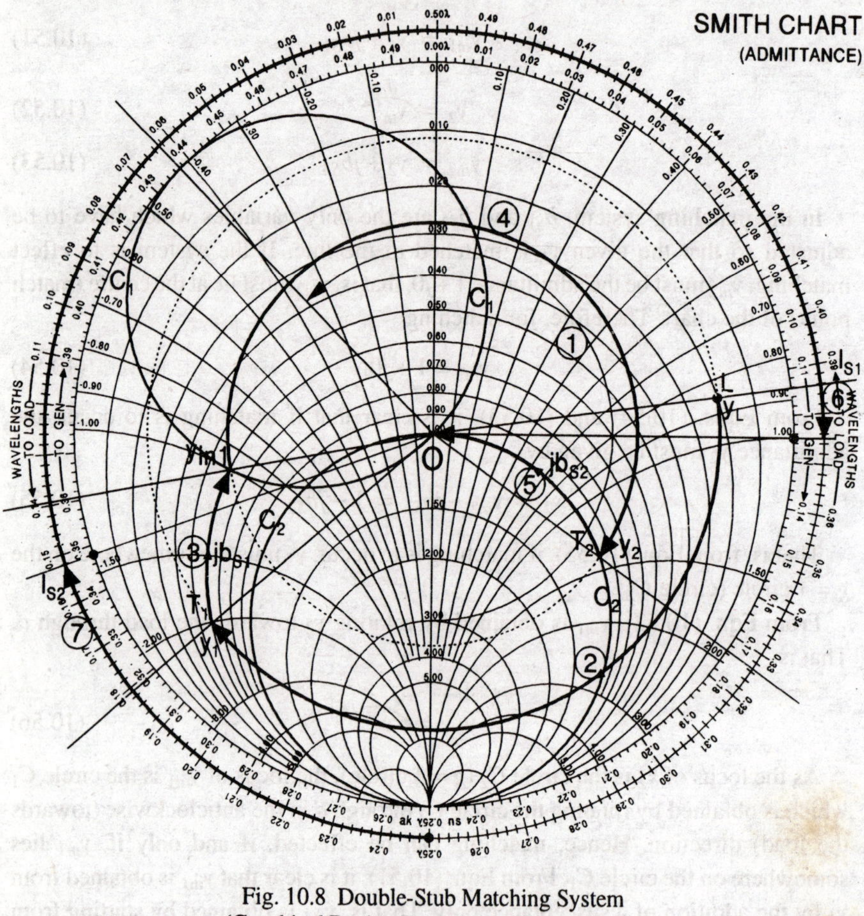

Fig. 10.8 Double-Stub Matching System (Graphical Design Procedur—Alternative 1).

Step 1 : Draw the circle C_1 of radius $\frac{1}{2}$ with its centre on the $S = 3$ circle and displaced from the centre ($g = 3, b = 0$) of the circle, C_2, ($g = 1$ circle) by the distance d, towards the load.

Step 2 : Enter the Smith admittance chart at the load point y_L and move on the constant-S circle passing through the point y_L a distance l towards the generator and arrive at the point y_1.

Step 3 : From the point y_1, move along the constant-g circle passing through the point y_1, and arrive at a point y_{in1} on the circle C_1.

Note that as the constant-g circle passing through y_2 cuts the circle C_1 in two points, any one of these can be legitimately chosen as the y_{in1} point. Therefore, two alternatives or two solutions are possible. [Ref. Figs. 10.8 and 10.9].

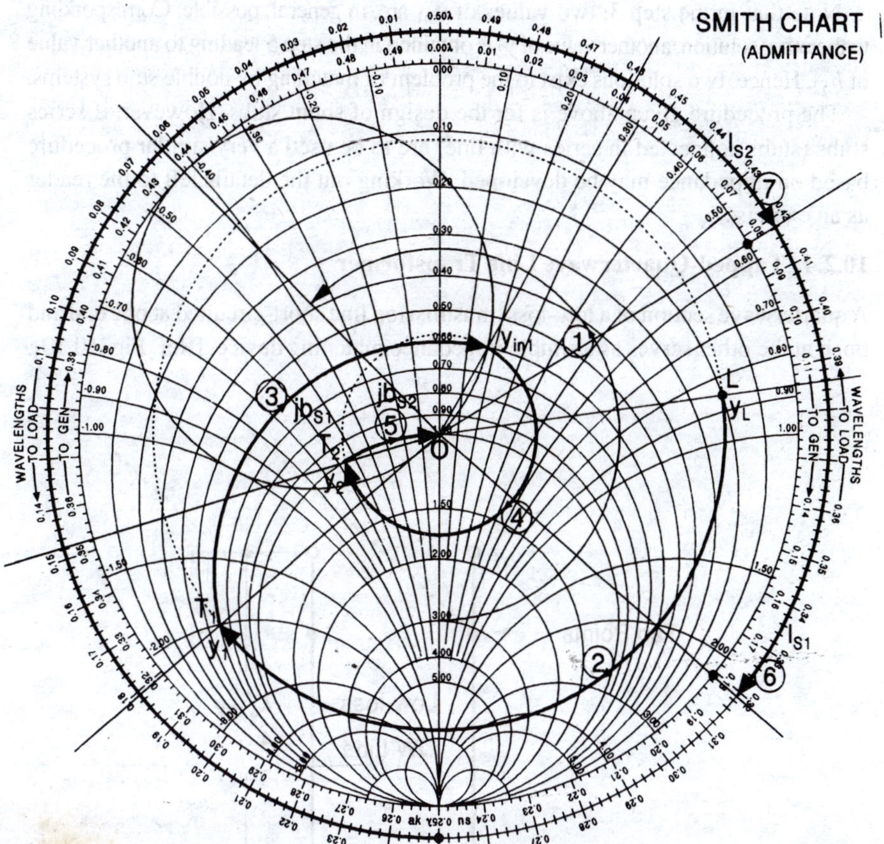

Fig. 10.9 Double-Stub Matching System
(Graphical Design Procedure Alternative 2)

Find out the susceptance, b_{S1}, that must be ***added*** to y_1 in order to arrive at the point y_{in1}.

Step 4 : From the point y_{in1} on the circle C_1 move along the constant-S circle passing through the point y_{in1}, towards the generator through a distance d and arrive at the point y_2 on the circle C_2.

Step 5 : From the point y_2, move along the constant-g circle, that is on the circle C_2 till the match point is reached. Find out the susceptance, b_{S2}, that must be ***added*** to y_2 to go to $1 + j0$.

Step 6 : Starting from the voltage minimum point ($g = \infty$) on the $S = \infty$ circle (periphery of the chart), move along the $S = \infty$ circle towards the generator to the point $0 + jb_{S1}$. The distance thus moved gives the length, l_{S1}, of the first stub.

Step 7 : Find out the clockwise distance between the $g = \infty$ point on the $S = \infty$ circle and the point $0 + jb_{S2}$ on the $S = \infty$ circle. This distance equals the length l_{S2} of the second stub.

Note that in the step 3, two values of y_{in1} are, in general, possible. Corresponding to the other solution, another value of y_2 is obtained in the step 5 leading to another value of b_{S2}. Hence, two solutions exist to the problem of matching by double stub systems.

The procedure given above is for the design of shunt stubs. However, if series stubs (stubs connected in series with line) are to be used a very similar procedure based on impedance may be developed. Working out the details left to the reader as an exercise.

10.2.4 Tapped-Quarterwave Line Transformer

A quarterwave section of a low-loss transmission line short-circuited at one end and open at the other serves as a simple impedance matching device. [Ref. Fig. 10.10].

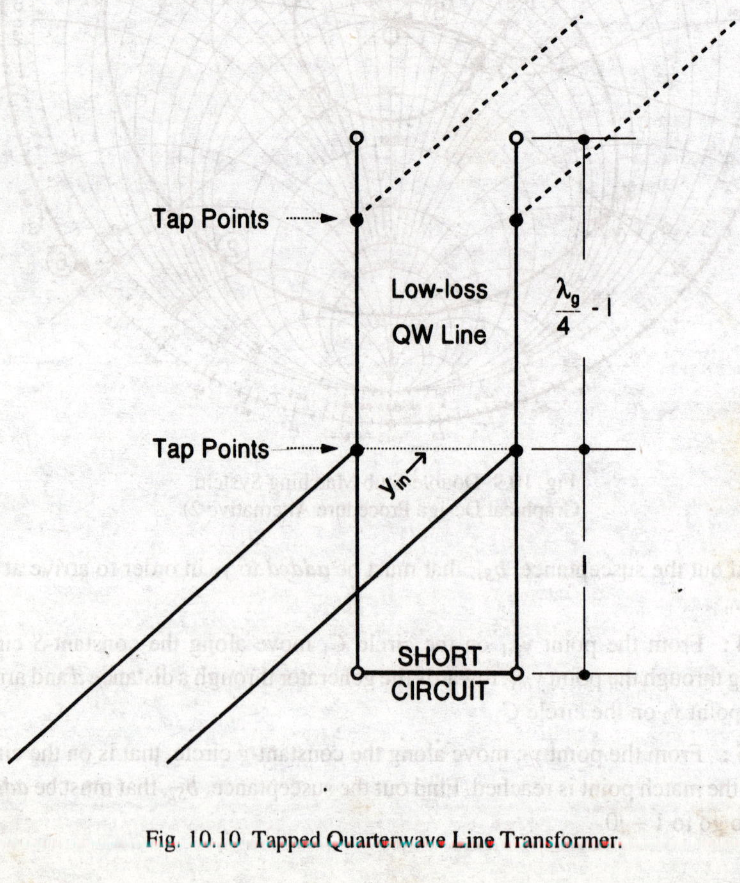

Fig. 10.10 Tapped Quarterwave Line Transformer.

The input admittance of this line as viewed from the tap-points a distance l from the shorted end is the sum of the input admittance of the short-circuited line of length l and the input admittance of the open-circuited line of length $\left(\frac{\lambda_g}{4} - l\right)$, that is

$$Y_{in} = Y_o \coth(\gamma l) + Y_o \tanh\left[\gamma\left(\frac{\lambda_g}{4} - l\right)\right] \tag{10.58}$$

where
$$\gamma = \alpha + j\,\beta_g \tag{10.59}$$

with α so small that $\cosh(\alpha\,\lambda_g) \approx 1$

and
$$\sinh(\alpha\,\lambda_g) \approx \alpha\,\lambda_g \tag{10.60}$$

Substitution of Eqn. (10.59) into Eqn. (10.58) gives

$$y_{in} = \frac{\cosh(\alpha\,l + j\,\beta_g\,l)}{\sinh(\alpha\,l + j\,\beta_g\,l)} + \frac{\sinh\left[\left(\alpha\frac{\lambda_g}{4} - \alpha\,l\right) + j\left(\frac{\pi}{2} - \beta_g\,l\right)\right]}{\cosh\left[\left(\alpha\frac{\lambda_g}{4} - \alpha\,l\right) + j\left(\frac{\pi}{2} - \beta_g\,l\right)\right]}$$

$$= \frac{\cos(\beta_g\,l) + j\,\alpha\,l\sin(\beta_g\,l)}{\alpha\,l\cos(\beta_g\,l) + j\sin(\beta_g\,l)} + \frac{\alpha\left(\frac{\lambda_g}{4} - l\right)\sin(\beta_g\,l) + j\cos(\beta_g\,l)}{\sin(\beta_g\,l) + j\,\alpha\left(\frac{\lambda_g}{4} - l\right)\cos(\beta_g\,l)}$$

$$= \frac{j\,\alpha\frac{\lambda_g}{4}}{[\alpha\,l\cos(\beta_g\,l) + j\sin(\beta_g\,l)]\left[\sin(\beta_g\,l) + j\,\alpha\left(\frac{\lambda_g}{4} - l\right)\cos(\beta_g\,l)\right]}$$

or

$$Z_{in} = -j\frac{4}{\alpha\,\lambda_g}\left[-\alpha\left(\frac{\lambda_g}{4} - 2\,l\right)\cos(\beta_g\,l)\sin(\beta_g\,l) + j\left\{\sin^2(\beta_g\,l) + \alpha^2\,l\left(\frac{\lambda_g}{4} - l\right)\cos^2(\beta_g\,l)\right\}\right]$$

$$= \frac{4}{\alpha\,\lambda_g}\sin^2(\beta_g\,l) + j\left(\frac{1}{2} - \frac{l}{\frac{\lambda_g}{4}}\right)\sin(2\,\beta_g\,l) \tag{10.61}$$

Note that the input reactance is very small compared to the input resistance and, therefore, it may be ignored. As the tap is moved away from the short-circuited end, the input resistance, r_{in}, rises from 0 at the short-circuited end to $\frac{4}{\alpha\,\lambda_g}$ at the open-circuited end. Further, as α is very small, $\frac{4}{\alpha\,\lambda_g}$ is quite large and, therefore, the range $\left(0, \frac{4}{\alpha\,\lambda_g}\right)$ over which r_{in} varies is very large. This property can be exploited for matching a resistive load lying in the range $\left(0, \frac{4}{\alpha \lambda_g}\right)$ to a transmission line or to any other resistance lying in the range $\left(0, \frac{4}{\alpha \lambda_g}\right)$. This device is frequently used for matching an antenna to its feeder.

10.2.5 Exponential Line Transformer

It has been seen that for a lossless uniform transmission line whose parameters L and C are constants along the length of the line, the characteristic impedance is equal to a resistance equal to $\sqrt{\frac{L}{C}}$ and is constant along the length of the line. If now the parameters L and C are made to change slowly along the length of the line that the theory of uniform transmission lines may still be applied piecewise then it would be expected that R_o varies along the length of the line. A finite length of such a non-uniform or tapered transmission line, as shown in Fig. 10.11, may, therefore, be used to match two resistances on an image-impedance basis.

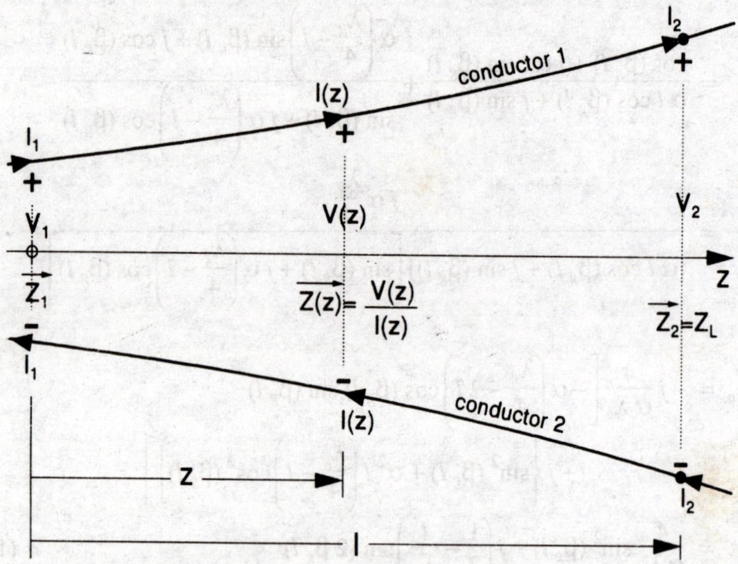

Fig. 10.11 Exponential Line Transformer.

To simplify the mathematical treatment of tapered transmission line, let an exponential taper, defined below be assumed.

Let $$L(z) = L_1 e^{2\theta z}$$

and $$C(z) = C_1 e^{-2\theta z} \tag{10.62}$$

where, L_1, and C_1 are the values of $L(z)$ and $C(z)$, respectively, at the input of the line, $z = 0$. θ is of very small magnitude and the taper is termed as positive or negative according to θ is positive or negative. Assuming the line to be lossless so that $R = G = 0$ and Eqns. (2.3) and (2.4) become for this case

$$\frac{dV}{dz} = -j\omega L I \tag{10.63}$$

and

$$\frac{dI}{dz} = -j\omega C V$$

Substitution of Eqn. (10.62) into Eqn. (10.63) gives

$$\frac{dV}{dz} = -j\omega L_1 e^{2\theta z} I \tag{10.64}$$

$$\frac{dI}{dz} = -j\omega C_1 e^{-2\theta z} V \tag{10.65}$$

Differentiating both sides of Eqn. (10.64) with respect to z and substituting Eqn. (10.65) yield

$$\frac{d^2 V}{d z^2} = -j\omega L_1 [2\theta e^{2\theta z} I + e^{2\theta z} (-j w c, c^{-2\theta z} V)]$$

$$= 2\theta \frac{dV}{dz} - \omega^2 L_1 C_1 V$$

or,

$$\frac{d^2 V}{d z^2} - 2\theta \frac{dV}{dz} + \omega^2 L_1 C_1 V = 0 \tag{10.66}$$

The solution of Eqn. (10.66) is

$$V(z) = e^{(\theta \pm \sqrt{\theta^2 - \omega^2 L_1 C_1})z} \tag{10.67}$$

In order that Eqn. (10.67) represents a propagating w(10.60)ave,

$$\theta^2 - \omega^2 L_1 C_1 < 0$$

or,

$$f > \frac{\theta}{2\pi\sqrt{L_1 C_1}} \tag{10.68}$$

Let

$$f_c = \frac{\theta}{2\pi\sqrt{L_1 C_1}} \tag{10.69}$$

It is observed that for $f > f_c$, propagation occurs along the line but if $f \le f_c$, no propagation occurs and the voltage amplitude is exponentially attenuated or the wave is ***cut-off***. Thus the tapered transmission line behaves as a high-pass filter with the cut-off frequency of f_c.

For the propagating wave, $\theta^2 - \omega^2 L_1 C_1$ is negative and may, therefore, be written as

$$\theta^2 - \omega^2 L_1 C_1 = -\beta_g^2 \tag{10.70}$$

or

$$\beta_g = \theta \sqrt{\left(\frac{f^2}{f_c^2} - 1\right)} \tag{10.71}$$

Substituting Eqn. (10.70) into Eqn. (10.67) gives the most general solution of Eqn. (10.66) to be

$$V(z) = V^+ e^{(\theta - j\beta_g) z} + V^- e^{(\theta + j\beta_g) z} \tag{10.72}$$

Therefore,

$$\frac{dV}{dz} = (\theta - j\beta_g) V^+ e^{(\theta - j\beta_g) z} + (\theta + j\beta_g) V^- e^{(\theta + j\beta_g) z}$$

$$= -j\omega L_1 e^{2\theta z} I(z)$$

which gives

$$I(z) = j\frac{\theta - j\beta_g}{\omega L_1} V^+ e^{(-\theta - j\beta_g) z} + j\frac{\theta + j\beta_g}{\omega L_1} V^- e^{(-\theta + j\beta_g) z} \tag{10.73}$$

Applying the conditions

$$V(z) = V_1$$

and

$$\frac{V(z)}{I(z)} = Z_1 \text{ at } z = 0 \tag{10.74}$$

to Eqns. (10.72) and (10.73) yields,

$$V_1 = V^+ + V^- \tag{10.75}$$

and

$$\frac{V_1}{Z_1} = \frac{j}{\omega L_1} [(\theta - j\beta_g) V^+ + (\theta + j\beta_g) V^-]$$

$$= \frac{j}{\omega L_1} [\theta V_1 - j\beta_g (V^+ - V^-)]$$

or

$$\frac{\omega L_1}{\beta_g} \frac{V_1}{Z_1} - j\frac{\theta}{\beta_g} V_1 = V^+ - V^- \tag{10.76}$$

Solving (10.75) and (10.76) for V^+ and V^-, gives

$$V^+ = \frac{V_1}{2}\left[1 + \frac{\omega L_1}{\beta_g}\frac{1}{Z_1} - j\frac{\theta}{\beta_g}\right]$$

and

$$V^- = \frac{V_1}{2}\left[1 - \frac{\omega L_1}{\beta_g}\frac{1}{Z_1} - j\frac{\theta}{\beta_g}\right] \quad (10.77)$$

If a finite length of this line is used for matching on an image-impedance basis, the reflected wave is absent which means that

$$V^- = 0 \quad (10.78)$$

Substituting (10.78) into (10.77) gives

$$1 - \frac{\omega L_1}{\beta_g}\frac{1}{Z_1} + j\frac{\theta}{\beta_g} = 0$$

or

$$Z_1 = \frac{\omega L_1}{\beta_g + j\theta} \quad (10.79)$$

From Eqn. (10.70),

$$\omega^2 L_1 C_1 = \beta_g^2 + \theta^2 = (\beta_g - j\theta)(\beta_g + j\theta) \quad (10.80)$$

Using Eqn. (10.80), Eqn. (10.79) gives

$$Z_1 = \frac{\omega L_1}{\beta_g + j\theta} = \frac{\beta_g - j\theta}{\omega C_1} \quad (10.81)$$

Under matched conditions, the voltage and current are obtained by substituting (10.78) into (10.72) and (10.73),

$$V(z) = V^+ e^{\theta z} e^{-j\beta_g z} \quad (10.82)$$

and

$$I(z) = \frac{\beta_g + j\theta}{\omega L_1} V^+ e^{-\theta z} e^{-j\beta_g z} \quad (10.83)$$

These give

$$Z(z) = \frac{V(z)}{I(z)} = Z_1 e^{2\theta z} \quad (10.84)$$

If the impedance to be matched to Z_1 is Z_L, then the length, l, of the exponential line required is given by

$$Z_L = Z(l) = Z_1 e^{2\theta l}$$

which gives

$$\theta = \frac{1}{2l} \text{in}\left(\frac{Z_L}{Z_1}\right) \quad (10.85)$$

Note from Eqns. (10.82) and (10.84) that

$$\left|\frac{V_L}{V_1}\right| = \sqrt{\frac{Z_L}{Z_1}} \tag{10.86}$$

This shows that a section of the exponential line behaves as an ideal transformer. It is usually used to match two lines of different characteristic impedances, to each other. Note that if θ is very small, a linear taper may be employed in place of the exponential.

10.3 BANDWIDTH OF MATCHING DEVICES

It may be observed that the transmission line matching devices are truly so only at one single frequency. However, in practice, a certain amount of mismatch or *VSWR* may be tolerated and this permits a matching network designed to work at a given frequency to be of acceptable performance within a range of frequencies called the ***bandwidth*** of the matching network. Usually, it is easy to obtain a narrow bandwidth of 1 or 2 percent by employing simple matching elements. To obtain larger bandwidths, extremely complex matching structures and cut-and-try methods may have to be employed. To illustrate these broadbanding techniques, a quarterwave transformer is considered below.

Consider a quarterwave transformer which matches a load of resistance R_L, assumed, without loss of generality, to be less than R_o, to a line of characteristic impedance, R_o. The characteristic impedance R_{om} , and the length l, of the quarter-wave line transformer are given by

$$R_{om}^2 = R_L R_o \tag{10.87}$$

and

$$l = \frac{\lambda_o}{4} = \frac{\pi}{2}\frac{1}{\beta_{go}} \tag{10.88}$$

Let the frequency change from f_o to $f = f_o + \Delta f$. This would result in a mismatch which may be analysed as follows :

At the changed frequency, the electrical length of the quarterwave transformer becomes,

$$l = \frac{2\pi}{\lambda}\frac{\lambda_o}{4} = \frac{\pi}{2}\frac{f}{f_o} = \frac{\pi}{2} + \frac{\pi}{2}\frac{\Delta f}{f_o} \tag{10.89}$$

The input impedance at the input port of the quarterwave line section, normalised with respect to the characteristic impedance of the main line, is given by

$$z_{in} = \frac{Z_{in}}{R_o} = \frac{R_{om}}{R_o}\left[\frac{R_L + j\,R_{om}\tan(\beta_g\,l)}{R_{om} + j\,R_L\tan(\beta_g\,l)}\right]$$

$$= \frac{R_L - j\,R_{om}\cot\left(\frac{\pi}{2}\frac{\Delta f}{f_o}\right)}{R_o - j\,R_{om}\cot\left(\frac{\pi}{2}\frac{\Delta f}{f_o}\right)} \tag{10.90}$$

Therefore, the reflection coefficient, ρ_1 at the input port of the quarterwave line transformer is given by [from Eqn. (10.90)]

$$\rho_1 = \frac{z_{in}-1}{z_{in}+1} = \frac{R_L - R_o}{R_L + R_o - j\,2\,R_{om}\cot\left(\frac{\pi}{2}\frac{\Delta f}{f_o}\right)}$$

$$= \frac{\dfrac{R_L - R_o}{R_L + R_o}}{1 - j\dfrac{2\sqrt{R_o R_L}}{(R_L + R_o)}\cot\left(\frac{\pi}{2}\frac{\Delta f}{f_o}\right)} \tag{10.91}$$

or

$$|\rho_1|^2 = \frac{\left|\dfrac{R_L - R_o}{R_L + R_o}\right|^2}{1 + \dfrac{4\,R_o R_L}{(R_L + R_o)^2}\cot^2\left(\frac{\pi}{2}\frac{\Delta f}{f_o}\right)} \tag{10.92}$$

If the quarterwave transformer is not used and the load directly connected to the line, the reflection coefficient, ρ_{mm}, and the *VSWR*, S_{mm} produced are given by

$$|\rho_{mm}| = \left|\frac{R_L - R_o}{R_L + R_o}\right| \tag{10.93}$$

and

$$S_{mm} = \frac{R_o}{R_L} \tag{10.94}$$

Substitution of Eqns. (10.93) and (10.94) into Eqn. (10.92) gives

$$|\rho_1|^2 = \frac{|\rho_{mm}|^2}{1 + \dfrac{4\,S_{mm}}{(S_{mm}+1)^2}\cot^2\left(\frac{\pi}{2}\frac{\Delta f}{f_o}\right)} \tag{10.95}$$

The maximum value that Δf is permitted to take is determined by the maximum value of the *VSWR* permitted or the maximum value of the reflection coefficient that may be permitted to exist in the main line. If ρ_{1m} is the maximum value of ρ_1 permitted, then Δf is given by

$$|\rho_1|^2 = \frac{|\rho_{mm}|^2}{1 + \dfrac{4 S_{mm}}{(S_{mm}+1)^2} \cot^2\left(\dfrac{\pi}{2}\dfrac{\Delta f}{f_o}\right)} \le |\rho_{1m}|^2$$

or

$$\cot^2\left(\frac{\pi}{2}\frac{\Delta f}{f_o}\right) \ge \left[\left|\frac{\rho_{mm}}{\rho_{1m}}\right|^2 - 1\right]\frac{(S_{mm}+1)^2}{4 S_{mm}}$$

or

$$\cot\left(\frac{\pi}{2}\frac{\Delta f}{f_o}\right) \ge \pm \frac{S_{mm}+1}{2\sqrt{S_{mm}}}\sqrt{\left|\frac{\rho_{mm}}{\rho_{1m}}\right|^2 - 1}$$

Hence, the fractional bandwidth of the matching system is given by

$$\text{Bandwidth} = \frac{2\,\Delta f}{f_o} = \frac{4}{\pi}\cot^{-1}\left[\frac{S_{mm}+1}{2\sqrt{S_{mm}}}\sqrt{\left|\frac{\rho_{mm}}{\rho_{1m}}\right|^2 - 1}\right] \tag{10.96}$$

Note that the factor $\dfrac{S_{mm}+1}{2\sqrt{S_{mm}}}$ is a monotonically increasing function of S_{mm} and has the minimum value of 1 at $S_{mm} = 1$. $|\rho_{mm}|$ is also a monotonically increasing function of S_{mm} and has the minimum value of zero for $S_{mm} = 1$. Hence, the quantity inside the brackets increases monotonically with S_{mm} and, therefore, as S_{mm} increases, the bandwidth decreases. That is, the greater the *VSWR* a quarterwave line transformer is required to match or tune out, the smaller is the bandwidth of the transformer. The permissible bandwidth of a Quarterwave Transformer, expressed as a percentage of the centre frequency is plotted in Fig. 10.12 as a function of the mismatch *VSWR* for three permissible values of the *VSWR* in the main line.

From the above discussion, it follows that the bandwidth of a quarterwave line transformer may be increased by reducing the *VSWR* it is required to tune out. For instance, if R_L is required to be matched to R_O, it may be done in two steps, that is, by employing two quarterwave transformers; the first designed to match R_L to R_{o1} and the second designed to match R_{o1} to R_o, where R_{o1} is chosen to lie in between the values of R_L and R_O, that is $R_L < R_{o1} < R_o$. This ensures that the *VSWR*s that the individual sections are required to match, namely, $\dfrac{R_{o1}}{R_L}$ and $\dfrac{R_o}{R_{o1}}$, are both less than $S_{mm} = \dfrac{R_o}{R_L}$ and hence the bandwidth of the match will increase. It is clear that if another quarterwave section of characteristic impedance, R_{o2} is added such that $R_L < R_{o1} < R_{o2} < R_o$, then the *VSWR*s that the individual sections match, namely, $\dfrac{R_{o1}}{R_L}, \dfrac{R_{o2}}{R_{o1}}$ and $\dfrac{R_o}{R_{o2}}$ would be less than $\dfrac{R_{o1}}{R_L}$ and $\dfrac{R_o}{R_{o1}}$ of the previous case. Hence, the bandwidth of the matching system increases further. Thus, by careful design, the bandwidth of the matching system increases as the number of the quarterwave

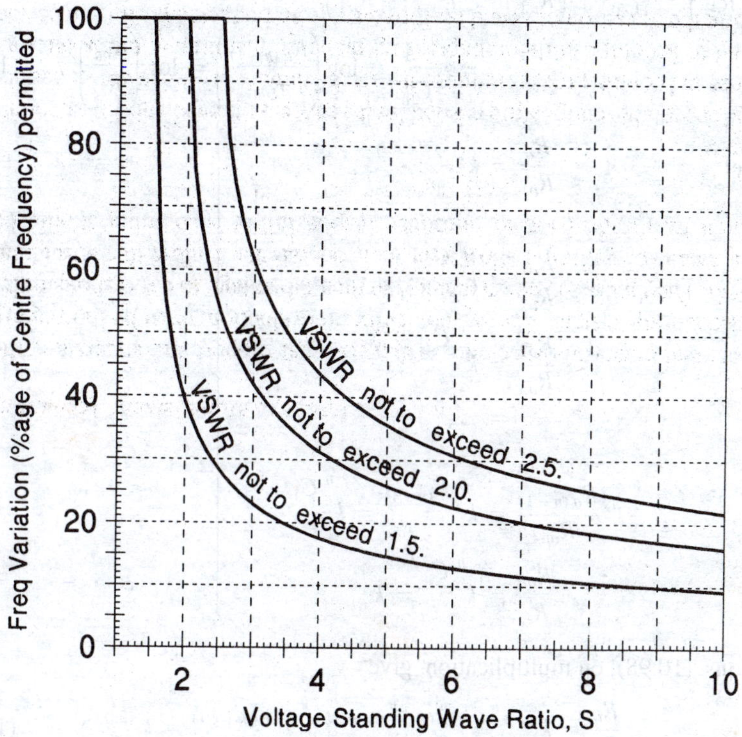

Fig. 10.12 Permitted Frequency Variation (percentage of Design Frequency) as a function of the Mismatch VSWR tuned out by the QWT.

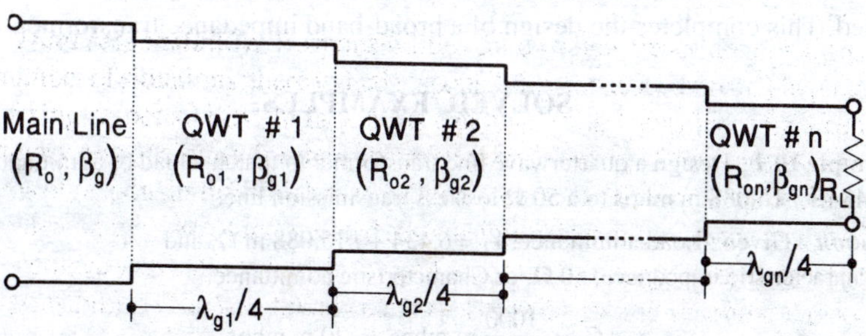

Fig. 10.13 Multi-Quarterwave-Section Transformer.

sections increases but the size and cost of the matching system also increase. The question then arises how best to choose the values $R_{o1}, R_{o2}, R_{o3}, \ldots R_{on}$ of an n-section matching system like the one illustrated in Fig. 10.13. Slater has suggested that the common logarithms of the impedance ratios at the junctions in the system follow the coefficients of the binomial expansion $(a+b)^n$, where n is the number of the quarterwave sections in the system. That is,

$$\log\left(\frac{R_o}{R_{o1}}\right) = \frac{1}{^nC_1}\log\left(\frac{R_{o1}}{R_{o2}}\right) = \frac{1}{^nC_2}\log\left(\frac{R_{o2}}{R_{o3}}\right) = \ldots$$

$$\ldots = \frac{1}{^nC_{(n-1)}}\log\left(\frac{R_{o(n-1)}}{R_{on}}\right) = \log\left(\frac{R_{on}}{R_L}\right) \quad (10.97)$$

Let $\frac{R_o}{R_{o1}} = k$

then $\frac{R_{o1}}{R_{o2}} = k^{^nC_1}$

and $\frac{R_{o2}}{R_{o3}} = k^{^nC_2}$

$\ldots = \ldots$

$\ldots = \ldots$

$\frac{R_{o(n-1)}}{R_{on}} = k^{^nC_{(n-1)}} = k^{^nC_1}$

$\frac{R_{on}}{R_L} = k^{^nC_n} = k$ (10.98)

Eqns. (10.98), on multiplication, give

$$\frac{R_o}{R_L} = k^{(1+{^nC_1}+{^nC_2}+{^nC_3}+\ldots+{^nC_2}+{^nC_1}+1)} = k^{(2^n)} \quad (10.99)$$

Therefore, k can be determined from Eqn. (10.99) and hence with the help of Eqns. (10.98), all the characteristic impedances, R_{o1}, R_{o2}, ..., R_{on} may be determined. This completes the design of a broad-band impedance transformer.

SOLVED EXAMPLES

Example 10.1 Design a quarterwave line transformer to match a load of admittance of $Y_L = 6.434 + j\,15.088$ m mhos to a 50 Ω lossless transmission line.

Solution : Given : Load admittance, $Y_L = 6.434 + j\,15.088$ m Ω, and
Characteristic impedance, 50 Ω, or Characteristic admittance,

$$G_o = \frac{1000}{50}\text{ m mhos} = 20\text{ m mhos}$$

For the solution given below, reference may be made to Fig. 10.14.

Step 1 : Normalise the load admittance, giving

$$y_L = \frac{Y_L}{20} = 0.3217 + j\,0.7544$$

Step 2 : Enter the Smith admittance chart by plotting the point (Y_L), (0.3217 + j 0.7544), using the (g, b) curvilinear coordinate system.

Step 3 : Draw the transmission line, that is, the constant-S circle passing through the point, (Y_L) and having its centre at the centre of the chart, O.

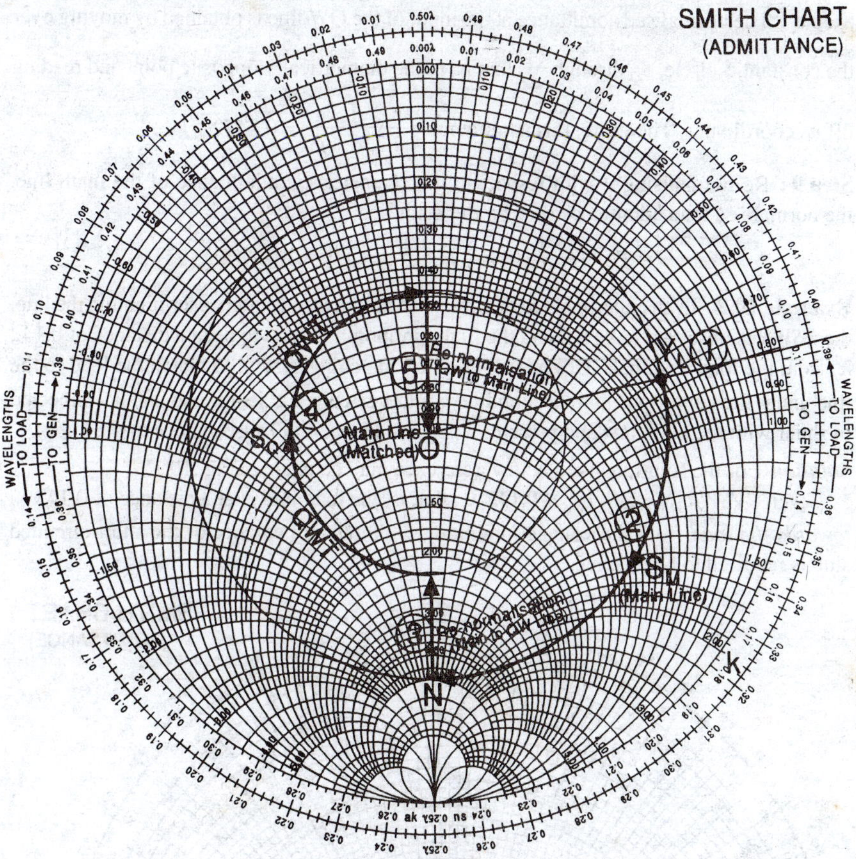

Fig. 10.14 Quarterwave Transformer.

Step 4 : Starting from the point, Y_L, move towards the generator (clockwise) along the constant-S circle, S_M, to the point of intersection, (N), of this circle with the lower vertical radius of the Smith Chart, is the voltage minimum. The admittance coordinates of N are (5, 0). Therefore, the *VSWR* on the line is 5.

Step 5 : The distance, 0.142 λ_g, moved through in Step 4, gives the distance of the Voltage minimum from the load-end, as read off the outermost circular distance scale. It is this point at which the load-end of the Quarterwave line transformer is to be located.

Step 6 : The characteristic admittance of the QWT reqd. is given by $g_{om} = \sqrt{S} = 2.236$

Step 7 : Before proceeding further along the line, as the characteristic impedance is different now (being equal to that the QWT), re-normalisation is necessary. Therefore, for the QW line, the normalised load admittance

$$= \frac{(S + j\,0)}{g_{om}} = \frac{(5 + j\,0)}{2.236} = 2.236 + j\,0$$

Step 8 : The normalised admittance at the input of the QW line is obtained by moving over the constant-S circle, S_Q through $\frac{\lambda_g}{4}$, that is, to the diametrically opposite point and reading off its coordinates. Therefore, this admittance $= \frac{1}{2.236 + j\,0} = 0.4472 + j\,0$.

Step 9 : Re-normalising it with respect to the characteristic admittance of the main line, the normalised load admittance seen by the main line

$$= (0.4472 + j\,0) \times 2.236 \;=\; 1 + j\,0 \;=\; \text{matched.}$$

Example 10.2 In an experiment to determine an unknown admittance terminating the line, the *VSWR* is observed to be 5.0 and the first minimum is found to be at a distance of 0.142 λ_g from the load terminals. Design a single stub matching system to match the load to the line under the conditions that the stub line is short-circuited at its far end and is connected in shunt with the main line and presents a positive (capacitive) susceptance to the line.

Solution For the solution given below, reference may be made to Fig. 10.15

Given : *VSWR*, $S = 5.0$; $l_{min} = 0.142\,\lambda_g$, that is, the load-end is at a distance of 0.142 λ_g *towards the load* from the voltage minimum; the input susceptance of the short-circuited stub is required to be positive.

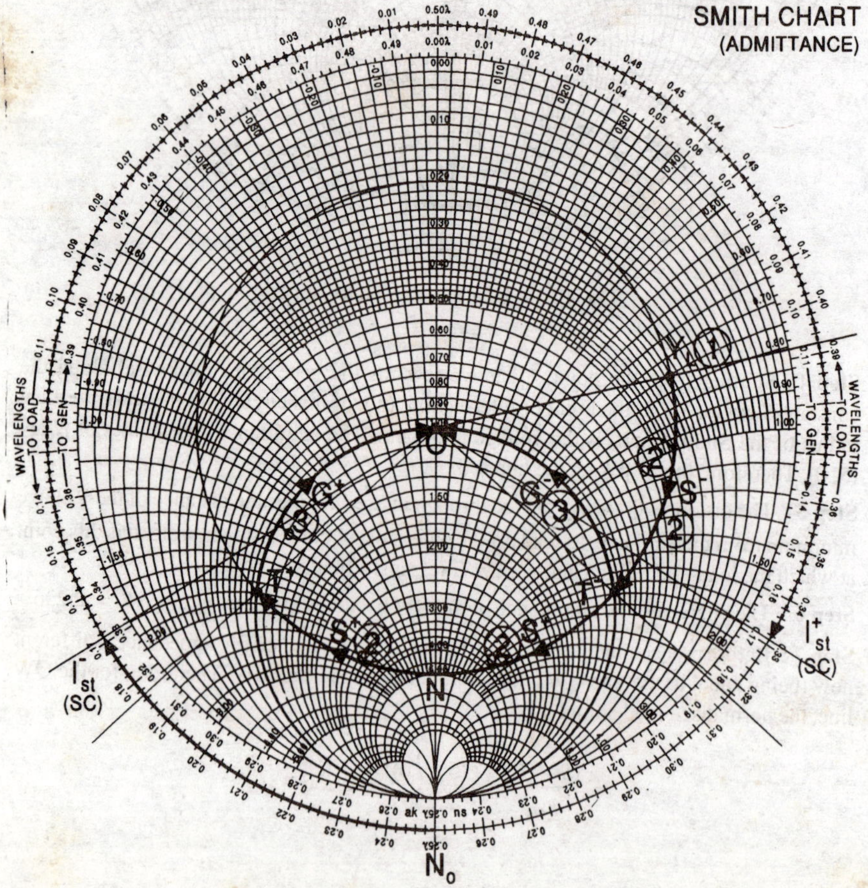

Fig. 10.15 Single Stub Matching : Shunt Stub.

Determining the Load Admittance :

Step 1 : Enter the Smith Chart by drawing the constant-S circle corresponding to $S = 5.0$. The point of intersection, N, of this circle with the lower vertical radius, $\overline{ON_o}$ is the voltage minimum.

Step 2 : To determine the load admittance, starting from the voltage minimum, N, move a distance of 0.142 λ_g along the line (along the constant-S circle) **towards the load** and arrive at the load-point, Y_L, where $y_L = 0.3217 + j\,0.7544$ (normalised).

Single Stub Matching :

Step 1 : Locate the load admittance as given above.

Step 2 : Starting from Y_L move along the S-circle, S^-, till its intersection, T^-, with the g = 1-circle. The distance moved (read off the outermost circular distance scale), (0.183 – 0.108) $\lambda_g = 0.075\ \lambda_g$ gives the distance of the stub from the load end. The line admittance at this point is $1.0 + j\,1.7889$ (normalised). Therefore, the stub connected at this point is required to present a susceptance of $-j\,1.7889$ to effect matching. Or, graphically.

Step 3 : From the point, T^-, move along the constant-g(g = 1 circle), to arrive at the **match point** or the admittance point, $1 + j0$, or the centre of the Chart. To do this, a susceptance of $-j\,1.7889$ is to be **added** to the admittance at the point, T^-. Therefore, this is susceptance to be provided by the stub at the point, T^-

Step 4 : To determine the length of the short-circuited line (stub) required to provide a susceptance of $-j\,1.7889$, start from the short-circuit point, N_o, on the periphery of the chart (the $S = \infty$ circle) and move towards the generator till the admittance point, $(0 - j\,1.7889)$, is reached. Draw the radius vector passing through this point and read off the outermost circular distance scale, the distance moved through, (0.331 – 0.250) $\lambda_g = 0.081\ \lambda_g$, gives the length of the short-circuited stub required for matching.

However, the above design does not meet the requirement given in the question, that the stub is required to present a positive susceptance to the line. Therefore, the alternative design may be examined.

Alternative Design :

Step 2 : Starting from Y_L move along the S-circle, S^+, till its **second** intersection, T^+, with the g = 1-circle. The distance moved, (0.317 – 0.108)$\lambda_g = 0.209\ \lambda_g$, gives the distance of the stub from the load end. The line admittance at this point is $1.0 - j\,1.7889$ (normalised). Therefore, the stub connected at this point is required to present a susceptance of $+j\,1.7889$ to effect matching. Or, graphically,

Step 3 : From the point, T^-, move along the (g = 1 circle), to arrive at the **match point**. To do this, a susceptance of $+j\,1.7889$ is to be **added** to the admittance at the point, T^+. Therefore, this is susceptance to be provided by the stub at the point, T^+.

Step 4 : To determine the length of the short-circuited stub required to provide a susceptance of $+j\,1.7889$, start from the short-circuit point, N_o, on the periphery of the chart (the $S = \infty$ circle) and move towards the generator till the admittance point, $(0 + j\,1.7889)$, is reached. Draw the radius vector passing through this point and read off the outermost circular distance scale, the distance moved through, (0.5 + 0.169 – 0.250) $\lambda_g = 0.419\ \lambda_g$, gives the length of the short-circuited stub required for matching.

Example 10.3. Redesign the single stub matching system for the problem given in Example 10.2 above but under the condition that the stub is to be used in series with the main line.

Solution : As the stub is to be in series with the main line, one must work with impedances and the Smith Chart is to be used as an impedance chart. Consequently, the voltage minima lie on the *upper vertical radius* of the Smith Chart. Except for these changes warranted by duality, the design procedure for series stubs is the same as that for shunt stubs explained in the solution of the previous question. However, the graphical solution is given in Fig. 10.16 which is self-explanatory.

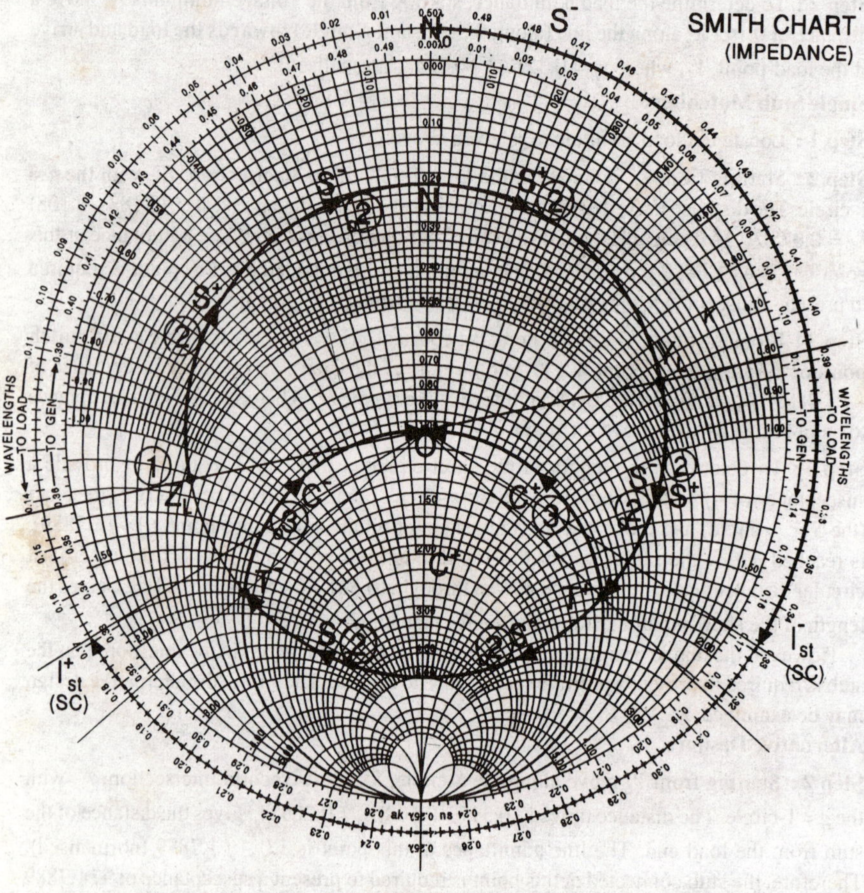

Fig. 10.16 Single Stub Matching: Series Stub.

Example 10.4 A load of (30 + *j* 44) m Ω admittance is connected to the load-end of a 50 Ω coaxial cable. Design a double-stub matching system for the given load. The first stub is to be connected at a distance of 0.168 λ_g from the load-end. The two stubs are $\frac{3}{8}$ λ_g apart and are to be connected in shunt with the line.

Solution :

Ref. Fig. 10.17.

Step 1 : Normalise load admittance, giving $y_L = \dfrac{30 + j\,44}{20} = 1.5 + j\,2.2$. Plot it on the Smith admittance chart and label the point, Y_L. Draw the *S*-circle passing through Y_L.

Step 2 : From the point, Y_L, move through a distance of $0.168\lambda_g$ along the S-circle, towards the generator and arrive at the point, T_1. Read the admittance coordinates, y_1, of T_1. $y_1 = 0.4027 - j\,1.0048$.

Construction :

Rotate the $g = 1$ circle (C_2) through a distance of $0.375\ \lambda_g$ towards the load giving the circle, C_1, in the following way : Draw the $S = 3$ circle. Draw a radius vector at a distance of $0.375\lambda_g$ from the lower vertical radius of the chart, towards the load. This operation may alternatively be stated as: Rotate the lower vertical radius of the chart through a distance of $0.375\lambda_g$ towards the load. The intersection of this rotated radius vector and the $S = 3$ circle, gives the centre of the circle, C_1. The radius of the $g = 1$ circle is $\frac{1}{2}$ unit, the radius of the Smith Chart being one unit. Draw the circle, C_1, which is nothing but the $g = 1$ circle (C_2) rotated through a distance of $0.375\lambda_g$ towards the load.

Step 3 : From the point, T_1, move along the constant-g circle passing through T_1 to its intersection, $T_1^{\,1}$, with the circle, C_1. Note that one could go to the other point of intersection, T^2, of these two circles. That is being called the 'alternative design' and denoted by the superscript '2'. The admittance coordinates of the point, $T_1^{\,1}$ are : $y_{in1} = 0.4027 - j\,1.8020$. Therefore, the susceptance required to be added for going from T_1 to $T_1^{\,1}$

$$= y_{in1} - y_1 = (0.4027 - j\,1.8020) - (0.4027 - j\,1.0048) = -j\,0.7972$$

Step 4 : Now, starting from the point, $T_1^{\,1}$, move through a distance of $0.375\lambda_g$ towards the generator on the S-circle passing through $T_1^{\,1}$, and arrive at the point, $T_2^{\,1}$ on the $g = 1$ circle (C_2). The admittance at $T_2^{\,1}$ is $y_2 = 1.0 + j2.9916$.

Step 5 : Finally, move from the point, $T_2^{\,1}$, to the match point, O, representing an admittance of $y_{in2} = 1 + j\,0$. In doing so a susceptance of $-j2.9916$ is to be added, which must be provided by the second stub.

Step 6 : Lengths of the short-circuited stubs required to present the respective susceptances, may be determined by following the procedure already explained above. The lengths required come out to be $0.1429\lambda_g$ and $0.0513\lambda_g$ for the first and the second (short-circuited) stub, respectively.

Note : An alternative double stub matching system results, if in step 3 above, instead of the intersection, $T_1^{\,1}$, the other point of intersection ($T_1^{\,2}$) of the constant-g circle with the circle C_1, is chosen to go to, starting from the point T_1. The rest of the procedure is the same as described above. In Fig. 10.17, the superscript '2' on the labels of points and other quantities and the subscript '*o*' on the circles circumscribing step numbers, have been used to identify the second choice.

Example 10.5. Redesign the double stub matching system for the problem given in Example 10.4 above but under the condition that the stubs are to be used in series with the main line.

Solution : As the stubs are to be in series with the main line, one must work with impedances and the Smith Chart is to be used as an impedance chart. Consequently, the voltage minima lie on the *upper vertical radius* of the Smith Chart. Except for these changes warranted by duality, the design procedure for series stubs is the same as that for shunt stubs explained in the solution of the previous Example. However, the graphical solution is given in Fig. 10.18 which is self-explanatory.

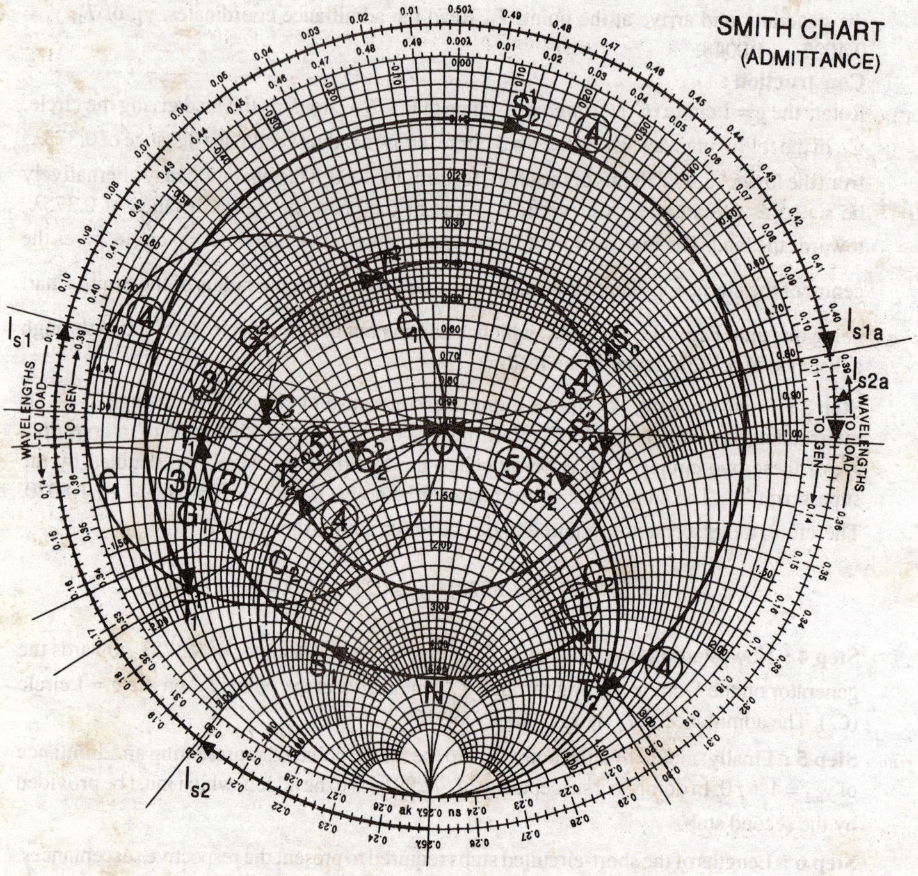

Fig. 10.17 Double Stub Matching: Shunt Stubs.

EXERCISES

1. Design a quarterwave transformer to match a load impedance of 70.7(1 + j) Ω to a 50 Ω lossless line at a frequency of 2.55 GHz.
2. Design a quarterwave line transformer to match a load of impedance $Z_L = 50 + j50$ Ω to a 50 Ω lossless transmission line. If the frequency is increased by 15%, what will be the *VSWR* produced ?

   [**Hint :** When the frequency increases by 15%, the normalised length of the quarterwave line used for matching would no longer be 0.25 but would also increase by 15%. The input impedance of the QWT would, therefore, be different from 1 + j0 and hence a mismatch *VSWR* would appear.]
3. A (210 + j 60 Ω) load is to be matched to a 300 Ω line to give a *VSWR* of unity by connecting a quarterwave transformer and a short-circuited stub directly at the load terminals. Calculate the length of the stub and the characteristic impedance of the quarterwave transformer required.

   [**Hint :** Design the stub to present at its input a susceptance which annihilates the load susceptance. Once that is done, the design of the QWT is quite simple.]

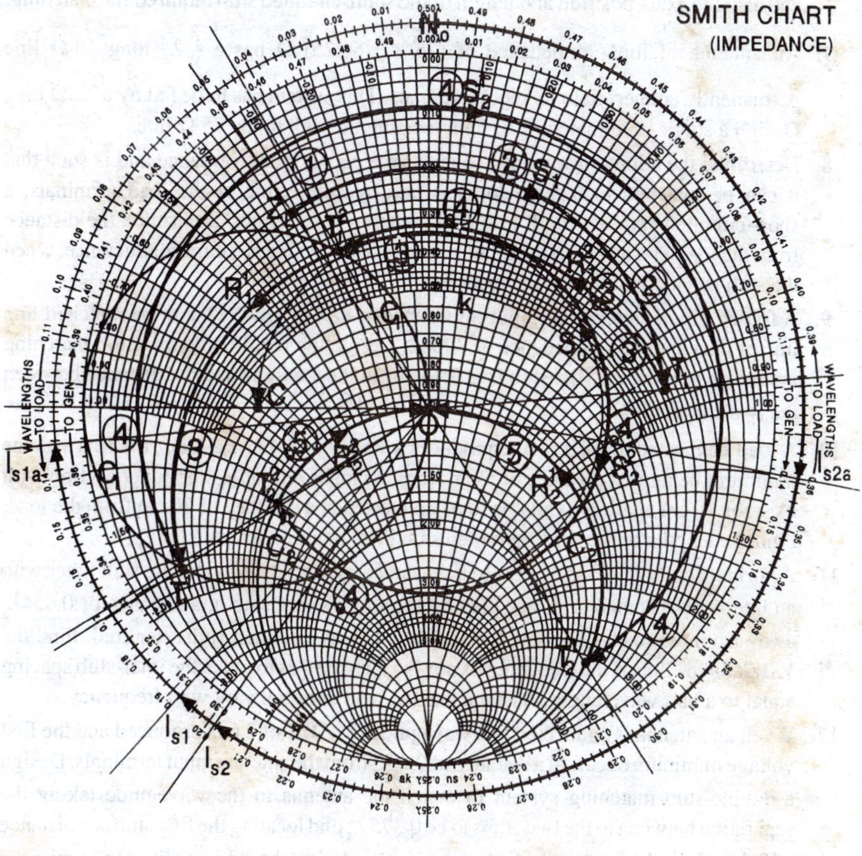

Fig. 10.18 Double Stub Matching: Series Stubs.

4. Design a single-stub matching system required to match a load of (24 + j5 Ω) to a 50 Ω coaxial line using a short-circuited stub of the shortest length. The characteristic impedance of the stub line is to be 75 Ω. Compare the frequency sensitiveness of this matching system with that of a similar one employing a 50 Ω stub line.

5. A resistance of 600 Ω is connected at the load-end of a 300 Ω line (lossless) at 100 MHz. Match this load to the line using a single short-circuited stub. Calculate the *VSWR* on the line when the frequency of operation increases by 10 MHz, assuming the load to remain constant. Re-locate the stub by moving it a distance of one wavelength towards the generator. Now calculate the *VSWR* produced when the frequency of operation increases by 10 MHz.
[Hopefully you would appreciate why it is recommended that the stub be located as close to the load as possible]

6. A 300 Ω lossless line feeds an antenna whose input impedance is a pure resistance. When the antenna is connected to the line a *VSWR* of 4 is observed on the line and further it is found that a short-circuited stub (connected in shunt with the line) has to be located within a distance of $\frac{\lambda_g}{8}$ from the antenna terminals to effect matching between

the antenna and the line, using Smith Chart. Calculate the input resistance of the antenna and also the exact position and length of the short-circuited stub required for matching.

7. An antenna of input impedance of $(75 + j\ 66.25)\ \Omega$ has a $\frac{3}{8}\lambda_g$ long $50\ \Omega$ line permanently connected to its input terminals. This system has to be fed by a $75\ \Omega$ line. Design a single-stub matching system locating the stub in the $75\ \Omega$ line.
8. Determine the admittance whose conductance equals its susceptance and is such that it can be matched to a $50\ \Omega$ lossless line by connecting at the load terminals, a short-circuited stub of length less than a quarter wavelength. Determine the distance from the load-end at which the input admittance of the line is a pure conductance, when the stub is not connected at the load-end.
9. A folded dipole antenna of input impedance of $(292 + j148)\ \Omega$ is to be matched to a line of characteristic impedance of $300\ \Omega$. Design the required double-stub matching system, given that the first stub is at $0.1\lambda_g$ from the load-end and the distance between the two stubs is $0.4\lambda_g$.
10. A load connected to a $70\ \Omega$ line produces a *VSWR* of 3.4 at 1.2 GHz, the first voltage minimum being at a distance of 5 cm from the load. A double-stub tuner section with its stubs 10 cm apart, is available to be connected at a distance of 20 cm from the load. Find the lengths of the short-circuited stub lines required for matching.
11. A normalised load $Y_L = 3 + j\ 3$ is to be matched by employing a double stub tuner with an inter-stub distance of $0.49\lambda_g$ and by locating its first stub at a distance of $0.054\lambda_g$ away from the load. Calculate the lengths of short-circuited stubs required. Find the *VSWR* on the line, if the frequency is now so increased as to make the inter-stub spacing equal to a half wavelength. Assume that the load does not vary with frequency.
12. When an antenna is connected to a waveguide, a *VSWR* of 2.6 is produced and the first voltage minimum occurs at a distance of $0.2\ \lambda_g$ from the antenna input terminals. Design a double-stub matching system to match the antenna to the waveguide taking the separation between to the two stubs to be $0.375\ \lambda_g$ and locating the first stub at a distance of $0.2\ \lambda_g$ from the antenna input terminals. The design should be such as to require the least lengths of the two short-circuited stub lines made of the same waveguide as the main line.

    If for some reason, the first stub (the one nearer to the antenna) gets detached from the waveguide, what will the *VSWR* in the feeder be ?
13. Design a quarterwave transformer which will match a $100\ \Omega$ load to a $50\ \Omega$ line at 10 GHz. What is the frequency band of operation over which the *VSWR* remains less than 1.2 ? Compare this bandwidth with that obtained for a matching system which employs two quarter wave transformers in tandem.
14. A purely resistive load of value $1200\ \Omega$ is connected at the load-end of a $400\ \Omega$ lossless line. Design a two section quarterwave transformer required for matching the load to the line at 1200 MHz. Find out the *VSWR* in the line produced when the frequency of operation increases to 1380 MHz.
15. Design a 3 quarterwave sections system to tune out a *VSWR* of 4.3 produced by a purely resistive load whose value is less than the characteristic resistance of the line. Use the Smith Chart to determine the *VSWR* in the main line when the frequency of operation decreases by 20%, explaining clearly the procedure you adopt.

Subject Index